AF765751

ADAPTED BRAINS AND IMAGINARY WORLDS

Adapted Brains and Imaginary Worlds

Cognitive Science and the Literature of the Renaissance

DONALD BEECHER

McGill-Queen's University Press
Montreal & Kingston | London | Chicago

© McGill-Queen's University Press 2016

ISBN 978-0-7735-4680-6 (cloth)
ISBN 978-0-7735-4681-3 (paper)
ISBN 978-0-7735-9852-2 (ePDF)
ISBN 978-0-7735-9853-9 (ePUB)

Legal deposit first quarter 2016
Bibliothèque nationale du Québec

Printed in Canada on acid-free paper that is 100% ancient forest free (100% post-consumer recycled), processed chlorine free

This book has been published with the help of a grant from the Canadian Federation for the Humanities and Social Sciences, through the Awards to Scholarly Publications Program, using funds provided by the Social Sciences and Humanities Research Council of Canada.

McGill-Queen's University Press acknowledges the support of the Canada Council for the Arts for our publishing program. We also acknowledge the financial support of the Government of Canada through the Canada Book Fund for our publishing activities.

LIBRARY AND ARCHIVES CANADA CATALOGUING IN PUBLICATION

Beecher, Donald, author
Adapted brains and imaginary worlds : cognitive science and the literature of the Renaissance / Donald Beecher.

Includes bibliographical references and index.
Issued in print and electronic formats.
ISBN 978-0-7735-4680-6 (bound). – ISBN 978-0-7735-4681-3 (paperback). – ISBN 978-0-7735-9852-2 (pdf). – ISBN 978-0-7735-9853-9 (epub)

1. English literature – Early modern, 1500–1700 – History and criticism. 2. Cognition in literature. 3. Emotions in literature. 4. Memory in literature. 5. Self in literature. 6. Cognitive science. I. Title.

PR428.P93B43 2016 820.9'353 C2015-906758-8
C2015-906759-6

Set in 10/14 Sabon LT Std
Book design & typesetting by Garet Markvoort, zijn digital

This one is for my beloved daughter
Sophie

I kept my promise – it took a while
Now you have to keep yours – no hurry

CONTENTS

ACKNOWLEDGMENTS

I wish to offer sincere thanks to all those who have tolerated my literary obsession of a decade, chief among them my colleague Grant Williams, who posed challenging questions and rebuttals; Richard Hillman and colleagues at the Tours colloquia on Renaissance drama, who heard me out on these matters on several occasions with skeptical good will; Massimo Ciavolella, friend and collaborator now into our fourth decade, for conference invitations that inspired me to think about conversion, nostalgia, and invective; Smaro Kamboureli, who arranged for the challenge of writing the Landsdowne lectures back in my cognitive salad days; Maurizio Ascari, for the invitation to Bologna that included a lecture on memory; Jim Davies of Carleton's Cognitive Science Department, who has given tips and encouragement; Joseph Khoury, colleague and conference associate, who shared in my thoughts along the highways of Canada and France; and the students at Carleton University who thought there was promise in such approaches to literature and joined in the cognitive sciences and literature seminars; the most recent among them are Patrick Juskevicius, Tyler Gogo, and Kevin Soubrian. I am grateful, as well, to Josh Elyea and Patrick Juskevicius for the attention they gave to the final manuscript. I am also deeply indebted to Jonathan Crago and all those at McGill-Queen's University Press who believed in me enough to see this manuscript through the vetting process and onwards into print. A major contributor to that process was Kate Merriman who brought greater brevity and light to many an errant phrase through her editorial expertise. To her my sincerest thanks. Similar gratitude is in order to those thoughtful readers for the Awards to Scholarly Publications Program who supplied their diverse and instructive evaluations.

ADAPTED BRAINS AND IMAGINARY WORLDS

INTRODUCTION

A major hypothesis lies behind the studies in Renaissance literature that constitute the present volume, beginning with a postulate that would seem entirely axiomatic: that the makers of narrative representations featuring human beings in social contexts achieve their ends by employing all the faculties of mind at their disposal for making those representations mimetically cogent, in accordance with their chosen literary purposes.

But then, what are the properties and conditions that define and produce narratives, human beings, social systems, mimetic cogency, and faculties of mind? How far back must we go to find an agreed upon beginning to the literary enterprise? Since the critical lens can properly focus in only one direction at a time, my request here is that we focus on the faculties of mind that shape the literary representation and interpretation of persons, much as they shape the understanding of persons in everyday life. These include the systems that produce weeping, or laughter, or jealousy, or falling in love, not to mention intimations of the self or the binary values of ethical evaluation. Elements in each of these modes seem to well up from platforms below the surface of active deliberation, and it would appear valid to ask how and why these things happen, and "who" initiates these behavioural patterns.

Inquiries into the nature and meaning of these affective and cognitive events in literary representations, whether in the minds of characters or the minds of their makers, are hardly new (think of Freudian criticism). But I am asking whether recent investigations into cognition, evolutionary psychology, and the neural networks that shape thought and emotion can fruitfully be included in the debate concerning the features of human nature represented in art. We may all agree that what makers produce is shaped by their intellects, their creative aptitudes, their conscious minds, in turn conditioned by personal experiences, unique memories, and cultural allegiances. We are less certain, however, about the modes of cognition and emotion readers apply and attribute to literary characters. And least certain for some is that those same faculties of thought and feeling arise from a material brain that was selectively engineered over evolutionary time according to the principle that best traits could be retained only through reproductive success, mate-selection preferences, or the luck of genetic drifting, and consequently reflect the values of evolutionary adaptation. Are there critical moments in literature as in life at which a consideration of our phylogenetic natures provides the best explanatory version of our predilections, choices, actions, and instincts? The studies to follow are out to test that thesis.

To be sure, a lengthy back-study into the properties of the adapted brain need not be invoked to explain all the dimensions of literary personhood. There are circumstances in which personal and cultural matters, the stuff of recent learning, conditioning, and the preoccupations of the historicized self (despite their dependencies on the enabling platforms of an evolved species) provide sufficient psychological contexts for critical judgment. We have, for a long time, known about memory and its limitations; about emotions, their sensations and correlations to social values; about the implicit associative operations required for the interpretation of analogies, symbols, riddles, and paradoxes; about self-awareness and our first-person point of view; about feelings evoked by the joys and sufferings of others; about computing self-advantage through deception or cooperation; about falling in love, mental obsessions, and so much more concerning the quirks and compulsions of the human self without entering into scientific backstories fit for Pandemonium. Yet each one of these properties of the mind is potentiated and enabled for mental recognition by our designed neural platforms, in accordance with the values and constraints that defined those production systems. Accordingly, there must be first-order human experiences, which in turn imprint their properties on all our infinite variety in matters cultural, institutional, personal, and behavioural; there must be genetic properties which define us as the homo sapiens species we are. It is what Jerome Barkow meant by the phrase, "Beneath new culture is old psychology," the Pleistocene brain that operates unaltered and undiminished beneath every expression of human intentionality and cultural conditioning.[1] Nothing emerges in the life of the mind that is entirely independent of the purpose-designed neural systems underlying all the operations of the human brain. Or, in the words of George C. Williams, "Is it not reasonable to anticipate that our understanding of the human mind would be aided greatly by knowing the purpose for which it was designed?" That is the simple hypothesis, but axiomatic as it may seem, it is far from having received universal acceptance.

The only alternative to the evolutionary history of that design is that human nature, with its innate characteristics, is acquired entirely after birth, entailing that the human brain, without any inherent knowledge tilting it in the direction of adaptive conceptualization, must wire itself within the single lifetime of each individual, exclusively through lived experience. But imagining that process is a philosophical koan; it makes demands upon early learning that defy possibility. This alternative holds that our natures must rely on imitation and cultural instruction, not only to equip the subject for all the vicissitudes of life but to design all the learning systems of a blank brain at the same time. This is a real egg-chicken dilemma. No learning platforms – no learning. And if the brain is not quite blank at birth, but is already wired up to learn, quickly and efficiently, then it already "knows" something, because even learning aptitudes can have come about only through the adaptive advantages of, well, learning. Is knowing how to suck and look for sustenance at birth a designed form of behaviour to purposeful ends? If so, the argument for the empty-at-birth brain is already lost. After that concession, we are seeking merely to define the list of all the many kinds of behaviours too vital to survival to leave up to fast-track learning through instruction, imitation, and heuristics. It is counterintuitive to think that humans, even as the smart and chatty primates that they are, do not have the same advantages of genomic learning that define the instinctual behaviours of our fellow mammals. Inconvenient as the fact may seem, there is no apparent moment in our developmental history at which that rich heritage of problem-solving platforms embedded in neural design was turned off. And if our brains serve up these computational and emotional biases as the raw potential probabilities of our behavioural patterns – patterns we would expect to encounter in honest literary representations of our kind – is there not hermeneutic value in rethinking the selves of literature in relation to the selves of reality?

But, alas, before we can get started, there is ungrateful polemical work to be done. Much of the analysis in the essays to follow is not directly involved with arguments pertaining to so-called "literary Darwinism"; nevertheless, by dint of the general reliance of cognitive philosophers on the principles of evolutionary psychology, they are subject to the comprehensive objections of the anti-Darwinians. Thus, it would seem appropriate to attend to these objections in the introduction.

Such objections fall into a few major categories, which can briefly be summarized as materialism, reductionism, determinism, genocentrism, bioprophetics, and the unreliability of provisional explanatory backstories, often dismissed as "just so" stories fancifully offered to explain phenomenological mysteries.[2] Inescapable, to my way of thinking, is that neural organization, cells, synapses, neurotransmitters, peptides, enzymes, myelin, glucose, and related materials enable, through a collaborative community of material operations, our remarkable ability to care, remember past events, feel pain, worry, plan, enjoy, believe, sense our freedom, and relive old memories. The materialism of thought, nevertheless, deposits a strange ontological paradox, because it ostensibly reduces the mind to mechanics, in opposition to the immateriality and freedom of its emergent productions. This is a well-established philosophical debate and continues to tease our minds like the two-dimensional Necker cube, which seems to open up and then down as our brains pop the figure into an illusory third dimension. The orientation of the opening becomes a figment of our Gestalt-driven reconstructive brains. How can brains be so substantial and minds so ethereal? There is something about the relationship between these properties that defies ready understanding, a philosophical crux known as the Mind-Body Problem.[3] For those seeking the soul in the psyche and the freedom of the will in pure deliberative thought, or the pre-eminence of culture formulated exclusively by post-natal human experience in a blank-slate brain, the war against the materialization of the mind is a sacred and self-evident cause.

But much depends on how the problem is posed. The emergent mental properties we call experiential thought and sensations – in all their nuanced richness – are precisely what the material brain was designed to produce, however the unique and free-formed qualia of mentation may seem to dictate their own essence. Nothing is mentally experienced without the material platforms that enable the production of those precise sensations. Conceding that, however, comes with intolerable consequences for anti-Darwinians, for it entails an endorsement of the principle that systems are selectively designed to purposeful ends, and those ends are remembered by the genome which replicates the most adaptive solutions to mental production through a long, step-wise re-organization of the material brain as its own inherent problem-solving instrument. For anti-Darwinians, all material explanations of characteristic behaviours are assaults on freedom of the will. Still, when you think about it, free will in absolute terms is under pressure whether the subject is conditioned by nature or nurture – by criminal genes or abusive parents. Genes constrain by design, but they also enable by design and are flexible in their operations. They may be tough on theories of the dematerialized soul, but they are not tough on the freedom of the will or the plasticity of the altricious or learning-adapted brain, both of which gained immeasurably through selective amelioration. We know from our own experience that we can plan, feel options, and reason out choices. The brain promotes these adaptive capacities, including our obsessive bent toward the reading of the intentional states of other minds – but of necessity in its own idiosyncratic and enabling ways. Thus, reasoning in biogenetic terms may impose the conditions of our nature, but out of that nature emerges all that we have ever imagined of freedom or fashioned into cultural variations.[4] The determinism that matters is not in the materialism of the brain; it is in having to choose among provisional templates and then live with the consequences. Just ask Adam and Eve. Mental freedom is to neurons what sunlight is to combustion. Yet laws and constraints apply which condition and circumscribe all experience. In the words of Jeremy Campbell: "Can an intelligence that is perfectly open make sense of the world? In a limited organ such as the brain, working under the constraints of space and time in a real world, reason cannot explore all possibilities, but needs to be guided by organized structures of knowledge stored in memory. As a consequence, the essence of organized knowledge is that it tilts the mind toward a particular interpretation of reality and tilts it away from other interpretations."[5] Such knowledge is divided between the genes and experience, and what we know defines us, whether from embedded or acquired understanding. Both forms of learning are destinies.

Our concern here, however, is with what the genome "knows" on our behalves about the world and offers to us as essential interpretational benefits. If such systemic knowledge proves to be the least bit true, it has to matter to the way we process not only the real world, but fictional worlds as well. There will be much more in the essays to follow on how the mind, by definition, sees the world through its own biogenetically programmed tendencies tilted in tell-tale directions – those which make us specifically human and organize all that we create, including culture itself and our innumerable imaginary inventions through which alone the genome expresses itself.[6] Hence the founding axiom: nothing happens in the mind, no emergent property is imaginable, without the corresponding enablement of the adaptive brain.

Critics have argued against reductionism by targeting careless analysts who overzealously account for recent cultural achievements in causal genetic values, eager to point out the absurdity of pairing up genes with goals specific to the modern world – and they are right. Those fallacies have become the hobby horses upon which to discredit evolutionary psychology *tout court* as a voice of material and biological determinism.[7] For careful naturalists, however, the remarkable feature of the brain is precisely its inherited latitudes of learning and considered reflection whereby it adjusts itself to new social and technological environments without the need for specific genes to explain our liking for string quartets, or psychological novels, or abstract impressionist art. But these concessions do not dismantle the powers of genes to frame and tilt our adaptive behaviours, powers expressed through the purpose-designed and systemic organization of the human brain. In this quintessential regard, we are inevitably constrained, for we can only invent things that fall within the scope of our cognitions and talents, or exploit the "affordances" available to us in our environments.[8] Moreover, our cultural creations become meaningful only when they address our human capacities for recognition and response, just as food is selected and prepared in relation to the physiological properties of taste. Such productions depend on faculties, faculties are related to the material brain, and hence art and its reception are aligned with neural design. That design was, in turn, confirmed selectively according to its operational or problem-solving efficacy throughout evolutionary time. How could it be otherwise?[9] Through the careful construction of that long chain of impingements, fictional worlds may be resituated in the productions of human nature, proof of which I have set out to demonstrate in the chapters to follow.

Nevertheless, for some opponents, splitting the difference between nature and nurture, between genes and cultural memes, has never been a satisfactory compromise, in spite of the fact that that relationship goes back to the foundations of evolutionary psychology, as in the writings of Lena Cosmides and John Tooby: "Every time one gene is selected over another, one design for a developmental program is selected over another as well; by virtue of its structure, this developmental program interacts with some aspects of the *environment* rather than others, rendering certain environmental features *causally relevant* to development ... Thus, both genes and the developmentally relevant environment are the product of natural selection (my italics)." There is great care in their work, when closely read, to show how human psychology uses genes, "not as implacable determinists of an inevitable human nature, but as subtle devices designed by ancestral selection to extract experience from the world."[10] Reductionism, along with determinism, is an accusation of philosophical malpractice on the part of the so-called literary Darwinists, but it does not apply to considered evolutionary thinking.

Further objections pertain to just how the human being's adaptive traits were established in relation to ancestral environments. How could the unified traits of human nature have arisen in such diverse settings? And how can we generate explanatory backstories regarding things that cannot be witnessed firsthand? Are these not just fanciful myths to account for how we are today, no longer in terms of tales of the gods, but tales of the ancestors? Answers depend on whether you are looking for justification for alternative explanations through culture and learning, or challenging explanatory models in order to work out the anomalies of a necessary truth. Concerning that necessary truth, as a species we are able to read the intentional states of other minds because we have interbred over eons to form one kind of mind, a oneness that depends on designs transmittable only through our genes, along with prehensile grip, canine teeth, and stereoscopic vision. If that homogeneity seems problematic, consider that we are dealing with huge stretches of time during which there has been a blending and averaging of traits through generations of cross-breeding and mate selection.[11] Then, consider the universals of the human condition which pertain under any and all environmental circumstances: the need to mate, bond, raise children, cooperate for scarce supplies, deal with cheaters, resist outsiders, protect wives, secure paternity, and fight for clans over hundreds of thousands of years. Cruel mortality made the choices, rewarding those with the mental responses which gave them the edge for survival while consigning the losers to early graves.

Evolutionary history is one sustained tragicomedy. Nature – combined with the speed-up process of partner selection whereby culture impinged upon nature – found the means to ameliorative differentiation. The winners in that contest slowly wrote our collective attributes and cognitive advantages in systemic and phylogenetic terms. The imaginative stories we tell are about how well or ill suited those latent and ancestral thought-production capacities are in confronting the exigencies of modern social environments. Tragedy sometimes ensues precisely at the crossroads where mental conflict betrays the rational person, when a stuttering Billy Budd, all love and good will, suddenly resorts to his fists over the indignity of his treatment, or The Husband in *The Yorkshire Tragedy* gives himself over to the murder of his own children to settle his rage against his collapsing social order. (Paradoxically, his emotional proclivities to react so decisively – adaptive in some circumstances – have just eliminated his contribution to the gene pool!) The diversification of ancient environments and the lack of living witnesses are objections in need of consideration, but from the perspective of a necessary truth. After all, here we are, with material brains controlling our conscious and volitional lives, brains which are specialized to do precisely what they do, in relation to all the other species we might have become. That has to matter to the way we do business with the world.

Another counter-approach, represented by Elliott Sober, is to grant the premise that all that is human must correspond to the evolutionary facts which pertain to the human, and with all that is entailed in being a creature made of matter, while denying that evolutionary sciences will ever tell us anything of interest or value about the ways in which we conduct ourselves in our environments.[12] This is a perplexing give and take-away argument which bows to the necessity of evolutionary reasoning in the making of homo sapiens, yet disallows any involvement of the developmental results in the shaping of human culture or behaviour. It is a little like agreeing that the world looks older than the Bible account allows, but that the evidence is invalid because God designed it that way to test our faith! Nevertheless, there is nothing in our natures that can have been selected without demonstrating its superior properties through behavioural efficiency. If there is something there that, by degree if not by category, differentiates us from our branch-off primate ancestors, we will be tempted to explain those degrees in terms of the adaptive distinctions that enabled the extension of traits and behavioural attributes, including our enhanced capacities to plan, desire, reason, choose, learn, adopt, imitate, override, laugh at irony, counterfactually imagine, and play according to complex rules. Sober may seek to confirm a truth in satisfactory terms by denying its consequences for human conduct, but it is only by modifying conduct that those traits received confirmation in the genome. Moreover, such traits constitute a form of learning, and this "knowledge" often frames most critically our reactions to the environmental prompts that attract our attention. I will be concentrating on these elements in the literary studies to follow.

Jonathan Kramnick, in his article "Against Literary Darwinism," advances a more comprehensive objection to the Darwinian rapprochement between science and the humanities, which he sees as entirely "misguided" even in its central arguments: "literary Darwinism fails to make its case."[13] He begins by framing the Darwinian critical enterprise through cognitive perspectives as an assault on humanism, drawing up the "we/them" camps which no thoughtful cognitivist could wish to endorse.[14] No one is denying the many cultures which have incontestably emerged from human intellectual activity. But Kramnick reasons that Darwinian-informed thinking is political in posing a takeover agenda to the study of the arts, thus depriving cultural studies of its rightful hegemony. His principal objection is to a certain aggressive manner in the categorical declarations of some cognitive philosophers, but that alone does not invalidate their claims. I confess, there is a kind of ebullience in the explanatory promise associated with recent investigations into the mysteries of the human brain as the instrument responsible for the production of all human meaning. If these hermeneutic insights gain a critical mass of cogency, probability, and reason, a new synthesis will indeed be in order. The studies to follow are a part of that bid, to be judged by their merits as arguments, not as sophistry on a calumniating mission against humanism and culture. Presumably Copernicus did not engage in his explorations merely to spite the Ptolemaic position, but to follow the evidence where it led. Self-evident, for Kramnick, is that the "humanities believe in an infinitely plastic human nature," and that such plasticity must be preserved at all cost.[15] But the infinite plasticity of human nature is not a self-evident truth, for to possess a nature that is human is to distinguish it from all other natures according to the constraints by which it was designed – a design which, inversely, must matter to the behaviours of the species. The question, rather, is what that species can possibly do that is not in keeping with the conditions, properties, and emanations of its nature. Round and round we may go, but every new science aspires to bring enlightenment to its domain of study, including human nature in its phylogenetic dimensions. To style cognitive- and evolution-based hermeneutics as a hostile takeover of the humanities has little bearing on the legitimacy of its claims. But true, validated claims do have a habit of taking over, eventually. Such are the hazards of learning and progress.

A subsequent tactic on Kramnick's part is to discredit even the most basic axioms of Darwinian criticism insofar as discussions and disagreements on many fronts persist among cognitivists, thereby proving that the current work on cognition remains controversial, thus unproven, thus premature, thus invalid for purposes of literary application. Divide and conquer. What can I say? That there are explanatory differences proposed by members of the academy is hardly tantamount to a proven discontinuity between evolutionary causes and their effectual relevance to modern categories of inherent behaviour. This is a variation of the missing link argument, that until every intermediary phase is not only visibly demonstrated but also agreed to by all observers, evolutionary processes can have no relevance to their living productions. But that is to mistake the give-and-take of scientific process with the incremental establishment of a common goal, which is to fine-tune our understanding of the development and state-by-stage influence of the human genome over time on the ethological development of our species. To be sure, there are those far more qualified to do the scientific fine tuning, but the implications of their research, its relevance and methodological progress, are not beyond an understanding that necessitates conviction. Among the most readable of these rationales is Luigi Luca Cavalli-Sforza's *Genes, Peoples and Languages*, a study in the evolution of our genetic history in which a leading geneticist links the implications of his research to human anthropology.

Kramnick also resorts to the argument concerning the absence of eye-witnesses reporting from ancestral times, an argument which he takes over from Gould and Lewontin. Without direct investigative access to ancient environments and the social organizations of our forebears, sociobiology can only be "an exercise in plausible story telling rather than a science of testable hypotheses," that is, a set of just so stories.[16] But as stated above, we are here, brains, natures, instincts, aptitudes, and all, and we got here by a process. We no longer agonize over the monkey to human development of the human body because the evidence is sufficient without eye-witnesses, but when it comes to the specialization of the brain through those same processes, the Scopes trial is yet to have taken place. What would we have seen had we been there? No one witnessed the "big bang," but there is little doubt among physicists today concerning the continuity in the behaviour of the elements from then until now, thus enabling a meaningful pursuit of their history. The emanations of design are with us now, and they have histories of continuous survival in prehistoric environments, whether or not we have living testimony to each developmental phase. What we have is inference and probability from a growing magazine of genetic, ethnographic, and closely reasoned data. By all means let us fine-tune our backstories in relation to who we are now, but not give up in despair over the speculative gaps.[17] Survival, food, shelter, tool making, caring for children, cooperation, mobility – the number of determinant pressures is finite and generic, and the effects of earth's varied environments will have created through genetic blending a collective inheritance pertaining not only to our tool cultures but to our mental capacities for meeting environmental challenges.[18]

These are not questions of specific environments, or specific social arrangements at specific times, but of the effects of mate selection among contiguous groups over vast tracts of time, a question of genetic transmission and massive statistical components behind the collective human story. That at one time in history we were nothing like ourselves today entails a story according to inherent conditions about how we changed and why those changes matter to our acquired performance capacities, instincts, and mental productions. We have little choice but to try to imagine that history if we are to carry self-knowledge to a new level, and that work can only be done speculatively through estimations based on the principles of adaptation. To worry over the presence of witnesses may seem reasonable on the surface, but dinosaurs did not have the brains required to write about their history, and neither did we during our most critical developmental phases. Yet the fact of our evolution according to material processes, together with surviving indicators, necessitates a history. That 98 percent of all aboriginal North Americans have type O blood, and that they branched off from a population with much lower percentages of that blood type, tells us that either fewer than five of them emigrated to North America, making the chance predominance of the blood type one in thirty-two, or that type O, in having resistance to syphilis, gave that blood type a massive advantage in fitness survival. Either way, in some such terms, there is a history to be imagined, much as there is a history to be imagined for the adaptive fitness of each universal characteristic designed by the genome.[19]

The evolutionary philosopher who wrestled most directly with the problem of ancestral stories without contemporary witnesses is Bernard Williams in *Truth and Truthfulness*. It is a subtle but compelling argument in defence of the genealogy of informing principles before they can be fully written according to the evidence necessitated by history. He argues that the story built by evolutionary thinkers need not be an observed and witnessed set of events over millions of years. Rather, it is an effort to explain our current habits, concepts, values, and institutions by revealing how, in simpler environments, they represented fundamental human concerns and capacities endemic to our natures. He illustrates how this is done in many other disciplines, as when political scientists speculate about the sequence of social pressures and political configurations that gave rise to the organization of states and nations independently from what is known about the careers, councils, and declarations that brought them into existence. To be sure, we can only imagine in generic terms how ancestral humans came to practise communitarian behaviour, extended their information sharing, and elaborated their technological cohesion, but given that all such advancements of necessity correlated to the collective capacities of human thinking, a dialogue opens up between pressures and enablement platforms, problems to be solved and the resources for solving them. A story emerges, of interest to our present literary purposes, only because that evolutionary story exists, whereby we are encouraged to imagine ourselves in our present social environments as animals with capacities and resources stabilized by the genetic congruencies which characterize our species at this moment in our development, and thus constitute the enabling architectures upon which all that we conceive and do is based. In these terms, we can, as Williams states, "get away from the preoccupation with reductionism."[20] Williams also concedes that not all cultural practices must be taken back to evolutionary explanations. There are even genetic mutations that are neutral and thus are not tested, eliminated, or confirmed by fitness. We can also invent to amuse ourselves in biologically indifferent ways. Nevertheless, in his words, "individual members of the species must of course typically, standardly, or in the right proportion have the psychological characteristics that enable humans to have this ethology, to live under culture."[21] That is not an ancestral, but a current fact, which evokes both cultural and genetic explanations.

Interestingly, Kramnick makes an about-face by stating that "literary Darwinism fails to make its case because it does not take the relation between the humanities and sciences seriously enough."[22] This is an intriguing concession that renders his global antipathy rather nugatory. Good Darwinian philosophy can be done after all if properly pursued. In what, then, consists the shortfall? It turns out to be in a specific and already familiar domain, the defining of artistic expression in adaptive terms – the unguarded linking of creative output with genetic causation. Such reductive associations seek to link the forms of art to the themes or tropisms of human nature and thus link art to adaptation, fortifying its usefulness to us as a species even to the point of positing the genes or brain modules devoted to its production. Should this error be proved upon literary and aesthetic evolutionists, revision is in order, but not the invalidation of all evolutionary psychology, as Kramnick intimates. After all, artistic expression, as with all forms of mental output, draws on the enabling platforms of the brain, and those platforms are selectively tilted to purposeful ends, even if they were not specifically designed to write novels. His target is principally a chapter in Brian Boyd's

On the Origin of Stories entitled "Art as Adaptation?" in which Boyd opines that "an evolutionary account of art, far from being reductive or deterministic, can do more than any other to explain art's force and freedom."[23] It all depends on what Boyd means by "an evolutionary account of art," for the point is indeed debatable whether art forms have been around long enough to have conditioned the genome expressly to the ends of specific artistic production. "Art genes" are to be doubted, attractive as they might seem as the scientific causeways whereby the human condition has been improved by the efficiency with which we share knowledge through narrative and other artistic media.

But if art as we practise it is too recent to have built the genes that build the production centres in the brain on which art relies, we still have the many adaptive platforms upon which all art is of necessity based. Something of value is still to be gleaned by assessing the degree to which art is parasitic on other-directed systems and incorporates, to its own expressive ends, the tilted emanations of those neural enablers. As Jim Davies states, regarding his book *Riveted*, "I have discussed the arts mostly as a by-product of other evolutionary adaptations. That is, we like art because it satisfies desires that were evolved for something else."[24] Storytelling, once in place, undoubtedly did become a medium for teaching through example and hence participated in a learning profile. In that regard, narrative is not unrelated to the informative powers of gossip to identify cheaters and cooperators in ways critical to collective order. The signalling and communication of value judgments on the actions of others is basic to any system of reciprocity (delayed gratification in search of future benefits). Arguably, then, the architecture on which gossip is based, in modelling literary imitation, indirectly expresses its evolutionary values. Envisioning art in the context of that backstory requires a disciplined form of analytical imagination. Many platforms were available for recombinant opportunities involving rhythm, voice, enthusiasm, chanting, memory, variations upon themes, information sharing, or cries to invisible forces. The creative imagination, as a form of planning, is free to build all that it can, based on our capacities for recombinant play in spatial, kinetic, and auditory media. In that regard, art genes aside, biology must go before culture, as Boyd himself has said: "there are good reasons to suspect that we may need biology as well as culture to explain art."[25] On that score, if some have gotten ahead of themselves with enthusiasm, we simply have to proceed with greater caution. Boyd wanted to explore the narrative arts as an extension of our inherited natures, but only in conjunction with both biology and culture. In that, he is still on track.

Kramnick would also fault much that follows as adhering to the old-fashioned sociobiology of Edward O. Wilson and the selfish gene era of Richard Dawkins (the late 1970s), hinting at their intolerable perpetuation of the world "red in tooth and claw," while himself misrepresenting the arguments concerning reciprocal altruism by Robert Trivers and others from those same years, which have modified or entirely displaced notions of cruel survival by solo fitness. Emotions may direct us toward aggression, but they also engineer our sense of guilt, of honour, shame, and troubled conscience – all of which are guides to social and cooperative behaviour. Evolutionary psychology is well beyond misguided Victorian brute psychology and deep into the consideration of bioethics – the evolutionary production of an ethical species.

Here again is a brilliant new analytical tool built on an evolutionary substrate which is likewise subject to the nips and bites of those who saddle it with a materialist reductionism never intended by its apologists, including the improbable existence of specifically "altruistic genes."[26] There is no backstory to accompany the putative emergence of a dreamy, generous mind that constantly denied its own benefits. There are no genes for altruism. But there is a most cogent backstory behind the computational logic that gave advantage to those who shared resources and delayed goals in expectation of reciprocal benefits to the advantage of both parties. This computational trick, an invention tantamount to the ethical wheel, is predicated on the complex of design circuitry that enables and favours the calibrations of cooperative exchange. Such exchange is computational because it requires not only the understanding of provisional scenarios as opposed to immediate benefits but also an application of close memory concerning those who are trustworthy and cooperative and those who give less than they get, or fail to repay debts altogether. Modules or genes are conveniently simplified explanatory vehicles, but good science requires more fully considered accounts. So be it.

Reversing the argument, we are a species capable of cooperation to our collective advantages, although we remain tribal in our reckonings.

Nevertheless, cooperation depends on basic provisional understanding and a kind of moral refereeing of all participants in order to guarantee the benefits to the deserving. Such reckoning requires a high order of cognition, and if we have such capacities and other species do not, the only story to account for that difference will be Darwinian. Arguably, learning how to compute these values had far more to do with the development of our reasoning brains than any other environmental pressure. How powerful are the tropisms of our natures, fearful as we are for our loss of honour as a trustworthy reciprocator, subject to guilt for defection, and to pleasure in supplying those in need within a community of trust. These things we perform, not like the eusocial ants or the antisocial bears, but like the humans we are, always calculating we/them strategies based on self-interested cooperation and considered deception when the absence of risk gives assurances of non-detection. I make a full story of this here to demonstrate how far we have come from raw survivalist sociology. Hence, the irrelevance of the sociobiological critique. The objection is based on a half-told story. We are fitted for survival by means other than distrust, combat, and claws, alone or through social organization. We also have caring brains.

Kramnick is correct in observing that the neurology of thinking is far from a finished science, and that the models proposed remain suggestive in their mechanisms, even though they seek to account in some operational manner for the ways our brains generate the categories of meaning so familiar to us as to go unobserved. On one side is the physiology of a thought-producing material brain, while on the other there are the analytical and modal parts of the emergent properties of the mind: the material systems of the brain and the parts of reason and reflection. The challenge in aligning the two remains constant, no matter the intermediary rise and fall of explanatory models. The brain has parts, functions, and areas of specialization, tempting observers to associate qualities of mentation with locations, pathways, thought-related material changes, and collaborations among brain centres. In this, Kramnick performs his own reductionist analysis by collapsing the entire research program bent on observing the brain in action by revisiting the critiques raised against modularity in the 1990s. Again he is quick to define the regionalizing of brain functions with one-on-one production centres: one module for this behaviour, another for that. Studies in cognition and the neurosciences have long since moved on.

But I do not wish to leave modularity maligned in the process. In fact, the subtleties of its provisional modelling of consciousness are instructively challenging insofar as they attempted to solve the blind operations which contribute to the seriatim singularity of conscious awareness. There are implicit impedance and channelling operations responsible for the prioritizing of the neural configurations which find sequential and hierarchical representation as conscious experience. These must be formulated and temporally supercharged in order to surpass, at specific moments, all contenders for conscious reflection. Hence, by what criteria do these operations make their sequential "choices?" The inefficiency of sheer randomness must be discounted, especially when, through introspection, we can identify levels of urgency and contextual obsession. Shouts of "Fire!" will cut into our ruminations with an urgency that is clearly survival related. If such adaptive overrides enjoy an attention priority, we may be well advised to back-reason the deliberations of consciousness on a scale of pertinence and urgency determined by the logistics of survival. Why would consciousness itself not be systemically composed according to adaptive values? Modularity was a dispersed system which, through impedances, sought to account for the diversity that comes to focus in sequential consciousness by imagining such a triage system in neurological terms. Dennett recognized his own model as nothing more than a provisional mechanism corresponding to a phenomenological reality. Such a system also entails thought clusters held in abeyance in sub-energized forms, some of which never rise to conscious recognition. Do our brains not hold at all times a plethora of potential thoughts, among which relevance-oriented triage mechanisms must choose? Is consciousness itself made up of the components most fit to survive the prioritizing operations of a systemic brain?

Modularity's fall from the analytical spotlight in favour of other models does not invalidate a pressing need to explain the operations of consciousness – a subjective phenomenon which provides all we can know of the self and its world. Consciousness may seem an everywhere, but it is narrow, sequential, and largely sub-directed according to design, and that design shapes the epistemological lens through which we experience our entire life. Understanding consciousness may well constitute the leading scientific challenge of the twenty-first century.[27] In this, even literary representations may play a part in the modelling of consciousness. The Renaissance soliloquy sought to replicate in words an equivalent of the thinking mind – a goal which re-emerged in the epistolary novel and came to an experimental fullness in the rambling, expressive, private, thought-paced, promiscuously searching, confessional mind of Molly Bloom in James Joyce's *Ulysses*. Kramnick's modularity assault, in any case, avoids far more recent work in neural network modes, connectionist, or parallel distributed-networks through which the algorithmic sequences of learning in vast networks are more approximately simulated. No one today is looking at localized information storage. That notion was put to rest in the 1920s when Karl Lashley trained rats to run through mazes and then extracted bits of their brains to see where the maze competence was stored, only to discover "that a rat's knowledge could not be localized to any single region."[28] Now scans and imaging, the roles of neurotransmitters, and the localized productions of specific operations such as memory formation have assumed priority. Such dialogic investigation does not render the entire discipline irrelevant or invalid. Kramnick, in fact, does not investigate the shortfalls of the modular brain, but falls back yet again on a rhetorical barrage against Darwinist criticism and its conspiracy to "begin its cleaning out of the stables of the humanities."[29] That polemical fear and recurrent salvo remains at the centre of his entire critique.

All these debates are tied to the ostensible contest between nature and nurture, toing and froing since the Renaissance, even though one of the earliest to recognize the contest in these terms saw them as working hand in hand. The Elizabethan schoolmaster Richard Mulcaster, in his *Positions Concerning the Training Up of Children* (1581), spoke of what was implanted in children by "nature" that required enlargement by "nurture."[30] It is a modest beginning for a passionate debate, the claims of which are falsified the more they approach the extremes. Fallacies of exclusion are tied up with notions of causation, that one system excludes all others, and that if nurture accounts for human behaviour then nature cannot, and vice versa. Such thinking characterized the nature/nurture debate throughout the twentieth century, with the social constructivists in control down to the 1980s. Thus the new balance has been slow in coming, even after the wholesale dismantling of the tabula rasa brain in such studies as Steven Pinker's *The Blank Slate*.

In effect, social and cultural constructivists, behaviouralists, cognitivists, and evolutionary psychologists have busied themselves with moving the finishing lines marking the contributions of ontogenetic and phylogenetic learning and their respective contributions to the building of the brain's architecture, for if the genome does not organize the brain, then culture must. But the moment we begin to look for archetypes, *idées forces*, patterns of consciousness, the social instincts of groups, the urge to personification, and why our emotional lives are structured the way they are, the genome supplies the best explanatory mechanisms; all human brains produce, in basic categories, the same classes of emergent properties. As a data processor, even with its remarkable plasticity and open-ended capacities, the brain has its systemic design which not only circumscribes the ways we process information, but the kinds of realities and social interactions we can pursue. It would stand to reason that the realities we produce are the best achievements in relation to the environment that millennia of evolutionary engineering have been able to achieve. The issue is how much that design imprints its own interpretations on the formation of values and volitions. That principle stands in apparent opposition to the view that "the mind of the child constructs itself in ontology,"[31] which, by definition, reduces the role of evolutionary design to nil.

Merlin Donald's lucid and enlightening study on "the evolution of human consciousness" just quoted takes up the torch on behalf of all cultural constructivists. He makes the best case there is, not only for the powerful influence of culture through the child's exposure to the select vocabularies of customs, beliefs, habits, and values obtained through parents, siblings, friends, and teachers, but moves on to his own conclusion that such acquisitions not only inform the brain but wire the brain, arrange its architecture, and alone teach it how and what to learn. Incoming data not only finds interpretation without a designed brain but instructs the brain on how to build the functional and computational faculties required for further interpretation. In essence, what is learned through directed attention, in the process of acquisition, organizes the brain's capacities to deal with that very data.[32] He provides an account of the famous statue of Condillac, an eighteenth-century thought experiment about how the human person learns and organizes her world through the senses, taking the story of Helen Keller as the clearest example of a person whose principal senses were shut down at eighteen months, but who, in spite of her handicaps, learned how to reason and reflect on the problems of contemporary society by building up her languages of communication through touch alone, by which she came to know the greater world. Step by disciplined step she constructed her brain to know external realities despite her limited access to experience.

The uncomfortable factors which go nearly unmentioned, however, include where Keller gained her intuitive sense of languages and grammar, where she derived her drive and curiosity, her will to know, her pleasures in accomplishment as well as her frustrations in failure. Could Annie Sullivan also teach her how to learn, desire, believe, and feel? Here we slip back and forth between explanatory paradigms. What does the brain know about responding to the world at birth, what did Keller achieve during her first year-and-a-half with her faculties intact, and how much of her future success depended on building the learning architectures of her brain through smell and touch alone? Of all the examples imaginable, Keller's story is a witness to the common intuitions she shared with all other humans on the basis of a common biogenetic heritage; she had to rely more essentially than anyone imaginable on the hardwired endowments of being human. Donald's point stands fast for him that, come what may, "we are our own conscious creations," and that all learning takes place in consciousness; through self-knowledge and plenary awareness we design our own brains.[33] In that blank slate vein, we are confronted with a neutral brain purportedly able to reinvent the operations of consciousness itself on the basis of an infant's first encounters with a mysterious environment. From a different perspective, there is in the approach a misplaced inclination to overrate consciousness as the only faculty through which we respond to our world, downplaying all those supporting faculties without metacapacities for self-knowledge that nevertheless contribute to our meeting with reality from our first moments of life out of the womb. How did those get wired? We are back to the elements of experience that consciousness alone cannot explain: how we feel, remember, perform analogy, respond to rhythm, achieve theories of mind, confabulate "reality" Gestalts, empathize, or comprehend the play mode (not just the rules). To underestimate the power of culture is blindsided, but to assume that culture alone designs our mental architecture is equally so, as Donald himself discovered in his passing acknowledgment of "blueprints" and "innate dynamism."[34]

We may just be negotiating finishing lines and diversifying contributions, but the brain is not a tabula rasa at birth or any time thereafter. We are simply too vulnerable to enter the world devoid of all systemic knowledge and instincts; we emerge looking for comfort and a source of sustenance and, untrained, we communicate a wish in that first postnatal cry. For the constructivists, that is, of course, merely the foundation for another slippery slope, both positions tending to excessive claims. The concern of my studies is not a denial of culture, but a reckoning with select literary challenges best explained through the frame of human universals, for literature in its insistent themes, designs, social preoccupations, and demands on readers often points back to those phylogenetic characteristics which define us as a species. The ensuing studies have been chosen and crafted to bring that perspective into bold relief, fully cognizant that not all literary questions are equally responsive to biogenetic analysis. Each will have to stand on its own merits.

Meanwhile, I am attracted to the kind of meeting of perspectives sought by David S. Wilson in his five graduated models, an analysis which begins with an exclusively genetic approach to human nature (nativist) according to which we are merely stone age creatures in a modern environment, and ends with a culturally constructivist position (empiricist) according to which we enter the world as blank slates to be wired and programmed entirely through postnatal experience, including all that pertains to sex as well as gender.[35] What matters for Wilson are the intermediary relationships between these uncompromising positions in the debate over nature and nurture. For surely, humans are not all genes, even though the genes are responsible for the learning platforms upon which all experience builds. The pinnacle of genetic design is a general and supple intelligence which allows us to play, invent, reason, project, choose, and sense our freedom in doing so, even though adaptive biases and systemic limitations remain in the background, colouring our experiences and conditioning our understanding. In these terms, matters of freedom and necessity will remain under consideration, but for Wilson, we must move beyond the if-one-then-not-the-other positions based on anxieties over determinism and freedom of the will.

On that score, Daniel Dennett, in *Freedom Evolves*, has made a powerful statement about the necessity for an integrated understanding based on a renewed scrutiny of the genetic within culture, and the cultural within genetics. Dennett concentrates on a keener understanding of the determinism pertaining to most lower life forms, but from which humans have evolved toward a remarkable sense of their own lived and experienced freedom as part of their mega-brain endowment. Freedom itself is an emergent property of mental experience which is as real as any mental formulation can be in relation to reason and perceived environments.[36] We cannot wish for states we are not equipped to know and enjoy, but freedom is not one of them, and our cultural values, legal systems, and religious beliefs have much to do with the framing and enjoyment of that mental potential. David C. Geary, in this tradition of careful reflection on the balance between genes and culture, cautions that "from an evolutionary perspective, the folk knowledge and inferential and attributional biases that define primary abilities are not sufficient for academic learning in modern society, but, at the same time, are the foundations from which biologically secondary academic competencies are likely to be built."[37] There are, to be sure, constant alterations in emphasis in trying to get the essential balance just right. Matt Ridley, in *The Agile Gene: How Nature Turns on Nurture*, states in parallel that "It is genes that allow the human mind to learn, to remember, to imitate, to imprint, to absorb culture, and to express instincts. Genes are not puppet masters or blueprints. Nor are they just the carriers of heredity." They respond to environments, construct the body and brain in the womb, "but then they set about dismantling and rebuilding what they have made almost at once – in relation to experiences ... Somehow the adherents of the 'nurture' side of the argument have scared themselves silly at the power and inevitability of genes and missed the greatest lesson of all: the genes are on their side."[38] Giving credence to these foundational properties of the genes, I would be hard pressed to accept Merlin Donald's cultural determinism, which results from the putative wiring of the entire brain through postnatal learning. The effects of both nature and nurture pertain to the genesis and constituencies of imaginative worlds and to the mental frames of readers dealing with the complexity of literary works and the mind's own reductive habits in the production of meaning. But in the following studies my preoccupation is with our genetic intelligence, our genetically proned aptitudes and biases, and the critical moments in literature in which they become the drivers and best explanatory systems behind human behaviour, culturally mediated to be sure, but shaped by systemic hereditary values. If genes design our brains, and they do, then the output of those designs must matter to the way we act and the kinds of stories we tell.

The following enquiries enter no further into this philosophical debate than is necessary to claim that, in some matters, going back to the genome is our best approach to explaining the transmission of innate knowledge. Insofar as we prioritize in our stories what nature prioritizes in our thoughts, my intent has been to state with all the cogency at my command what I think the implications are for the study of literature in redefining its themes through the constitution of the human brain. Restated in the words of David Geary, "the content of many stories and other secondary activities reflects evolutionarily relevant themes," which, at the same time, constitute the "themes" embedded in our mental productions which guide our responses to real and imaginary worlds in identical ways.[39]

I am the first to confess that an integration of biogenetic principles into the critical act may require adjustments to a lifetime of habits concerned with language, conventions, and the artifices of the imagination.[40] Professional investment is difficult to renegotiate for us all, and in direct proportion to the degree of that investment. But to my mind, we find ourselves on a new threshold, for while we have a discipline that is based on principles of structure and language, we also have a discipline that is involved with the representation of the human in all its cognitive and affective variety. That is hardly a new idea, but what constitutes the nature of the human both in characters and readers is under constant reassessment. What it means to cry or fall in love has been hardwired since Palaeolithic times; cultural exfoliations must make do with these universalizing properties and build their institutions around them, because whatever cultures may teach, it is stimuli from the social environment that produce crying, and largely without conscious participation, just as they incite fear, anger, or desire. This is a first order of determinism at the level of the species because biogenetic design aligns limbic readings with environmental changes. If the materiality of the brain holds us at times to ransom through its hedonic productions, must these correlations not be displayed with fidelity in our storytelling? If we cannot prevent ourselves from crying, and we cry when select circumstances are present, we are, in those moments, socially modelled by our emotions arising from an ancient brain system shared with all our mammalian progenitors. To resist this deterministic moment is to resist our membership in communities also defined by emotions. In such a manner, art determines its genres by working through the deterministic platforms of the limbic system by evoking emotional contagion. We cannot ignore how these systems work in the name of autonomous self-fashioning.

Let me introduce some of the topics to come up for further investigation.

Alan Palmer reiterates in his own way the program already stated, but adds the troublesome and longstanding question concerning the ontology of fictional minds. It is a critical first principle, for if fictional minds do not participate in the same psychological values whereby we evaluate real social life and assess other minds, fiction belongs to alien worlds beyond our capacities for evaluation. What are characters if not representations of persons? Palmer stated axiomatically that we should not make any final pronouncements about the minds of literary characters if we remain "in ignorance of the rich, insightful, and exciting, but also bewildering, arcane, and difficult debates on the nature of real minds."[41] That is a verily-I-say-unto-thee pronouncement insofar as it joins the fictive world to the real in bedrock psychological terms and asks how they can possibly be differentiated, apart from their respective mental modes: imaginatively composed reality; perceptually composed reality. This perspective originates in the two modes our minds possess for generating worlds: those which arise from within the mind, and those which are taken in directly through the senses. Rarely do we confuse these modes, for to do so would be to live in a maladaptive state of perpetual hallucination.[42] Yet the modal distinction between the imagined and the perceptual does not extend to the phylogenetic properties of the brain by which human experience is defined. From that single estimative faculty arise the reality principles of both worlds. The ontology of the fictive emerges from the same design features of the brain that determine our unique way of seeing the world: the common ontology of the human, both perceived and imagined.

The implications for literature are huge. Arguably, the validity of the fictive mind is drawn directly from the properties of the real mind; its only substance is found through the mirror held up to the operations of cognition and emotion by which the human brain does business with its world. Thus, for Palmer, to read fictional minds means knowing our own minds first. How do we theorize other minds in reading, and how do authors do so in writing? Palmer's position is that fictive minds, although created through the artifice of authors, differ in no significant ways from those of living subjects because authors, like their readers, of necessity create characters through their own categorical assumptions about personhood largely through introspection. Like God, they can create other minds only in their own image; readers, in interpreting such minds, can only do the same. That has to matter to any study of character. The debate about the nature of fictional minds will come up again and again, and seminally so, because the ontology of fictive minds forms the basis of their interpretation.

Another topic is intentionality. Consider the salvo launched by Lubomir Doležel, that "in the philosophy of action the problem of intentionality is at the center of interest," but that nevertheless "empirical studies of acting, including narratology, have hardly noticed its existence."[43] To speak of narrative plotting, we should think less in terms of structure, function, and circumstance, and more about how minds establish what Daniel Dennett calls intentional stances, insofar as meaningful action is only that which the decided mind sets in motion, once it is screwed up to the volitional point.[44] Plots that are less than pure flights of fancy are shaped by the qualities of volition apt for actualization according to the themes embedded in mental structures. Such is the only order of plotting that ultimately matters, and that takes us back to the psychology of intentionality. An act is that which measures the degree to which agency can alter conditions in the environment according to an imagined and desirable state. That definition is challenged by acts which to others appear neurotic or pathological. Planning is a cognitive business, which Doležel makes axiomatic to the hermeneutics of narrative. Plots are always concerned with the motivations behind actions, but there are plots in which intentionality becomes openly thematic, as when our folk psychological capacities for theorizing other minds fall short and we find ourselves entirely perplexed by the actions of others. We crave knowledge of intentionality, indulge cagey plotters who wish to keep us in the dark for a time, but censure those whose explanatory powers fall short of the actions they have created.

One of the essays to follow takes up the case of The Husband in *The Yorkshire Tragedy*, whose actions have been explained in terms of melancholy, or demon possession, in keeping with sixteenth-century perspectives. But in a theatrical study based on history purporting to reveal the processes of mind whereby a father comes to murder his own children, we are invited to assess his intentional stance by what we know of human nature unconstrained by period science. Hardly would the case matter to us if The Husband were deemed and perceived merely as a theatrical invention. He is, of course, but he is also historical, a factual performer of the deed, acting on intentions generated in a real human brain, through design components much like our own. Our inquiry is not only aesthetic; it is posed in hard causal, psychological, and legal terms, working back through melancholy as depression, or demon possession as compulsive behaviour, to the biogenetic potential of a distraught mind capable of actual infanticide. Both tragedy and justice require a precise understanding of how, in the human mind, acts originate and the extent to which they enter consciousness as volitional states. Nowhere is this more critical than in cases of ambiguously motivated murder. What are we to think of those for whom we feel genuine pity because their actions have momentarily escaped moral invigilation, but whose deeds call for justice unmitigated by extenuation, as in the case of this play? The author may be aiming for an aestheticized closure in the tragic emotions, but the play is about crime and justice, precisely as The Husband, himself, insists when his reason returns at the play's end. It takes us to the core of narrative predicated on and guided by the intentional stance of the protagonist. What analytical perspectives should be barred to humanists in dealing with the tragic conditions that arise in what it is like to have a human brain? If intentionality is the essence of plot, and plot is critical to the unfolding of tragic events, can the genre also become an art form about the errors and destruction that arise with the "accidents" of the volitional brain? In that regard, tragedy is brain science before its time in dwelling on the mental operations which beget distraught and counterintuitive deeds as well as errors in judgment through a limited understanding of options. We may resign ourselves to the mysteries of a cathartic genre, but the tragic sense of life inheres as profoundly in the features of our natures that brought those events to pass. The most troubling and compelling revelations of destiny are those which arise from within. What constitutes a man that he can will so much havoc in the world as to kill his own children? The answer will never be final, but gains are to be had in considering the evolutionary backstory to the intending mind, the precondition to action and narrative order.

By a different optic, we cling to the mysteries of the mind because we need them. Sometimes we want to believe that experience is invisible and immaterial, that mind can be detached from matter, and that what we know of our experience is reliable, even infallible.[45] Well-intentioned overconfidence or self-deception can also be the substance of literary themes and actions, and the psychology of our self-confabulation and incorrigibility also has its place among the traits of the adaptive mind. These are robust convictions and deeply cherished. Hence, one of the less attractive aspects of the cognitive enquiry into the brain is the exposure of these deceptive garrisons of well-being. Studies in cognition tell us just how messy, approximate, and confabulatory we are as a thinking species, albeit largely to our benefit.[46] Too much obsession with dreamy detail, methodical one-item-at-a-time data inventories, and related computational distractions would have been fatal in ancestral environments. We compute far faster than computers through massively redundant systems and data blending, and this for reasons of efficiency in the environments that threatened us, but at a cost to our infallible accuracy.[47] We have certainly made ourselves interesting in that regard; many a plot has been constructed around the kinds of mistakes we are inclined to make on the basis of subjective incorrigibility. Think of Austen's stories in which snap judgments of character require complex social processes and painful personal reflection to put right. Memory, in other contexts, in following its methods of expediency and efficiency, amounts to an uninvited editor of every flash of consciousness – misconstructions which, as false memories, play havoc with our lives and those of others. More conflict follows. Even our first-person accounts are far more skewered by rationalization and interestedness than our third-person accounts.[48] So by these studies, we may find ourselves diminished somewhat as truth-generating beings by the constraints of our cognitive operating systems and the biases of self-orientation. Such factors of the brain/mind construction are more than germane to our assessments of human nature in social settings. Having a self has been enormously adaptive because we can meta-reason about our goals and desires, but it has also made us devious, rhetorically biased, and infinitely political, a problem not experienced by termites and worker bees. Stories are made of this, including Theodore Dreisser's meditation on the eusocialization of the ants.

Then there are the traits that have been ratcheted up to extraordinary levels of acuity by ambient pressures. Other minds are, in fact, our most menacing environment, thereby launching an arms race between our decoding powers of alien intentions and the ability of radical replicators to cheat and deceive us through trickery, lies, and misrepresentations of themselves. This is a generic plot grounded in biogenetic attitudes – the tales of cooperators and tricksters, the former best characterized by gossip (the exchange of mutually beneficial social information to catch cheaters) and the latter by the antics of those who seek to exploit groups by taking more than they contribute.[49]

That incentive to plot finds its origin in the genome, which makes the contest altogether more "archetypal." Much of our intelligence is social and thrives on information exchanges about other people whereby our salient interests are protected from cheaters and defectors. As illustrated above, these invigilating talents serve us well in evaluating literary relations by treating representational societies as real.[50] Literary criticism, in this regard, is an extension of the psychology of gossip. Lisa Zunshine's most recent study explores in some depth the principle that we are readers because we crave such social information and take incessant delight in practising our expertise in interpreting other minds in novelistic and related literary environments. In her words, "we have the hungry theory of mind that needs constant input in the form of observable behaviour indicative of unobservable mental states."[51] Not only are we adept at it but we seek out this kind of information with gusto, in keeping with the urgent obsessions of a socially designed brain. Donald Symons posits rightly that the brain is not a promoter of general goals such as striving for survival, but a community of mechanisms independently contributing to inclusive fitness. Among them are the mechanisms that urge us to remain strategically alert to those who contribute to the communal good and those who hoard resources, especially surreptitiously. In these matters the computational mind, in conjunction with our hedonic prompts, runs overtime, particularly in "detecting cheaters in social exchanges."[52] Moreover, attentive readers make such calibrations systemically as the basis for all forms of moral judgment. Fictive worlds, like games, allow us to practise our survival skills in precisely these terms. Within the safe rules of play and the distancing of fiction we gain experience without personal exposure to the risks of actual worlds. That is what storytelling is for, inter alia. These worlds are fictions representing realities, but the mental performances aroused by those representations are real.

Another of Doležel's axiomatic nuggets is his declaration that all mental faculties "operate between the poles of intentional acting and spontaneous generation."[53] In the perspectives raised thus far, we have spoken of grounded intentionality and systemic mental states. Yet concessions must sometimes be made to subconscious behaviours and instincts. Although we have remarkable, design-constrained powers to reflect, plan, and execute, we also rely on systemic forms of mental activity which rarely if ever feature in conscious awareness. This is not the Freudian unconscious, but the brain that drives the car, stops at red lights, and turns onto Bronson Avenue while the driver daydreams his way into work. This is the brain that rides bicycles, trains to play entire Rachmaninoff concertos, and tells us after a time just how badly pews are designed to accommodate the human posterior. This is business as usual until a crux emerges when the deliberative and the systemic enter into conflict, only partially under the scrutiny of moral reasoning and judgment.[54] Literature tells these kinds of stories with relish, when decisions seem to arise from inscrutable urges or the emotions take control of the mind and turn us into momentary neurotics or berserkers. Hamlet's arm thinks faster than his cortex when he puts his sword through the arras, but somewhere in his brain that action was cogitated, drawing the constitution of that instrument into the moral/volitional program of his life and his destiny, once Polonius is dead. The backstories may not be extenuating, but they are explanatory, and contribute, again, to the tragic sense of life. Laertes carries mental obligations to his father in parallel to Hamlet's. It is the collateral damage of revenge ensuing from a problem in cognition and how the complete brain makes up its mind.

Emotional sensations and expressions are a matter of daily occurrence, and our mastery in interpreting them we may well take for granted. They are an intimate part of experience, somehow distinct from reasoning, generally held to be less trustworthy, arriving as sensations to indulge or escape, to override in the name of reason, even as they supply efficient prompts to immediate action. Characters caught up in ambiguous emotions are hermeneutic challenges to us as well as to themselves. We know that Renaissance lovers can pass through such fields as quickly as they can say the words, "I burn, I freeze." We try to imagine a quality of love through such emotional equivocations. We know, too, how long-term emotions become moods, and finally temperaments, from the sardonic (Thersites) to the melancholy (Jacques). Such critical matters relate, in turn, to the default methods whereby we understand the human economy of limbic values. If literary meaning includes the emotions of characters reacting to the provocations of their imaginary worlds, such interpretations might incorporate an interest in what the emotions are, where they originate, and what they are for. But criticism rarely looks into such matters, tending, rather, to favour the explanatory value of theories of the passions contemporary with the work under study. Whole conferences have been held on the "literary" emotions, often seeking explanations in terms of early modern science as the only valid means for understanding the nature of early modern emotions. But the pressing question is whether early modern homo sapiens can have differed from postmodern homo sapiens in any respect whatsoever concerning the emotions which, again, received their production mechanisms in evolutionary time.

I drag in Aristotle on this score in more than one argument because he so famously argued that tragedy, as a literary genre, was nothing more than the calibration of a story type involving specific plot and character formulae apt to enjoin upon as many spectators as possible at any subsequent time an emotional response combining both pity for someone and fear of something. What did he have in mind in defining the genre in terms of a collective emotional response and how do the emotions formulate that unified community of interpretation? That was Aristotle's gnomic definition of tragedy, the value of which can be determined only by the way in which emotions achieve collective expression according to collective values. Implicit in his definition is that the genre remains the same for all time only because the emotions of the spectators in relation to stories remain the same for all time. Yet there is a study to be made – a more challenging one – to determine how the mirror authors held to their own natures and those around them in order to validate the emergent properties of their characters' psyches were somehow fused with the vocabularies of contemporary analysis. There are negotiations to be made between that which is valid for all time and that which points toward era-specific explanatory frames. This is critical in light of the long, slow process of evolution, which disallows any measurable change in the constitution of the emotional mind within historical time. All this is to say, by way of prelude and apology, that the cognitive sciences offer valuable perspectives on the feeling brain, perspectives which, through hermeneutic extension, reveal much about the emotions of imaginary persons and the feelings they arouse in readers throughout all literary history.

As a foretaste, the circuitry in our brains which imposes felt experience, and thereby creates much of what it is like to be ourselves, is truly ancient, a paleo-mammalian feature, shared with all our warm-blooded, mostly four-legged, mostly furry ancestors. The cortex is a neo-mammalian add-on, which makes its judgments of the environment in entirely different terms. In essence, then, we have two apparatuses for knowing and judging environmental information, and both make compelling evaluations: feelings in the form of pleasant or unpleasant (hedonic) sensations; and thoughts made available to all the computational capacities of reasoning and memory. In daily life, these are so intermingled as to seem mere nuances of a single experiential instrument, until emotional drives enter into direct conflict with reason in relation to more complex and strategically oriented social objectives and cultural duties.[55] All the social narratives that come to mind are in some way concerned with the emotional lives of characters and the crises they face in managing these often conflicting elements of mental production. Authors may task themselves with mastering a descriptive language of inner experience of such complexity as to profile the limbic system itself as a weapon of mass confusion, insofar as literary expression is preoccupied with the interiority of subjects and the felt qualities of experience that constitute the gamut extending from pleasure to torment. Genes build the limbic system, for a start, and feelings continue to impose hedonic evaluations on our experiences, no matter how much we may seek to contain them culturally. We are pleased to anatomize the behavioural repertories of chimpanzees or bonobos, for example, hardly imagining that some distant future observer might visit the favour upon us in terms of the evolutionary tropisms of our natures.[56] But these properties pertain, nevertheless, to the purposes for which the brain was designed.

The chapters on laughter and crying will take up these issues in relation to emotional cruxes in literary situations. What may be ventured here by way of orientation is that all attempts to name and prioritize the emotional states as goal-directed and felt information is an implicit recognition of "all the hidden wisdom and know-how that nature embodied in innate, homeostatic dispositions."[57] These are Damasio's terms for profiling the emotions as systemic centres of ingrained guidance and instruction. Emotions are knowledge and know-how because they make sub-cortical judgments of our circumstances. Any attempt to qualify and name them is simultaneously an attempt to define the salient categories of change and impingement arising in world conditions; kinds of emotions, after all, were designed by the kinds of iterative challenges posed by ancestral environments. Clearly, emotions give powerful instruction in such matters as mating, parenting, protecting the brood, fighting, and mock fighting, and they shade and nuance our readings of situations involving cooperation, guilt, empathy, loss, felt injury, and victimization, which in turn give structure and content to our moral systems. Emotions, moreover, have their profiles of waxing and waning, and in these values determine who we are during successive moments of rage resolving to reason and calm, or the inverse. Aligning these hedonic states with moral values on a binary scale is critical to our conscious social understanding and is central to our storytelling. The origin of binary morals can have arisen only through the polarized understanding of experience imposed by positive and negative hedonic states. It is a complex but necessary backstory, because such binaries do not really exist in nature. It is our brains that conceive of relations as polarities.[58] At this juncture, Damasio speaks of primary emotions – the classic five or six – followed by secondary or social emotions, and finally by the background emotions which, when prolonged into moods, temperaments, or dispositions, constitute part of the destiny of personality. This brings us full circle, in a manner of speaking, to the Renaissance humours: the splenetic, the phlegmatic, the melancholy, and the sanguine. But what about jealousy, guilt, pride, fatigue, edginess, discouragement, or enthusiasm? These are hedonic qualities associated with particular provocations, each one requiring an explanatory narrative. In this way, the emanations of the limbic system urge us to alter our environments to the maximum possible to reduce pain or to enhance pleasure and well-being. This, for characters, can become their driving imperative – to escape the unpleasant emanations of their own minds ceaselessly exacerbated by unrelenting circumstances or sanctified goals. The feeling brain looked after our mammalian ancestors and still seeks to do so through its unabated repertory of sensations. Imaginative literature is made of this.

The relationship between feelings and consciousness is tricky. It would seem that cognition anticipates feelings by conceptualizing the stimulus passed along for hedonic evaluation. The thinking brain first identifies what the feeling brain appraises. Moreover, that which arises in the imagination is also subject to limbic assessment, arguably tricking the more primitive brain into believing in our mind-generated images as vitally as our lived experiences. This is how we get "gut feelings" about plans performed provisionally by the imagination in ways that guide decisions. But this cannot be the entire matter, for when a person strikes out in anger or leaps in fear before the provocation has been fully comprehended or even recognized, we begin to appreciate the autonomy of the emotions. There is benefit in celerity, as in fight or flight circumstances in which the emotions, without consultation, recommend whether to cut and run or stand the ground.[59] Our conscious minds must then struggle to understand what our emotions are telling us. We might call it the Billy Budd syndrome, when thought is too slow to contain the surge. Thus, cognition is not the sole reader of the environment, and not the first. Can the emotions, then, bypass our invigilating systems pertaining to goals, beliefs, and moral judgments, however fleetingly? Is tragic destiny sometimes linked to the breakdown in this community of the mind? It is a compelling question and the jury is still out, but it reveals again just how much the design of the data processors in our heads is involved in the cruxes of culture and experience.

One of the curious characteristics of the emotions is that some may be hidden, while others are written all over our bodies and faces, making them entirely communal in effect and value. It is an obvious point with less obvious implications. The backstory suggests that these giveaway traits are part of our membership in communities since time immemorial, and that pre-verbal communication in groups was vital to our collective success. Exposing our tears, our laughter, and feeling the vestigial raising of the hair on the backs of our necks are all part of an autonomic signalling system that passes important messages by which groups detect not only the moods of individuals but the symphony of oneness and understanding in relation to common threats and pleasures; we rely on laughter and crying as vital signals of the interior states of other minds. Moreover, group members constantly spy on each other to ascertain whether the expected emotional responses are present because they signal felt allegiance to the interests of the group. They are biogenetic lie detectors. Exile may be based on the failure of these hedonic tests insofar as spontaneous, hard-to-fake emotions are treated as basic data in the regulation of social worlds. Suspense arises with our well-wishing for the imperilled, which we feel even in the privacy of readership. Contagious weeping or laughter in the theatre is equally a builder of interpretive communities. Embarrassed we may be in weeping over a soppy romance, but at least we are signalling our empathic concern for others and participating in events pertaining to fertility and the future prospects of the race. Inferences are drawn in the studies to follow from these simple premises about felt learning, literature and the common good, the constitution of genres, the formation of communities, and the ontological substance we grant to fictive worlds. The emotional brain, as it defines us, often impinges on all these matters, and no less so when we are tricked into wishing even our enemies well.[60]

One inevitable consequence of these perspectives is the issue to be taken with the widely espoused doctrine that characters are mere patterns and constructs, and that it would be nothing less than neurotic or infantile to treat them as real persons.[61] We are back to the problem of the ontology of fictive persons with which we began. In relation to this axiom of the New Criticism, David Herman remarks that even those interested in a grammar of narrative have been willing to "shift attention from characters as 'beings' to characters as regularly recurring typifiable 'participants' in the syntagmatic unfolding of the narrated action."[62] But there is an insufficiency in such approaches because characters are constructs of the imagination based on categories of being and the attributes of personhood; their ontological substance includes all that authors and readers supply of what they know about human values and behaviour as lived experience (back to the leading axiom of the first paragraph). Together, they grant to characters all that is required for them to function as plenary persons; we are liberal, after all, in extending such attributions in accordance with our acquired schemas of knowledge. We are wired to read persons according to our folk psychology. Through such ascriptions we find ourselves confronted with agency much like our own, permitting all the calibrations we make in daily life concerning beliefs, desires, and probable goals. Function must therefore be measured by the operations of mind responsible for self-actualization through the limited ranges of agency enjoyed by each character. Dennett applies that principle widely in stating that "when one deals with a system – be it man, machine, or alien creature – by explaining and predicting its behaviour by citing its beliefs and desires, one has what might be called a 'theory of behaviour' for the system."[63] If that becomes our default mental mode for dealing with social realities, our psychology for decoding volition and action in causal terms becomes paramount in fictional worlds as well; we cannot help ourselves. We begin by attributing rationality and purpose to other minds, and by assuming that all actions arise with desire to realize hoped-for ends. Any creature achieving its plenary being through an evolutionary history is a creature guided by the values that enabled all its direct ancestors to survive, thereby lending predictability to its behaviour. Our problem is that not all humans are completely rational, and thus not all ascriptions of a purely logical nature should be assigned to them; plots are the means for finding out how those ascriptions should be modified. There is more to be known about literary agency than plot roles and literary functions, whatever the conditions imposed by conventions.

Consciousness remains a mystery if we attempt to anatomize it in neurological terms. It is at the centre of the Mind-Body problem concerned with just how these blind neurological patterns and chemical transmitters, systemically synchronized, manage to produce the wonderful mental picture of an orange, which I can turn in my mind as well as in my hands, texture at will, and attempt to smell and feel with all the qualia of orangeness. It is a miracle and an explanatory challenge. But it is also yet another design system which has been engineered by selection to produce precisely the kinds of thoughts and images we have – sensations so omnipresent to us as to constitute our universe and to pass for absolute perception and knowledge (linking to arguments above about determinism and plasticity). Paradoxically, our assessment of mind "has implanted powerful, but potentially misleading, assumptions that we can only begin to question once we realize we possess them."[64] Just because we live in our minds does not mean that we have an accurate understanding of what mind is like, particularly if we have no notion of the design perimeters that define its operations. And when we do come to that analysis, we have only metaphors to fall back on. We talk of cross-wiring, depth, layering, who knows how, through which the margins of our computational powers and habits have been extended to include meta-capacities such as an ability to think about thinking, and to reflect provisionally about what we do, about who we are, where our life has taken us to date, all from our first-person point of view. It is a miracle to our understanding and yet it remains a property made possible by the organization of a physical brain, like the picture produced as the emergent property of the electrical circuitry in a television. This is, likewise, the instrumentality of the reading brain, concerned as it is with the ontologies of counterfactual but possible representations.

Consciousness is a terrific parlour trick and has become very good at distilling compound impressions of attended things. A dimension of this reflexive awareness allows us to perceive that we are perceiving, to choose in part what we want to think about, to organize long chains of thought, to juxtapose our values and goals with others, and to associate all these things with the organism to which the self-managing instrument is attached. Fiction is created in the image of these specific capabilities, as when writers call upon the limits of our recursive computational powers to keep track of stories-within-stories and characters talking about other characters talking about still others. We have our limits, and while cultural creations can toy with those margins, playfully overwhelming demands will not teach us to remember more than the average five recursive layers which the gods of evolutionary creation thought germane to our survival and embedded in our mental architecture.[65] All storytelling makes its calibrations in accordance with those latent capacities; in brief, we can compute the limits of literary production in terms of our mental design. Concomitantly, there are the calibrations to be made by authors in relation to the perceived limitations of the mind of the "common" reader. By these same self- or author-directed powers of attention, we can point our thoughts toward emotions not formerly noticed. So versed are we in managing this spotlight of attention and calibrating our interests in relation to goals, modified according to our schematic and embedded knowledge of the world, that we barely perceive the limited terms of our world-making. Imagination itself is a salient mode, and one so potentially all-absorbing that we could become dangerous to ourselves in relation to the unattended world of reality, were it not that we snap between modes with a hierarchy of attention that is virtually absolute. "Fire," or "you there" will call us out of every reverie we can imagine. We are here today because not one of our thousands of ancestors ever became irrevocably distracted (at least not before they managed to reproduce themselves). But the modalities of the brain from dream to hallucination are reminders of our diverse engagements with mental sensations in a real world. This is but a *soupçon* of a vast and growing literature on the nature of consciousness and the self. We do not even know whether the Einstein of brain systems has already spoken through many voices or is yet to pronounce on what consciousness and the self really are. But I predict that best answers will look nothing like a Newtonian law. The brain is so agile that we will never be able to systematize all the qualities of emergent thought according to principles of design. And yet to ignore those questions may constitute the limit of our hermeneutic potential and leave literary scholars far in the rear of human understanding.

Hence, that the final word on consciousness and the emotions has not been written should not encourage us to ignore the progress that has been made. That the mind is tilted to the designed categories of its experience is a principle as old as Kant, and Darwin necessitated that the tilt be framed in relation to our long evolutionary history. That is a beginning. Thinking through the nature of fictive ontologies built up through our default mental processing takes these questions into the world of reading, narrative, and the projections of selfhood and subjectivity upon the characters within those representations. Language is the medium that inaugurates and patterns literary worlds and cannot be overestimated in that role. But readers are interpreters through the cognitive and emotional resources at their disposal that condition and qualify meaning formation, and those resources, in their modal properties, define both the perceiver and the perceived. Characters are actualized Gestalts according to the criteria of the real so that we can participate meaningfully in their worlds and invest our emotions in their destinies. Such are the preconditions for social learning through imaginatively lived experiences.

This activity may be far more basic to our natures than we are at first willing to admit. Children do these things at tender ages without benefit of books or theories. Their brains allow and perhaps even induce them to treat sticks as horses, to make them neigh and buck and chase bad guys, and they tell adults without any hesitation whatsoever that the horses are real in the way of serious play. That we *can* do this is a suggestion in itself that the capacity is phylogenetically beneficial, if only because refusing agency to a volitional entity is a far more dangerous strategy than over-attributing agency to entities that lack it.[66] Thus, we became an anthropomorphizing species. Not surprisingly, then, our ancestors enlivened their worlds with mythological creatures by animating deities, wraiths, trees, animals, or the dead, and then came to love or fear the beings they created and fortified their ontologies by telling stories about them or even putting them in charge. Religion is all about making attributions through personification to the point of losing that vital distinction between fact and fiction. Reading fiction may call on that same gift under different contractual circumstances. These questions I find most fascinating.

That we imprint our categories for seeing the world on the perceptions we receive brings us to the matter of forms, schemas, scripts, and frames, each word expressing the magazines of information that constitute our expectations about the world from physical substances to social rites and customs. Conformity to those frames and scripts permits ready comparison and ready ascriptions; they are shortcuts to mental response. This applies equally to the imagined as well as the perceived. Roger Schank and Robert Abelson made big strides in describing these "scripts" and "frames" which enable the mind to coast through familiar surroundings while alerting it to matters novel or dissonant.[67]

Yet even then, the mind seeks to regularize. Read an odd and elliptical story in an eccentric setting to a group of auditors and ask them to reconstruct it with as much remembered detail as possible. Invariably that which was strange, if mentioned at all, will be normalized to the world of the teller, even without awareness of departure from the original story, because best-guess scenarios have been good enough for ancestral purposes. Scripts pertain to events in familiar sequences. Plans are about achieving goals. We make use of frames to interpret fictional minds and to pad out imaginary worlds.[68] In the words of Alan Palmer, "because fictional beings are necessarily incomplete, frames, scripts, and preference rules are required to supply the defaults that fill the gaps in the story world and provide the presuppositions that enable the reader to reconstruct continually conscious minds from the text."[69] There are, in fact, surprisingly basic ways in which systemic cognition, rarely considered or noticed, contributes vitally to the construction of literary worlds. Just as the visual cortex supplies data based on schematic estimations in order to complete deficient visual representation, similarly data based on probability is added to the limited verbal information on the page in the construction of literary settings.[70] Just as adaptive design fostered visual white lies to repair the gaps in nature, it fostered the systemic construction of equivalent-to-real places to contextualize the stories of equivalent-to-real persons. Arguably, as planning became more vital to our survival, those imagined futures gained in value the more our schemata supplied them with richly provisional contexts or settings. We could "live" in that future the better to assess its usefulness. This default procedure entails an ancestral backstory whereby the planning mind became a maker of fictional worlds. For readers, the mystery of those mind's-eye creations is immediately evoked by their sufficiency and amplitude in light of the paucity of detail in the text. Such matters are easily investigated by comparing the saturated mental pictures to the actual words. Why can we not read without confabulating these settings? Can we instruct our brains to forego them? Are they mimetically essential to our belief in the characters whose lives are at the centre of our attention? "Who" assembles the materials? This innate property of a designed brain is a prime example of the mind's phylogenetic part in the making of essential fiction.

This is where we began – with the confabulating brain, but confabulation based on experienced probabilities concerning the nature of things and reliable general information about the world, all of which expedites the absorption of novelty and the promotion of social norms.[71] Our brains are like that. The fascinating element is the guiding template that determines the density and specificity of the imaginary world built up from such residual information. In sum, scripts and frames are preconditions to our experiences by enabling us to assimilate them.[72] So much of what we take from reading is, thus, not linked to language; linguistic prompts merely set in motion the probes, associations, analogies, memories, schemes, and multimodal images through which we generate meaning and emotionalize experience.[73] The *de dicto* of reading pertains to the words, but the *de re* pertains to the actualized qualities necessary to the genesis of concern and learning.[74] They are very different.

Criticism, in that regard, may be more than the distillation of meanings through words and formal considerations; it may be the unique understanding that comes with communities of readers sharing the same experience through their collective repertory of frames and schemas. That we have the capacity to compose these fictive worlds to felt and educational ends is not only the biogenetic substrate of all literary exegesis but also the link between stories and an adaptive species. Narrative actualization and assimilation is both a feature of a capacious mind and a feature made possible by the adaptive and enabling architectures of our own brains. Literary criticism begins here, with the propositions we can formulate about the prompted worlds of our own generation. We become so transported by this collaborative process in the formation of fictive ontologies, according to Richard Gerrig and David Rapp, that the beginning of interpretation is an ability, not to imagine, but to pass judgment on our ungrounded attributions. An effort must be made not to believe, but to "disbelieve" in the overstuffed roundness of our fictive worlds.[75] Characters, above all, are "constructed according to some sort of theory about how people are,"[76] namely, the frames and scripts that pertain to living persons, for only as such can we come to share with them our concern. But there is, perhaps, a line to be crossed when living vicariously through ascriptions amounts not only to a state of absorption but a state that Melanie C. Green describes as "transportation." At that juncture, we may well wish to call to account our own fabulous and evocative attributions by recalling our evaluative faculties.[77] (I remember distinctly a professor telling our undergraduate class that anyone who gets lost in fictive worlds is not only neurotic in wishing to identify the self with another, but immoral in closing down the active faculties of judgment. Discuss.) Literary theorists may wish to pursue just how we might train for an ideal level of attention between disinterest and self-escaping neurosis. But that we seek absorption in order to accumulate the most lived and vital information we can is one of the magical legacies of our brain design. We can imagine alternate lives and then meta-map ourselves into the picture for purposes of comparison and experience, all in the safety of a winged-back chair.

This introduction could be extended in many directions, but enough is as good as a feast. Each of the following essays sets out to bring a relevant cognitive perspective into the spotlight – memory, the emotions, the self, intentionality, the narrative brain in perceptual and imaginative modes – now extended from theory into practical hermeneutic approaches to specific literary texts. My intention has been to make the explanatory bridges from brain science to textual interpretation as direct and readable as possible, despite the complex logical and explanatory preliminaries that must be kept in view. I am cognizant of the long swatches of cognitive theorizing deemed essential to the success of these arguments, which some readers may find prolix. Yet I am equally aware of the roughshod trampling through the cognitive sciences in trying to keep up momentum. I have sought to make partial amends to the cognitively curious by filling otherwise suspect notes with the names of my sources and collateral readings, along with brief discussions of some of the classic debates among specialists. I am mindful, too, of the complex syntax sometimes necessary to deal with compound matters. At my back I felt the dreaded imputation of reductionism in skimming over the tacit middle elements of the causal chains connecting biogenetic conditions to social and cultural effects. In those moments, syntax can be pushed to the limits of its additive and subordinating capacities in accommodating those critical middle sections. To this I cry *peccavi* with explanation. Moreover, cognitive philosophers have struggled to find a precise vocabulary for expressing the properties of mental activity, one that does not misrepresent their qualities and roles. What constitutes the phenomenology of precepts? What is an idea? What is an emotion in relation to stimuli, sensations, and biogenetic purpose? How does one deal with the reality of things imaginary in origin? What are the best words for expressing the configuration of systemic aptitudes that constitute the human brain? How does the mind perform justice and how are we to imagine the origins of the systemic sensibilities and computational operations involved? What should we call the computational instinct that informs collective social exchange on a costs-to-benefits basis? How can the selfish brain care about the fortunes of strangers? There are terms such as "reciprocal altruism" or thoughts conceived as "emergent properties" of the brain – terms as carefully conceived as language allows in order to avoid errant connotations or phenomenological imprecision. Of their use I too am guilty, yet fully mindful of how even a careful set of terms representing complex operations can cloy like verbal silt. But cognitivists are most assiduous, on the whole, in avoiding jargon, hectoring rhetoric, or strategic obfuscation, and I hope to have upheld that tradition.

In matters of content, I have sought to develop the literary and critical potential offered by recent investigations into the human brain, its computational and emotional properties, its powers as a meaning-making instrument in the reading process, and the way authors intuit and reproduce its signature properties, often to critical and dramatic ends, in the representation of fictive persons. Brains are the vehicles that necessitate this superimposition of science upon literature by dint of their designs and the way they build our primary versions of reality. This, in turn, suggests critical perspectives yet to be explored, through which literary personhood may be more closely aligned with the generic values embedded in human nature.

Only one further point of clarification (or apology) remains. Admittedly, I have relied rather heavily on texts from the early modern period to illustrate this analytical rapprochement between literature and the cognitive sciences. Arguably, those choices may be justified by dint of the intrinsic value of each work for illustrating select properties of the human and for posing instructive challenges to the acumen of its readers. Nevertheless, those specializing in literatures from other periods or languages may feel somewhat excluded. In that regard, they need only consider that choices of any kind entail exclusion, and that choices from a remote period are no more disadvantageous than the choice of an unread author from their own period of specialization. Of far greater importance, whatever the period, is the conscientious and transparent mediation of all features of the work relevant to the critical understanding of the precepts under study. This I have sought to do, in essence leaving no reader inadequately informed, aware that some of my selections are unlikely to be within the common ken even of Renaissance specialists. They will enjoy little advantage regarding *The Dialogue of Solomon and Marcolphus*, *The Yorkshire Tragedy*, or Robert Greene's *Menaphon*, for that matter – which are far from being among the elect of the early modern canon. Yet as epitomes, respectively, of the proverb form, the dramatized criminal mind, and the characteristics of romance, no better choices could be imagined. They are of value both in themselves as forgotten worthies, and as representations of cognitive principles and potential hermeneutic methodologies apt for transfer to works in nearly any period, language, or genre.

I cannot shy away from this double focus in the chapters to follow; literary examples are essential and the Renaissance period is rich in texts in which the exemplification of human nature in challenging social environments is deftly revealed and problematized. But I am also keen to expose the hermeneutic usefulness of cognitive and evolutionary perspectives in the study of literature in far more general ways, and to that end I have implicit confidence in the analogical powers of my readers to convey these critical memes into their own periods of specialization and hermeneutic vocabularies.

CHAPTER ONE

On the Obsessions of Selfhood

Doctor Faustus and the Dramatization of Consciousness

There is nothing like the final soliloquy of Marlowe's *Tragedy of Doctor Faustus* to focus the attention of the literary interpreter. Protagonists have abandoned their lives on stage before, and have addressed the audience as they did so, thereby representing a consciousness in its final moments. The *memento mori* is, by definition, about the transition from awareness to oblivion or some form of trans-biological consciousness, and the quality of death will, in some sense, be relayed through a mind-scape characterized by hope, resignation, or despair.

But Marlowe takes the dramatized moment to new thematic heights. Faustus the apostate lapses back into the theological frame of the Christian world order as the only cultural system through which he can imagine consciousness after death – a troubling acquiescence after a life of denial that not even Mephistopheles could maintain. Thus, in his final minutes of life, he attempts to negotiate the security of his soul through an act of faith, but fails to complete the transaction before the critical moment passes. That is suspense in a new key, emanating from a free mind which accepts its damnation before the fact. Even during this crux, his mind wanders to collateral questions: why did he even have a soul (a spiritual identity), unlike other creatures; why could he not simply implode into oblivion; or why had he ever been born? The Faustian mind has been foregrounded throughout the play, but nowhere so dramatically, as the protagonist becomes all thought, wrestling with unsettling speculations arising from the core of his subjectivity (a word I have never fully endorsed, perhaps because it carries so much critical baggage), or better, from the centre of a self which has been informed and shaped by the conflicting values of the Renaissance world.[1]

So expressed, perhaps no reader would object to calling *Doctor Faustus* a play about the crisis of the self, now confronted by a cumulative life record that spells damnation according to terms he must struggle to comprehend as emotional convictions. From the outset, the protagonist has been self-aware and motivated by mind states calling for such labels as ambition, curiosity, daring, boredom, disillusionment, fear, and despair. In saying only this much about a character imbued with all the appurtenances of a full interiority, we have already granted to him much that is implicit in being a person.[2] He has a thinking mind unique to itself – yet potentiated by a brain common to the species – partially known to him through introspection, promoting an "I" stance based on a substrate of self-awareness; these would appear to be reasonable beliefs and conditions for conducting a characterologically centred reading of the play. Moreover, if Faustus is an Everyman figure, that Everyman theme requires his participation in the mental activities that are common to the species and provide the common denominators to that which is universal in human experience.

Yet what can we hope to gain by submitting the concept of selfhood to further philosophical investigation, including from a post-Darwinian genetic perspective? Perhaps the inquest should not be undertaken. Perhaps our intuitions about subjective being, whatever those may be, will prove fully adequate, obviating the need for an approach to the self through evolutionary psychology. Basing the formation of identity on a biogenetic substrate can only lead to a critical *mise-en-abîme*, in any case. The widely favoured approach to the dialogic self, constructed entirely through life experiences, may suffice. Why re-evoke the insoluble nature vs nurture debate through this new philosophical frame of reference? Even the compromise artists in this field such as Richard Dawkins have struggled to maintain a balance satisfactory to both sets of claims: "By dictating the way survival machines and their nervous systems are built, genes exert ultimate power over behavior," yet Dawkins qualifies immediately that "the moment-to-moment decisions about what to do next are taken by the nervous system … brains are the executives."[3] For Michelle Rosaldo, culture is far less prescriptive than has been thought in shaping the self, but not to be excluded: it is "a matter less of artifacts and propositions, rules, schematic programs, or beliefs than of associative chains and images that suggest what can be reasonably linked up with what: we come to know [the self] through collective stories that suggest the nature of coherence, probability, and sense within the actor's world."[4] On the nature/nurture front the teeter-totter is still in motion and likely to remain so.

But however the balance is to be settled, the contributing parts remain essential. The self that matters is both that which the genes have designed and that which spins stories about itself with full confidence in their seamless continuity and integrity. We are indeed unique in our own stories, yet we are one in the design of the biological instrument that interprets the world and unites us in a phylogenetic praxis through which the self phenomenon becomes a marker of our species. The story of Doctor Faustus is likewise unique among the sagas of the psyche found in the dramatic record, yet the "coherence, probability, and sense" by which we recognize Faustus's obsessions as falling within the range of human cognitive production invite further analysis. Thinking through the biogenetic conditions of the centralizing self – in the present case as they pertain to one of the most interiorized representations of the human to cross the boards – becomes a thought experiment worth attentive exploration.

This, in effect, amounts to a triune question, for it pertains, first, to the default understanding we have of the self through the common parlance of "folk psychology"; second, to the increasingly complex debate among cognitive philosophers concerning the nature and production of the self as an emergent property of the brain; and third, to the degree to which fictional characters can be said to function on the basis of any such considerations whatsoever. This third issue calls for initial consideration because if characters cannot have selves of a kind that pertains to philosophical analysis, then all subsequent reflections on the subject are in vain (this includes all psychological criticism and all literary-critical considerations of the self). But to admit the self as a literary topic is to invite the best reality checks there are concerning the nature of selfhood. To propose literary selves is to subsume all the conditions and restraints pertaining to the neurological equipment by which the self experience is defined – but that in due course.

Suddenly, we find ourselves in a preamble to a study of Faustus which is, in fact, one of the knurliest of problems in mental processing – the ontological status of imaginative persons (indeed, anything produced by the imagination). Thus the starting gate for a study of the self must be set back just a little farther. In pretending characters are real, how much do we pretend? (The answer is far less than we think.) The literary criticism of recent decades, in general, would have us keep our distance, but from a cognitive point of view, this position may require some serious revision. We have to take characters seriously, in some experiential sense, in order to care about them at all. As a tentative axiom, fictive minds do business in fictive worlds, but in terms we recognize as all too characteristically human. If they did not, we would have little interest in them, for perhaps the most important form of verisimilitude is that which pertains to characters who conduct themselves within the perimeters of the psychologically possible, even in the most eccentric of ways or in the most fanciful of contexts. What they say, will, and do must make sense in relation to our lifelong training in the reading of other minds – and who would admit of deficiencies in that most highly prized skill? Arguably, such evaluations are conducted automatically whenever persons are represented, whether in life or in fiction.

Alvin Goldman calls the procedure "mindreading," which begins with what we systemically attribute to other minds by way of attributes, beliefs, and motivational latitude, and concludes with the way in which we represent those minds to ourselves.[5] We do our best to identify the intentional states and feelings of other persons as a first order of survival activity, and this we perform on the basis of deeply generic psychobiological mechanisms. "Paul Ekman created a furor in anthropology by showing that isolated New Guinean highlanders could recognize the facial expressions in photographs of Berkeley students."[6] Emotions, for a start, are not relative to cultures; they are phylogenetic to the brain of the species. People everywhere are design-equipped to deal with emotional communication and to presume knowledge of the mind states of others through just such appraisals. It may prove that what we find compelling in the Faustian experience arises with the degree to which we seek access to his mental experience through the values of our default psychological understanding of other minds. We want to know what it is like to be Faustus. To that end, we must allow that fictive characters reflect on their own feelings, suffer from defeat, feel pain from their wounds, and so much more that pertains to human experience, because they are designed by human artists who, with similar brains, spontaneously recognize the universalities of our nature and employ their skills to represent them in social environments. By extension, characters are represented as having all the components of the human brain by which those mental emanations are generated. Stated otherwise, if we are concerned with the "Everyman" elements of the human experience, can we avoid attributing to literary characters all the material mechanisms whereby those universals might be instantiated in explanatory terms? More troubling, for some, we may have our only access to the nature that we call human through such analysis.

By now, no doubt, some readers may find themselves formulating objections: that behaviour and social understanding are not determined by genes; that authors are not scientists and characters are not case studies; or that characters are not alive and hence have no brains or psyches as such. They are agents. They function in plots. Their imitation of human nature pertains only to what they do, and not to what they might do if they were living flesh and blood. But no one really believes that once they begin to describe the algorithms of learning through experience. Can we take away the self and the soul of a Dr Faustus – or the presentiments of a brain disposed to produce them – given his obsessions with both the actualization of a protean self in life and the destiny of his soul after death? We have hit an impasse. We are not certain that characters can and should be treated like persons, that we should be concerned with their psyches, or that there are means for knowing other minds. The argument to follow is that we do anyway, and for just cause. It is bedrock to our natures to project volition even upon the winds that yearn and selfhood upon every creature demonstrably capable of self-awareness (as well as many that are not, but that is another matter) by reason of the adaptive traits of consciousness.

In an effort to quicken our experiential interest in literary characters as representations of the human, we clearly invest something of ourselves emotionally and analytically on entering their worlds. What that "something" is pertains to the arresting question of the ontology of fictional referents.[7] There are two ways to approach the problem. The first pertains to what is represented in characters as real, and the second pertains to readers and what they must believe about characters in order to gain full access to their intellectual and emotional lives. It would seem excessive to argue that Faustus, as a fictive character, constitutes a complete human being. And yet, how do we define the partially human or prevent ourselves from treating him as a plenary person? In nature there are no incomplete humans; the character package comes not only with arms and eyes and senses, and a cerebral capacity to interpret those sensations, but with memories and the ability to recall them, not to mention felt and interpreted emotions entailing a brain with a generic design potential for eliciting all the emergent mental states common to the species. There is nothing in the list that does not also apply to literary characters because, in processing them as persons, all of these features pertain, ipso facto.

The case of Dr Faustus puts the question to us directly because it matters that his motivations are perceived as self-directed, that his life events are mentally experienced, and that the meaning of his life is determined by states of thought. To fear death and punishment, he must have consciousness, and thus, we reason, he has a mind imbued with a sense of time, of former commitments, and moral sensitivities. Being conscious in agreeable ways is all that matters to him – or to any member of the species.[8] That is the essence of his life's choices; they are strategized in those terms. This premise entails that his personhood be constituted of mental experiences characteristic of biogenetic design, because what can the thoughts of a person mean that are not generated according to the design constraints that pertain to our inherited natures? These are matters of representation, to be sure, and we can hardly deny the writer's craft in creating such illusions, any more than we can deny the brain's design in creating the illusions we are pleased to call reality. Both instruments intend for us to experience seamless Gestalts: equivalent-to-reality representations, which include all the properties necessary for the ontological actualization of the entities named, including the human. That is the point, namely, that our minds have their ways of looking at representations of persons in accordance with what matters to us as an interpreting species through the architecture of our own brains. And what matters to us about others are their powers of agency, will, and volition in relation to self-benefiting goals; we can interpret characters as persons in no other terms. The point is unexpectedly pervasive, for it would seem that personhood forms a grammar that makes experience itself possible, because all human qualities of life are formulated in terms of individuals among their own kind.[9] Therefore, the deconstructive challenge to that Gestalt-formation posed by the legacy of new criticism must, of necessity, be at cross purposes with the systemic intentions of those who create characters. We have no qualitatively independent psychological frames for dealing with imaginative representations of minds. Reading is simply an extension of our lifelong interaction with other people through social interpretations predicated on a working understanding of the nature of other selves as ontological entities.[10]

Stating the problem of literary personhood differently, the question is how many of the features of the phenomenal brain we are willing to include in a cognitive approach to imaginary minds. When Freud came along with the unconscious and the superego, we did not hesitate to attribute these elements to characters, or to credit their stories with modern insights far beyond the theoretical comprehension of their authors. Inversely, no matter how great our interest in early period science, we are hardly willing to limit our interpretation of these characters to the theories of mind and brain current only during the ages in which those characters were conceived (although we often pretend to do so in new historicist ways). Humoral medicine presents a fascinating explanatory system for understanding how Renaissance thinkers interpreted human nature, and authors may even have designed characters according to those principles. Nevertheless – and these are troubling words for some – insofar as these same characters reflect the universal traits of the species, the origins of which can be traced only to the common genetic heritage that makes us all human, they reflect the emergent properties of biologically plausible brains in terms not confined to period science, thereby necessitating a universal scientific vocabulary for interpreting human behaviour. There is no way out, really. Either characters reflect what brains typically allow members of the species to do as persons throughout all the periods of recorded history, or we enter the realms of critically arbitrated quasi-personhoods by criteria I leave the reader to imagine: that medieval characters have medieval brains, etc. *Vive la différence*, except that their brains are structurally identical to ours.

The point is so important because much of what we understand the self of characters to be is what we categorically attribute to them psychologically as persons. Arguably, we have always granted working souls, selves, emotions, planning, memory, desire, and volition to literary characters. Psychology has long been a part of literary studies. The only question is what further information we are willing to tolerate in assessing the emergent properties of mind, and whether we should think about characters in relation to the baseline traits of the species, of which they are inevitably tacit representatives. Any of a number of scholars might be called to bear witness. Leslie Brothers reports that, "because of the way our brains work, our perception of 'person' is automatic. It is an obligatory part of our experience of others – and ourselves," and the critical part of this process is endowing such persons "with mental life."[11] So is it really only the naive readers who skip the hermeneutic hoops by entertaining such characters as ontological entities in all but name, aligning their own minds, in the present case, with the mind of a man terrified of death and lingering consciousness? If so, how naive are they as readers?

This brings us back to the all or nothing argument, now from the point of view of the reader. For as Peter Unger has insisted, "a subject is in no way a matter of degree … the subject himself either exists in full or else he fails to exist completely."[12] As a person, Faustus too must have emotions and desires, a self, an identity, a body, and a temporal space for his unfolding biography; he must express emotions according to meaningful contexts, seek to advance his status against conflicting volitions or obstructive conditions, and above all, think about who he is in the world and suffer in his mind the consequences of his decisions; without these abilities he is not a person – the argument is circular. The attributes of characters must chime with the attributes of persons, including ourselves, and they must remain within the potential of the biogenetic design of the species or not be representatives of that species. Can there be an alternative?

Opposition, nevertheless, has been stout and deserves further articulation. Again, witnesses abound, and one must be chosen to speak for many. Stephen Orgel pointed out that "characters … are not people, they are elements of a linguistic structure, lines in a drama, and more basically words on a page. This argument had, for some graduate students at Harvard in the 1950s, a wonderfully liberating effect."[13] His point is well taken in a historical sense, for it is good to remind students of the medium behind the literary message: the words, the patterns, the simulations, the ambiguities, the creative input, and what plays say in contradistinction to the characters within them. Moreover, the new approach gave professors a lot more to do in an analytical way, thereby launching another dimension of the "new" criticism. To be sure, characters are crafted by other minds than their own, language has its own confining and conditioning properties, and plots do assume iterative shapes to which characters, in ostensibly following their own minds, make their contributions.

But if students found this perspective liberating, what was it liberating them from? While calling attention to the many formal conventions of literature, this position drew back from the fundamental convention of simulated reality, pronouncing readers uncritical if not duped or insecure who allowed themselves to suppose that authors aimed at representational Gestalts based on the inner selves of their characters. The thoughts, motivations, and agency of literary humanoids had seduced them into critically pointless assumptions about literary experience, identification, and uncritically vicarious thrills. Readers should seek to learn from these characters only in terms of the dynamics of the social situations in which they function, but which are by no means generated by intentionalized minds based on the emergent properties of consciousness. That characters can be interpreted, according to Orgel, "through notions of psychology, of stimuli, acculturation, development, [or] childhood trauma is defeated at the outset."[14] We know what he is saying. But authors who create interiorized characters need not be setting up as psychoanalysts; they simply have their eyes and ears about them as well as access to their own minds as models upon which to base imaginary minds, and to ground them, no matter how eccentric, in all that pertains to the generic and phylogenetic properties of the human. Let it be confessed that such studies as these become supererogatory in the cases of characters who are all about manners and social puppetry. We attribute to them the same operational bases in personhood, while laughing at the degree to which their interiority is compromised by their superficiality or social pantomiming. But in the instances of characters whose existential preoccupations epitomize the common elements emblematic of the human condition, arguably the construction of mind itself becomes the "meaning" of the dramatized events.

In life as in reading we make it our business to read the signs, gestures, and words of others as emanating from transparent or hidden intentional stances; to deal with other minds, we must rely on the theories of mind which our own brains are biogenetically designed to produce (our so-called "folk psychology"). By these means we come to the reading of other minds, particularly as those states impinge on our own well-being and security. It is the very essence of social consciousness – one of the most honed of human specializations. As we are compelled by instinct to do in life, we are compelled to do in reading fiction. It is the way *our* brains address the world, and the precondition is that we bestow an adequate if not plenary personhood upon characters so that we can put our minds to analytical work in the same ways that pertain to real social environments in order to interpret and learn from them.[15] Such folk psychology is binding on our cognitive processes; we have no other ways of computing the social. Or, in the words of Rom Harré, personhood is part of a conceptual system in which "one can present oneself as agent, exercising one's personal powers and be fully responsible for what one has done."[16] The liberating effect of the Empsonian "eureka" – that characters are not living persons but merely functional entities within plots – has been good for pedagogy, as stated above, but I cannot imagine a single reader actually putting it into practice while reading social fiction. In the words of Blakey Vermeule, "Now, after decades immersed in this sin [the effects of new criticism], literary theory has scoffed the question of why we care about literary characters into irrelevance and spun the question of how we read fiction into a topic of nearly theological complexity."[17] Her argument centres on brains and gossip and the way we are deeply programmed to seek social information from characters by treating them on an ontological par with all living bearers of such information.

That Orgel chose the theatre as the medium for making his point about character is paradoxical, for actors are a de facto reminder that the characters they play are not performing themselves. They are the words on a page brought to life in the body and brain of a borrowed person. Plenary human agents employ their living faculties as the means for "bringing the words to life" (Orgel). The actor places at the disposition of the character his own capacities for thinking, feeling, and advancing the self in a social environment – all of them attributes which spectators shift with ease to the characters. Even Orgel found himself describing the theatrical mission in terms of "adapting the text to the demands of psychology," on the assumption that actors know what to do with the words by way of actualizing the personhood of the representans. Why would actors want to do that if characters, by rights, should be seen "simply as part of the text"? That actors cannot stop themselves establishes the emotionality that consolidates entire audiences into communities of common feeling.[18] We can hardly imagine the actor in the role of Faustus not giving his full human resources to the interpretation of a mind in the throes of inexpressible anxiety. An actor must seek to embody the plenary selfhood of the play's protagonist as the epitome of his art. Theatre is perhaps not the best medium through which to frame characters as mere constructs and functions.

To equip characters with selves as a categorical part of the personhood we attribute to them by default habits now points back to a truly challenging mission: to know what it is for real persons to function according to a subjective point of view. Folk conceptualizing will place the notion beyond question, but the working parts of the self are not easily anatomized by casual introspection. The evolutionary and neurological backstory of this emergent component of consciousness – the presentiment of having a self – seems remote from the matter in hand, and yet it offers many vital points of clarification. That backstory, in brief, builds on principles of adaptation, fitness, and genetic design; it involves the environmental pressures, as best we can retro-construct them, which destined our ancestors to host an expanding brain that organized itself increasingly around a new dimension of reflexive awareness that included notions of identity and deliberative agency. Language reflects this state of mind through the global employment of the various forms of the pronoun "I," which is used to express the implicit agency invested in this first person perspective, even if it is but an effective illusion for enhancing our operational focus.[19] Self is a mental vantage point from which the world is perceived and a centre of action which generates "person-oriented discourses."[20] As a universal, the self is a reductionist concept, a mental illusion composed of its discrete patterns of awareness held together by relational criteria, yet it is also a kind of bedrock mental stance through which we engage with all aspects of the environment, including our own bodies.[21] This wraith of thought is precious to us because it organizes our ambitions, ratchets up our sense of honour, and holds our identity both socially and metaphysically.[22] Moreover, it is that property of consciousness that we are least inclined to relinquish when the body reaches the end of its biological life. These are mere hints, but insofar as *Doctor Faustus* is a play about the self damned by the conditions within which it seeks completion and enjoyment, it is also a play about how the generic mind works to advance its interests through the production of a personal point of view, and thus a cognitive investigation into the hallmarks of being. That is the Doctor Faustus we want to reach: the man who arrives at his tragic plight through the systemic mental engagements that construct the Faustian self.

The foregoing on the nature of the self may suffice for readers eager to move on, but for the curious, here is the heart of the matter about the origins of the phylogenetic singularity of the phenomenon. This emergent property of the mind has several operative dimensions, all of them the products of the mind states fostered by problem solving in relation to the environment. A self allows the organism to be aware of its world from a conscious, unified point of view, thereby giving an interested grounding to all forms of data and information.[23] The self, illusory as it may be, has a notion of its identity, beginning with its attachment to a single body, and of occupying a place within its physical and social surroundings.[24] Through the aid of memory, the self becomes the focal point of a narrative which incorporates the sense of events in time that constitute an autobiographical story, with its own confabulated sense of consistency and continuity – a Faustian preoccupation to be sure.[25]

Faustus, in a sense, completes this task within the play, for the action begins with a post-adolescent scrutinizing of options in fashioning a unique and efficacious self in the humanist world. Faustus must establish control over what he would do, study, enjoy, desire and become – his choices daringly calibrated according to the novel options of the Renaissance world order, including necromancy. There is the self of the *Bildungsroman*. The play overall is an autobiography – the plot is tantamount to the shaping of a life – a life of special interest because it represents a lifelong crisis in the design-compulsions instructing the management of experienced selfhood. The action of the play reveals the confrontation between inner instruction and environmental encounters with their varying levels of impingement. In the disciplinary debates of the opening scenes we see a constructed, cultural, and moral self, which is also dialogically produced through contact with other minds including, ultimately, those represented as speaking to him from within his own brain – including Mephistopheles.[26] The self, in seeking to achieve complete executive control over the deliberative and volitional faculties, then encounters fear of punishment when the "robust self" meets threats of retribution followed by agonizing intellectual and emotional vacillation between secular ambition and regret.[27] Faustus's soliloquies are psycho-biographic in effect and ideographic in substance, making the entire play a study in the narrative of the self and its struggle for affirmation in conflicting worlds.[28] Life stories not only serve to unify the many mental frames which contribute to selfhood, but become habituated and iterative performances of rationalization and diminished suffering.[29] The self, meanwhile, gives definition to the nature of experience, of what it feels like to exist, to seek pleasure, and to shun the painful and unpleasant; these matters, too, are expressed through autobiographical narratives.[30] The self, moreover, seeks an alignment with the emotions in protecting the body and its interests. It is the vantage point through which the mind expresses honour, pride, ambition, humility, love, friendship, or hatred – all of them constituting mental-stance strategies for looking after the interests of a self-actualizing organism.[31] The design-structured self, in relation to literary characters, is a critical opportunity.

Through the play's concern with salvation and damnation, Doctor Faustus becomes a variation on the late medieval tradition of the Christian Everyman. This typology asserts, as a human universal, that all must die and confront a binary afterlife, anticipated in this life by the terms of the Christian order whereby the soul determines its post mortem condition according to an ethical grid.[32] That formulaic mission for the achievement of eternal consciousness becomes the one common denominator of the species and, itself, a theory about the life of the mind, insofar as the Everyman archetype entails the reduction of all significant experience to the life journey of the moralized soul – the Christianity of reward and punishment. Selfhood, thereby, becomes not only a universal property brought to attention by the values of a self-actualizing protagonist, but a prerequisite to the existentially Christian, insofar as the soul is the spiritualized and transcendental extension of the self.

The relationship between Christian universals and biogenetic universals now becomes a mental tease, because the representation of Everyman, the generic Christian protagonist, involves all the requisite features of personhood to find the path to salvation. I have been arguing for a protagonist of interest to us because we accord to him plenary personhood as the foundation for his actions and beliefs. Paradoxically, this same protagonist, through a life of conscious reflection, reconfirms the universality of the medieval tradition. It is one of the most brilliant and troubling aspects of the play. It is as though Faustus rediscovers the Everyman complex as the psychological tyranny of the natural brain.[33] Selfhood exists in its own economy of guilt and frustrated deliverance projected upon the external world by one of the cognitively most self-made and auto-determined protagonists on record.

Following up that prospect may point in the direction of one of the most pervasive binaries to preoccupy human consciousness, the nature of good and evil. Arguably, Faustus was in full flight from the menace of evil as a deterrent to the secular will by befriending its representative and adopting its powers, yet he must endure a lifelong anxiety over the option for good, inflicted on him by inner voices. A long debate might follow concerning the origins of binary thinking and the absolute hold it has taken on social and ethical considerations. Faustus does not live as bare, unaccommodated man in a pre-cultural world. He is a prodigal son of the Christian West. Yet, that theologically constructed perspective coexists with and emanates from the hedonic binaries of an ancient midbrain, binaries that generate a graduated set of feelings tending to pleasure and pain through which we evaluate the changes in our environment. As conscious creatures, we can track benefits and depravations, fairness and unfairness, daylight and darkness, delight and suffering, rewards and punishments, in subsequent times to be abstracted and socialized into moral categories. Arguably, such matters take root in the sensations biogenetically designed to generate approach and avoidance as part of our success as replicators.[34] Binaries are an emergent property of consciousness and a human way of seeing the world. They are also generic to the organization of medieval values, driving Everyman along the road to salvation. These values internalized are both cognitive and cultural and almost beyond categorization, yet they are at the forefront of the final Manichean and Faustian experience.

Among the captivating features of this play is the devilish pact, seen as the restructuring of the choices that predestine the life profile of the protagonist. That bargain is emblematic of the deals we strike with destiny in the form of incremental commitments that delimit future choices. The signing with one's own blood is a starkly emblematic representation of the end of choice through a demonstrative act of volition. The boldness of the Faustian bargain with time elicits a particular *frisson* not only because it sets a biological limit on the duration within which he must achieve all the quests of an experientially ambitious self, but also because it openly and baldly negotiates away all possibility of the extension of identity and consciousness beyond the conclusion of the pact. As a legal instrument, it epitomizes an irrevocable belief, made early in life, concerning the most insistent of intimations: that the life narrative, moved into the realm of pure thought, may or may not be concluded at the moment of death. The young Faustus, on that score, decides to hedge no bets, because his absolute control over the natural world entails an absolute commitment to the terms of his empowerment; that after death there is only hell, or that all post-mortal conditions are but projections of fantasy and fear. Little did he imagine the future torment that would seize upon his conscious mind when self-generated binary evaluations reappeared as imperative conditions. In this, the experience of the self is everything. The pact, for these reasons, becomes the central *donnée* of the play.

Theoretically, the pact could not touch the freedom of the soul, yet it remains the correlative of a bounded consciousness; it is internalized as a script that monitors and evaluates the facts and events critical to the organization of life, whether in a stultifying or liberating fashion. The crisis that ensues arises from over-commitment.[35] Faustus had settled on a course with such determination that he locked out the introspective dialogue that had constituted his scholarly quest.[36] The challenge of the play is to bring readers and viewers to grapple with the mental correlatives implicit in that contract as a metaphor for a quality of self-definition and experience, as a "commitment script," in Tomkin's words, whereby matters indeterminate can be swept away in favour of a narrowed purpose, a quality of self-homeostasis achieved by diminishing the self as a "continuum of relationships" which must constantly negotiate who is close and who is held at bay.[37] What does it mean to live according to terms that constrain the free operations of the soul by eliminating crucial areas of reflection – thoughts that can never be expressed or acted on? The pact is profoundly indicative of the conditions within which one knows the self, a centring of interest and an increasing emotional duress couched in ominous expectations that can ultimately render toxic the enjoyment of one's own mind.

Time, initially, had been a calculated window of opportunity – the time at the disposal of any one subject to bring to the self those things that it can imagine possible and to enjoy them as achievements. Selves have desires, imagination, conditional agency, strategies for implementation, and expectations of the experiences that accompany the realization of planned exploits by which influence and control in the world are measured. But there is the other side of the formula measured out by social disapprobation, the limited tolerances of competitors, the circumscribed reach even of the necromancer, the invasion of binary thinking that reduces all stratagems of the self to the discipline of a moral order, or the need for cooperation and community in order to reach personal goals.[38] Magic, at whatever price, was the skill set that gave Faustus tools, a means for deceiving others, ultimate mobility, powers of self-metamorphosis, control over nature, and a modus for exploring intellectual puzzles. It was a moment of cognitive epiphany – the moment in which genius seizes upon the field of enquiry that will orient all future investigation, the beginning of a lifelong quest deemed worthy, rewarding, and necessary to his intellectual stature. In that regard, the moment at which necromancy became a life commitment and an intellectual imperative chimes perfectly with such moments recorded in the lives of many leading thinkers, who "can usually identify a situation or even a moment when [they] first fell in love with a specific material, situation, or person – one that continues to hold attraction for them."[39] Faustus's exploits as a practitioner of magic correlate to the richness (or poverty) of the imagination and thereby become the measure of the self in the world. Ironically, however, the mastery of those arts, given the impermanence of his achievements, found its only reward in the quality of what it was like to experience these ephemeral operations in sequential moments of biological time. Faustus had selected his life's medium by excluding all others according to contract, limiting his self-repertory to that of an itinerant trickster and necromancer.

Worth a further moment's reflection are the mechanisms of mind that orient such goals. There is something perceived as the highest good in being Faustus that provokes him to declare: "Had I as many souls as there be stars / I'd give them all for Mephastophilis!" (I.iii.102–3; 1604). That is a steep price, if hollow in its hyperbole, for what his imagination has come to crave: "a world of profit and delight, / Of power, or honor, of omnipotence" and having "All things that move between the quiet poles" at his command (I.i.54–5; 57; 1604). This self-orientation has everything to do with interestedness in the world for the benefit of the organism: power, pleasure, prestige. What sustains the craving in excess, however, is not mere biological success, but what the mind enjoys as goals realized and experienced, thereby bringing us to ponder how brain design makes us addicts to the measures of pleasure which that same design encourages us to seek out.

To pursue this nostrum in relation to the self is to bring the tragic sense of life back to the paradoxes of our evolutionary inheritance. The compulsive desires which habituate in Faustus's mind and blind his moral judgment to the point of making him a gambler with his own soul have not escaped the attention of other scholars, who sense the need to define the dictatorial appetites that polarize his life. To see him as a victim of addiction is not to dismiss him as a man taken over by a foreign substance and dependence, but to profile him as a man given to anticipated pleasures with an analogous drive. Deborah Willis made that argument in an attempt to profile the curious commitments of a mind that constituted its own damnation.[40] It is an incomplete analogy, of course, as Carol Margaret Davison points out, because Faustus is under his own intellectual instruction, while addicts are physiologically driven. Mephistopheles, meanwhile, is only figuratively a drug dealer, although he caters to appetites in exchange for a soul.[41] Yet it is a sound attempt to profile a state of selfhood in which voluntary and involuntary factors have been juxtaposed, revealing a subject still possessed of choice, who, nevertheless, is systemically ill disposed to avail himself of choices he recognizes as beneficial. "Now go not backward: no Faustus, be resolute. / Why waverest thou?" (II.i.6–7; 1604). Under ancestral circumstances, the Faustian cravings were adaptively incited by scarcity, but the Renaissance supply of profits and delights reversed those conditions. Early modernity brought choice, and choice pits judgment against the psychology of entitlement through self-actualization.

Throughout the play, Faustus builds his life story by enacting a series of "I-positions" tantamount to an identity narrative.[42] His attempted escape from restrictive ethical frames is paramount, yet his peace of mind – his enjoyment of the self – is disturbed by unsolicited voices representing ethical polarities. They may be explained away as allegorical trappings, but that is to hollow out the representation of the dialogic self – a psychomachia of sorts arising in the deepest antinomies of his consciousness. Hypothetical and alternative thinking is endemic to the sapient brain, the mental equivalent of our nervous sensory scrutiny of the extended world. But persistent instruction in matters antipathetic to the hedonist and libertarian becomes a source of torment. The play thus fosters a thought-odyssey that is entirely modern in spirit and effect, for it delivers a story written by the psyche immersed in its own subjective experience. This moves the ethical fashioning of the self back to the productions of a brain designed to predetermined ends. As a necromancer, Faustus initially escapes, in the words of G.S. Gregg, the "empirical conceptualization of positioning in terms of constructivist, binary scales, with the idea that positioning takes place in a space defined by the moral order."[43] There is initial freedom and exuberance, but as early as the beginning of the second act, the angelic and demonic voices, with their vacillating homilies of hope and despair, begin their aggravating whispers. They represent the values pertaining to an external, culturally generated, religious system, to be sure, but the universalities behind them are phylogenetic; they are cognitive properties intrinsic to the self. At these moments in the play, we are most immediately aware of the dramatic exteriorizing of a mind present to itself in thoughts conditioned by the economy of mental production.

Faustus's head might be said to resemble the community of voices described by Hubert J.M. Hermans through which the self is constructed, but that community has hardened into absolute positions.[44] Good and bad angels link their exclusionary positions to a moralized order, namely the capacity to assign moral evaluations to all aspects of decision-making, and to suppose that those decisions will be monitored by success in relation to goals. This dimension of the human has been thoroughly sifted in studies concerned with the dialogic, negotiated, impaired, or polyvocal self.[45] Through intense reflection, Faustus may still entertain the overhaul of the self through an act of conversion, in some senses tantamount to a kind of self-therapy. P.B. Baltes asks whether the self is multi-directional, or whether development through life is unidirectional, irreversible, and universal.[46] There is potential in the question, whether a systemic pursuit of coherence and uniformity in the presentation of the self becomes its own uni-directional destiny. In Faustus's case, it is just such habits of self-confirmation that exclude the possibility of God's grace, the denial of which has figured so prominently in the Faustian identity. Minds are designed to be nervous, even promiscuously exploratory in terms of cognitive options, much as consciousness itself cannot dwell on single thoughts for more than a few seconds without revising them, adjusting them, or moving away from them. But this nervous self also negotiates with the pact-oriented, reputation-centred bias of the identified self, advancing from a base of known and reliable markers, while shutting down the voices for change. The historical self is a slow and cumulative destiny.

This is the Faustus of principled resolution. Identity is a personal achievement, according to the analysis of R.F. Baumeister, a unity forged from conflict and resolution, which, for the sake of effective agency, creates firewalls against ongoing contingency and choice. Yet the Faustian mind, although settled by pact, remained tormented within, still lacking the integration of a composed self.[47] To recapitulate, in the words of Peter Raggatt, his identity had become "dispersed in a moral landscape defined by often conflicting narratives."[48] There are two templates for being, each with its "own constellation of attachments," and the Faustian brain has given each a *force majeure*: for easy labelling, the secular humanist and the Christian. For this play, the tragic sense of life resides in the production of a self that is powerless to consolidate the appetite for limitless worldly influence with the contingent order of belief through which the psyche finds relief from the unbearable fear of eternal torture. The ordering of selfhood is the context for this all-too-human contest, the substance of the Faustian tragedy.

Faustus's only escape is rationalized denial. That too is a human universal: to rehearse and relate a justified version of the self to the point of comprehensive self-deception. When Faustus fails to complete the scriptures which offer a means of hope by way of justifying his apostasy, we are right to divine an act of basic mental prevarication. Faustus is fully engaged in converting the inconvenient dissonances of mind into confirmed intentions; he has come to believe his own lies by resolving indecision into self-affirmed truths. Robert Trivers has concentrated much of his research on the many forms of mental slippage whereby we reduce mental conflict and low self-esteem by constructing matters according to self-interest – constructions then rationalized as necessary truths. The self carves up the world in terms of justified "I' perspectives, often by construing others as selfish and abusive.[49] Faustus, of necessity, must harden himself against contending ideologies by immersing himself in the "world" of necromancy, hoping to escape his nagging conscience. The protean Faustus emerges with the enhancement of the self through confabulation.[50] It is a piece of how the mind works, and in the words of the man who made that phrase current, in his section on "Kidding Ourselves," Pinker states that "the mind has many parts, some designed for virtue, some designed for reason, some clever enough to outwit the parts that are neither." Nevertheless, while "one self may deceive another ... every now and then a third self sees the truth."[51] Faustus in a nutshell.

He is attracted to power of a kind that comes not with mere knowledge, but with knowledge that has efficient causal control over sectors of nature and the material world, operations which, in the absence of advanced chemistry, physics, and mechanics, might be achieved only through agents who, by their very natures, held sway over a set vocabulary of physical and pseudo-physical operations permitted to them as supernatural beings. Such agents might be looked upon as mere natural operators and catalysts if they could be divorced from the moral order which defines them – the essence of the entire Renaissance debate over white and black magic. Once again, self-deception alone enables Faustus to domesticate Mephistopheles into a mere hidden scientific medium. Constructed truth is the necessary cost for the integration of the self around exorbitant goals, a stance that seeks mastery by discounting the incomprehensible or the inconvenient.[52]

By related forms of self-justification, Faustus likewise discounts the goals and dignity of those who obstruct his will to power. His indifference to others constitutes a "deanimation" of the people he victimizes by essentially "excluding them from the circle of personhood"; he treats them merely "as means to some end."[53] The trickster mentality a substantial part of the Faustian experience of the self – is a complete psychological position in itself. What we should call this perverse but fully adaptive human proclivity is difficult to say – in one sense moxy, but in the Everyman context of the deadly sins, pride. That too begins with the subjective incorrigibility of the constructed self, namely the systemic and designed predilection to overrate the reliability of our beliefs and memories. Now claiming as his own these forbidden powers binding upon features of the natural world, Faustus could exercise his imagination seemingly without limits as an extension of self-experience – still untutored in the boundaries to be set on his fancy. Yet there was latitude enough to bring awe, admiration, and fear in all the social circles he entered. He could quell gossip and complaint through his secret powers and ignore the eye of God. In this, we see the autonomous self in its kinetic phases. Everyman, now contractually bound to the fallen world, is still in the formula, for there are the flights of the imagination that persist with their provisional glimpses of spiritual hope.

In due course, Faustus finds himself following the life of the itinerant necromancer and trickster, the triviality of which, by comparison with the inaugural vision of the man seeking world reform and unspeakable delight, has by no means escaped critical notice. Nevertheless, there is something that it is like to be a trickster – something that clearly fascinated the German imagination; it is the mentality which animated many of their folk heroes from Marcolphus to Tyl Eulenspiegel. The point is, after the signing of the pact, that the Faustian life profile becomes attached not only to the life of the necromancer, but incrementally to that of a prankster and mere entertainer. To ask what that means to the Faustian self is to anatomize the pleasure of trickery and the means whereby the art of magic enabled the practitioner to produce the cognitive shocks and obstructed expectations in others that satisfied his own will to power.

We may ask, as well, whether the architecture of trickery is best defined in terms of a psychological addiction, a means to advancement in a competitive world, a fascination with the strategies of deception, or the quest for status as a worker of wonders?[54] We need to know in order to get at this very particular experience of the self for which, ultimately, Faustus has traded his soul. There is first the commitment to a vocabulary or repertory of ploys and practices that demand careful attention. The mind becomes absorbed with matching strategic advantages to human vulnerabilities in a richly computational domain. The successful trickster must read the propensities of other minds and draw them, by their own misplaced desires or self-blinded enthusiasm, to participate unwittingly in their own deception and defeat – he must be a superlative mentalist. Trickery thereby becomes a plot scenario conceived in advance and actualized by a combination of victim susceptibility and a perfect understanding of the effects at one's command. The significance of such activities in the world of human affairs may appear bankrupt, yet the performative aspects of the consummate trickster constitute a world in which the practitioner may, for a time, lose herself. Dominance, or the illusion thereof, through each completed operation is the trickster's reward, and as such, occupies consciousness and drives the imagination forward. In trickery, the pleasure principle is paramount in rendering a projection into an actuality which entails the unwitting cooperation of a victim hoodwinked and bested, often by her own blinding desires. To understand Faustus, this is something we must fathom, because he had traded for it all the conventional disciplines of the university curriculum. Faustian amorality emerges in conjunction with the trickster view of the world, which he had chosen as the theatre for the Faustian self. The Faustian trickster thus combines contradictory roles in his making of deception from travelling entertainer to master mage, culture hero, and intellectual wizard, all within that single operational mindset.[55]

Paul Radin, in conjunction with his investigation of the Winnebago trickster cycle, perceived in the trickster archetype an "archaic *speculum mentis*" because this personality type reflected the human condition of mind at some phase in its pre-socialized history.[56] It represents a strategy of self-advancement which is pre-empathic and unqualified by the self-censuring of reciprocal altruism. This may be closer to myth than to science, but it intrigues, nevertheless, in suggesting that as an evolving species, we went through divers and partial conditions of socialization, to which we may sometimes revert. Carl Jung concurs, that given the "crude primitivity" of aboriginal trickster stories, "it would not be surprising if one saw in this myth simply the reflection of an earlier, rudimentary stage of consciousness."[57] Faustus is by no means as paradoxical as the fool-savant of these ancient stories who, like Wakdjunkaga of the Winnebago cycle, is far from controlling the contingencies of his world. Yet he belongs to the same mental order in the extent to which he is energized by the experiences of the trickster as practitioner, a figure largely indifferent to considerations of good and evil. Faustus owes his repertory of ploys, in fact, to the tradition of German tricksters, each one versed in problem solving through an employment of fanciful projections. They represent an obsessive exploitation of what it is to be "survival machines which can simulate the future" and thus be "one jump ahead of survival machines who can only learn on the basis of overt trial and error."[58] Faustus's paranormal powers gave him every advantage in defeating the default expectations of his audiences or victims. The trickster's reward is largely paid out in baffling or dazzling others, seeking revenge, or collecting gratitude and gratuities – a far cry from the Faustus who took up magic to complete his intellectual journey into the arcana of the universe for the sake of knowledge.

At the core of it all is the computational obsession with the trick as a Gestalt, an intentional structure described by Erving Goffman as the "effort of one or more individuals to manage activity so that a party of one or more others will be induced to have a false belief about what is going on."[59] As a mindset, however, it is about the powers of provisional thinking, which is a critical part of the arms race by natural selection that gave the more advanced computational brains their competitive and reproductive edge. Faustus is absorbed by one of the tools of human cognition – that which enables his own quest for absolute fitness and dominance as a trickster. His psyche, in that regard, may be reduced to a twenty-four-year elaboration on a single frame of activity. In those terms he defines himself and in those terms we assess him through the projections of our folk psychology – our default modes for the reading of other minds. The trickster, for a time, becomes a complete template for the profiling of the Faustian self.

With Faustus, however, given his anti-Papist instincts and Protestant geopolitical bias, we may also want to join trickster to homo politicus. Necromancy, for him, was initially an instrument for activating his intelligence in the world – and the intelligence of a communal animal includes the organization of groups and the wielding of power. In the words of Andrew Whiten, "primates often act as if they were following the advice that Niccolò Machiavelli offered to sixteenth-century Italian prince-politicians to enable them to socially manipulate their competitors and subjects."[60] He called this universal feature of human nature the "Machiavellian Intelligence Hypothesis" when it is extended into the contexts of more elaborate political behaviour. It is all about the negotiation of social competition through brokering, exclusion, and the building of alliances. For a time, at least, it appeared as though

Faustus's acquisition of demonic powers would extend his range as a political broker and manager of scarce resources in accordance with his political ideas. An entente with the Devil is by no means the classic form of Machiavellian political opportunism, but Faustus momentarily envisions it as a Renaissance option. In that regard, Vermeule reminds us that "there is a strong correlation between high Machiavellian intelligence and broad literary appeal. Any characters atop a list of the favorite literary characters of all time would score extremely high in Machiavellian intelligence."[61] Faustus had his chance to enter politics, but the motif was underdeveloped in his book of life and Marlowe did not extend his range beyond that textual authorization. Even so, the Machiavellian intelligence of the trickster is by no means a misplaced modifier of the Faustian experience. It was more than a latent part of the man.

More of the Faustian self is to be located in what it was to live in a state of damnation, perhaps reflecting the atheistic leanings reported of Marlowe himself.[62] The Faustian paradox has often been remarked, that while he denied, as a good atheist, the existence of a divine order, he nevertheless employed powers originating in the spirit world. Marlowe makes much of that irony, that lapse in the framing of Faustian reality, as the basis for his dying anxiety. But there is no consistently formulated rationale for his working ontology. If he could negotiate with the forces of evil in material terms, how could he discount the forces of good? Or had he simply escaped this binary understanding of the world through flat denial? Upon that aspect of the Faustian self we are free to speculate, critical as this discrepancy remains in the ensuing psychodrama over salvation and damnation. The play challenges us to learn what we can about the tyrannical inclinations of consciousness concerning the soul and the mental discipline apt to cross the finishing line of salvific sanctification.[63] This property of his being is juxtaposed with a mind awakening to its own sense of frustrated ambition and broken promises glossed over by the rhetoric of demonic commitment. Gathering up the salient mental themes that constitute the Faustian identity is work for the general critic, but attaching those themes to the way in which the brain is constituted to deal with its many environments is altogether more far-reaching, because that which was once made generic by archetypes is now made generic by the genome.

That connection is reaffirmed by dint of the Faustian obsession with positions of good and evil, a piece of the binary order of cognition which constructs our perception of much of the world experience in oppositional terms. In the image of that default mode, Marlowe personifies those positions as dramatic voices and thereby links the vacillating conscience of the protagonist to the religious systems which have carried that oppositional orientation of the mind into the structures of the spiritual universe. Heaven and hell are merely spatial materializations of absolute pleasure and pain, which find their origins in the human limbic system as the guardians and instigators of self-preservation. The dramatic action becomes the correlative to a mind in the throes of indecision, thereby creating a tableau of the protagonist's interiority. In that sense, being in the absence of self-enjoyment forms the substance of the play's tragic vision. Why, we may ask, did a man of such cognitive autonomy allow such voices to invade his own consciousness? We might speculate in practical psychological terms: guilt, advancing age, latent knowledge of Christianity, disappointment, Mephistophelian innuendo, or fear of death. Or, again, we may look to the existential conditions of the self, which are grounded in the binary models of the brain's architecture and its capacities for self-reflection in relation to the passage of time. We may speculate that he felt cheated; he did not get his questions answered; Mephistopheles became less friendly and more threatening; many legitimate pleasures were unavailable to him; and biological time was running out. Perhaps the mind simply dwells on things forbidden, bows to suppressed voices, or rehearses the positions associated with bad conscience, each of them disquieting components in an imperfectly integrated self. Or perhaps, with approaching death, the brain goes into overdrive in planning what it can of the future, or in avoiding the torment that, in former times, seemed safely remote. What Faustus found, as the crisis mode increased, were the sole conditions known to the Renaissance mind for securing the soul in the afterlife: reorganizing the mind itself in pursuit of the precise formula which constitutes genuine and efficacious faith.

This mental exercise, to borrow a few words, entails an implicit belief in the "unseen agents who can monitor our behavior and administer punishment or reward – the stories we call religion – [which] permeate and persist partly because they offer such powerful ways of motivating and apparently monitoring cooperative behavior."[64] Different periods in cultural history have settled on diverse belief systems for dealing with consciousness and identity after the death of the body, typically offering escape for the soul through obedience, service, and ritual in the context of a faith community. Christian soteriology, in sixteenth-century Europe, remained in the ascendant as the means of choice through which the self might achieve immortality through the sacrificial death of Jesus of Nazareth. But the transaction necessary to achieve that state of grace is a unique mental narrative that must be formulated by a self in dialogue with its own experiential potential. This play, more than any other in the Everyman tradition, equates salvation with a controlled and emotionally sustained reformulation of the self. St Paul did what he could in his writings to reduce the mystery of that procedure by using notions of law and contract by which God through His grace was bound to honour the faith of the believer and grant justification. Faustus clearly had intimations of what that faith state must be like, yet he was powerless to actualize the requisite cognitive stance. Present to his mind were the traits of a religious belief system, the conditions of reward and punishment, the sense of a soul with eternal potential, and the fear of death. But damnation comes in the end to a self that is incapable of completing this thought transaction for reasons as constitutionally deep as the reader might care to plunge. Salvation is yet another mental finishing line entailing the alignment of attention with requisite features of memory, belief-formation, doubt, and truth. If ever there was a play about universals and the propensities of human cognition, this is it.

The final crux of the play, therefore, becomes the transactional aspects of Christian conversion, involving what it is for the self to imagine its composite identity in provisional and projected terms. There is decidedly something it is like to undergo a full conversion experience, much as there is something it is like to be a wonder-worker and trickster. That which detains the protagonist's mind is critical, leading to a variety of readings centred variously in the nature of reprobation, doubt in the efficacy of prevenient grace (the unforgiveable sin), fear of demonic threats, intellectual pride, or, as stated above, the incapacity of a distracted mind to compose a state of consciousness acceptable to God. The play has been a sequence of belief states that have composed the narrative of the man and built his identity. He had chosen a profession, embraced necromancy, abjured God, pursued his trade with a modicum of imagination, and confronted the passage of time. He had lived for a time an "absoluteness of experience," nourished by a sense of "subjective incorrigibility" which induced him to act with greater confidence than his command of "truth" warranted.[65] We may backtrack on the literal to see the Faustian anxiety in terms of old age, depression, or the loss of the limitless possibilities of youth.[66] But in actual dramatized terms, the finale is a soliloquy concerned with perfecting a ritualized moment of consciousness.[67] Faustus begins this mental voyage, only to become repeatedly distracted by the "demons" of the mind. He knew the appropriate procedures and was near the collapse of his former identity – a collapse requisite to the act of spiritual transformation. He was in an eleventh-hour way of rediscovering the medieval Everyman, now in a Protestant mode, struggling to complete the thought formula for salvation.

The play thereby alludes to the psychology of conversion, just as it investigates the experiential nuances of the trickster mind.[68] Faustus's thoughts had been troubled by the promptings of a Good Angel and an Old Man, both recommending a comprehensive act of self-denial and self-renewal.[69] Conversion entails belief in historical promises, as when Faustus peers into the sky to see Christ's blood, amounting to an emotionalized divorce from his past intellectual commitments. But Faustus had denied the truths of scripture (I.i.39–46; 1604); he had preached to Mephistopheles the foolishness in thinking there could be pain after death: "Tush, these are trifles and mere old wives' tales" (II.i.134–6; 1604). He had been a teacher in his day on the fallacy of the soul. He would have taught that the transformation of matter into spirit was beyond demonstration and that mind could never know the soul. Yet even for him, the possible soul lived on, not only through religious belief systems but through intimations arising from the imagination dealing with possible futures.[70]

The thematic significance of that increasing presentiment is central to the play. Milton's Adam, in a suicidal mood, confronts the dilemma: "Yet one doubt / Pursues me still, lest all I cannot die, / Lest that pure breath of life, the spirit of man / Which God inspired, cannot together perish / With this corporeal clod; then in the grave, / Or in some other dismal place, who knows / But I shall die a living death? O thought / Horrid, if true!" (X. 782–9). Scholars of cognition have studied the kinds of fears we are least likely to shed, among them that consciousness and its executive functions will end with death, thereby confirming the concerns of both Adam and Faustus. The unimaginable impermanence of selfhood has been a powerful motor to myth. We are that kind of animal. Yet intellectual interference can also cripple the formation of mythic belief, like the failure of the folk hero described by Max Lüthi, who experienced "a reactionary impulse in his own soul which is not yet the equal of his own insight."[71] The self, by its very nature, tends over time to habituate its belief systems.[72] "To have a self is not a passive event."[73] It is a hard-fought construction for some, as the opening of the play demonstrates, and thus the less amenable to redefinition, given the "psychological organization which produces the experience of a stable substantial self."[74] Or, in the thinking of Harry G. Frankfurt, that which most pertains to the self is what we have given our assent to, following our desire for it.[75] Faustus had achieved what he desired: a plenary self liberated from the master narrative of the Christian religion by the signing of an irrevocable pact with anti-Christian forces. The play invites us to take the full measure of that assent as an expression of the elective self. It is tantamount to a conversion in its own right, not to join but to abandon an inherited religious order. That choice entails a heroic measure of tolerance for what A. Giddens called "ontological insecurity" – the precondition to any radical overhauling of the self-narrative through the adoption of comprehensive belief paradigms about the nature of the world. The call to a conversion experience at the end of the play must be evaluated psychologically in relation to the conversion at the play's opening and sealed by the new dispensation of the twenty-four-year pact.[76]

That courage to face "ontological insecurity" is simultaneously the substance of a fool and a hero: a fool if there is security for the self in faith, and a hero if the pluralities of that world necessitate an open engagement with the diversity of knowledge and being. In the latter regard, Faustus is an early champion of modernity, fully cognizant of the necessity to settle himself in a world of contending choices and belief templates. Peter Berger has called this "the heretical imperative" of the modern person, reminding us that the Greek verb *hairein* meant simply to choose (*hairesis* thus meaning "the taking of a choice").[77] In the early Church Councils provisional doctrines and indefinite truths, through strong-armed negotiations and factional bullying, were wrestled into the received dogmas of the faith, but at the cost of turning all dissenters within the group into heretics (those who contended with orthodoxy). The Renaissance constituted another age of politically negotiated truths among even greater choices, including the existential obligation of Everyman to make a religious choice and hunker down within a like-minded faction, or hide behind a personal palisade of lonely but determined conviction. Self-made protagonists confront those choices full on, as Faustus did in mastering all the disciplines of the medieval university before registering his disenchantment with their limitations. His purpose was to escape the tyranny of a pre-modern belief system and to live the freedom of modern plurality. The paradox of that freedom is manifested in the play, particularly in the craving for community and assurance that comes with age – a nostalgia for community. The ultimate irony for moderns is the revolt of their aging psyches against incertitude, as though a latent platform initiated a longing for the comfort of universal truth and deliverance as surely as newborns come into the world in search of sustenance. Faustus lived the heretical imperative until it became a menacing destiny. It began with the university and the pride of skepticism. He became a loner separated from a community of consensus. The heretical imperative had created the dimensions of the Faustian self, but contending conditions of consciousness gave him increasingly diminished pleasure in his brazen commitment to a heroic freedom.

Selfhood can be proudly residual, stubborn, and conservative, because the illusion of persistent and unified identity is foundational to efficient agency and mental focus. The final soliloquy unfolds and Marlowe aligns dramatic and rhetorical timing with a mind running frenetically through its options for altering the destiny of the self; then there are but eight more lines of compressed panic in which to utter a final request for more time, a flashing thought in the direction of God, a state of torment and regret, a promise to burn his books, and a nuanced reflection on Mephistopheles. His name becomes the final word. We are gripped by this mental vortex, struggling to experience what it is like to understand the nature of tears and repentance (V.i.35–41; 1604), to comprehend the power of Christ's blood to ransom (V.ii.93: 1604), and yet to lack the concentration and mental fixity required to conduct the thought-states of conversion. The Faustian mind ranges freely between pleasure, contemplation, planning, and spirituality, featuring the equivocally dialogic mind of a man who enjoys nearly universal knowledge, only to confine itself, with the passage of time, to a confrontation between master discourses and the necessity of choice. William James talks about this darker side of the Everyman in the limiting nature of choices.[78] Not all minds are constituted to produce those states that ostensibly guarantee the full approbation of the divine. Such is damnation in cognitive terms; not all selves can make themselves fit for the presence of God. The self, after all, can be constructed independently of religious orders, yet that self to its owner is, as a point of view in the world of self-empowerment, just as precious. The bad death, however, is that in which judgment destroys the repose of a unified and achieved self, and a worse death because the self assumes the responsibility for the repercussions of that judgment.

What, then, do we make of this play in relating all that transpires to the operations of the self belonging to and animating the protagonist, and the degree to which the making of the self is shaped by the dispositions of a genetically and adaptively organized brain? Ultimately, it all must come down to a richer hermeneutic perspective in following the Marlovian leads concerning the dramatic externalization of a reflective mind, richly endowed, in relation to selfhood, identity, and the crisis of being arising from a brain designed to negotiate its own status in the presence of conflicting cultural options. The Faustian protagonist, obsessed by the choices available to the constructed or elective self, takes us all the way back to the fictive representation of interiority, the evolutionary components of human nature, the tilting of biogenetic design, and the computational terms of self-representation. The self in literature is a relatively benign topic until we look further into what we mean by identity and the ontological status of the minds of fictive persons. Suddenly the self becomes a philosophical abyss in need of grounded definition. By restoring the ontological substance of characters through the attributions granted to them by our reading brains, storytelling, as never before, becomes a remarkably privileged vehicle for exploring the conflicting properties of the meta-conscious and self-reflective brain. In that, literature may regain some of its "legislative" power.

Doctor Faustus has an external social dimension, but the action takes place within the material soul constituted of the material brain and the emergent properties it is biogenetically designed to favour. Damnation is a condition of that soul emanating from the binary modes through which social experience is measured. Among the features of that brain is what Raggatt has called "religious authority structures,"[79] which have been suppressed but never obliterated. We are amused by the hedonist's response to the intellectually and historically enhanced kiss of Helen of Troy, and the lovelorn projections of eternity even as a devil putatively sucks forth his soul. And we are abashed by the urgency with which the self, terrified by endless time, begs to hide itself in atomic oblivion. The human emotions are closely calibrated to the passage of time, both with the shadowy antiquity of the humanist schoolboy and the time without end of the religious imagination. Emotions abound in this play, which Joseph Ledoux defines as a colouring if not a transformation of the mind that feels. In assigning such mental states to a character, observers must deal with the something it is like to be that person.[80] By the play's end, insofar as "people are their emotions,"[81] Faustus has become his terror and that terror sends his mind scrambling through all the magazines of his intellect for an interpretation and a final plan of action, therein reflecting Damasio's core self in emergency mode.[82] In such ways, the play evokes the society of the mind conducted in the name of the self.

We have come full circle to the awareness of the reading self that meets the awareness of self in characters, and ultimately to the vast literature pertaining to the analysis of what it is to be and have a self by recent scholars of the human cognitive processes. Volitional selves are at the centre of social history and there are questions that may be asked about how the mind makes up its mind, systemically, computationally, rhetorically, contextually, genetically, and emotionally. The Faustian experience dwells on the nature of decision-making in relation to the design structures that constrain thoughts and condition mental events – design structures that define what it is to be human. Among their emergent properties is the phenomenon of selfhood and the strategies employed by the species to affirm this valued point of view – one that is ready to place high wagers in the pursuit of its interests. The imperatives of the Faustian self produce in sequence the intellectual gambler, the fanaticized broker of world power, the social trickster, the hedonic escapist sending for Helen of Troy, the depressive neurotic, the victim of voices, the unrepentant seeker of eternal escape, and the default apostate surprised by damnation. The self is a point of view for reading the play; the play is a point of view for revaluing the place of the self as one of the thematic centres of dramatic representation.

CHAPTER TWO

The Biogenesis of Ethics and the Challenge of Shakespeare's *Measure for Measure*

Wittgenstein, like Dr Johnson before him, took exception to Shakespeare's excessive verbosity and ambiguous word play. He associated such indulgent imprecision with a lack of rigour in matters ethical and juridical. This appears to be particularly true of *Measure for Measure*, which, for many, comes to a close without the restoration of an acceptable moral order. Christopher Norris, elaborating on Wittgenstein, speaks of the Bard's "deplorable lack of standards when it comes to apportioning credit or blame among his *dramatis personae*, or distributing rewards or punishments in line with a due sense of moral worth."[1] This complaint runs high even among the most admiring of Shakespeareans when it comes to the so-called problem comedies with their messy applications of what has been called "poetic justice" actualized as rewards or depravations (material, social, or political) for each character in putative accord with his or her merits. The question is how we know this without doing some rigorous imaginative apportioning of rewards and punishments on our own in relation to norms of moral worth. But then, "who" does the ratings, and who sets the norms whereby we find Shakespeare deficient, despite the play's reassuring title? Could fairness be built into the very design of the brain like a Kantian category? And if it is, how can the results, as in the case of *Measure for Measure*, be so mixed? How do our brains calculate the morality of persons, and how do they compute social justice?

Measure, by its title, is precisely about justice, computational proportions, and calibrated merit. Seminar students are obsessed by little else in discussing this play and enter into the most subtle rationales not only for the work's entirely unsatisfactory closure, but for the unethical stances and contradictory natures of the characters: that Duke Vincentio abuses his powers, especially in assuming the disguise and privileges of a man of the cloth; that Angelo is too wicked to escape the full rigours of the law; that Lucio is a mere fly swatted by an elephant; that Isabella is too arch to arouse empathy and too rigid to survive marriage; while a dissolute, even sociopathic, prisoner named Barnardine is released just to make a point about pardon, and hence, that our expectations of justice have not been met. Laurie Maguire points out Isabella's discourse of mercy until she is corrupted in her thinking by the Duke in ways as "nefarious ... as Angelo's attempts to interfere with her body."[2]

Right or wrong, these are fairly sophisticated computational results, derived from the kaleidoscopic emergence of information about the characters rated by norms pertaining to goodness and badness, fairness and justice – all of them arrived at, presumably, in generic ways similar to those exercised by Wittgenstein and Dr Johnson. We score them along the gradations of an ethical grid, entailing a form of cognitive processing that may prove to be largely systemic. Ethical scoring appears to be simply a part of the way our brains are designed to process social information, given that, without conscious instruction to do so, we are constantly on the lookout for cooperators and cheaters, good behaviour and bad, rating characters as we go, employing capacities in which we place implicit trust and which, in *some* contexts at least, we apply with considerable finesse and accuracy to our unquestionable advantage. (We will come to the more complex ethical cruxes which place stress on these default systems.) This entire component of consciousness is so familiar as to seem commonplace and self-intuitive: that we approach interpersonal relations preoccupied with matters of fairness for ourselves and for the groups within which we function, as well as for the fictional characters and societies which compel our attention in imaginary worlds. Ethical judgment is a reality we impose on fictional narratives. "Nowhere is our concern for how others treat others more apparent than in our intense engagement with fiction."[3] That such a capacity depends on the competencies of our brains is likewise self-evident, so why even go there?

Here is why: if we are all obsessed with ethical scoring based on urges as deep as the construction of consciousness itself (a point to be further elaborated), the play epitomizes the need to distinguish between the architectural, emotional, and computational instructions behind such scoring and the diversity of ethical views based on those same cognitive mechanisms. Brains clearly provide the platforms for doing the "mathematics" of ethics, the computational capacities, and they supply the emotions, the reward/punishment systems, for guiding ethical choices, but there are no innate answers to specific ethical questions. In the words of Donald Symons, genes merely "make possible novel behavioral means to the same old specific ends."[4]

Measure for Measure has been sealed, probably forever, into the cluster of so-called problem comedies precisely because of the scoring discrepancies in ethical terms among readers.[5] "Shakespeare did not 'solve' the problems of moral justice he apparently set himself in his problem comedies."[6] Plays in this group may even be defined as works bent on confusing and disorienting our moral judgments – as ethical knots we can only attempt to disentangle. Richard Hillman notes how this play subjects the romantic elements to a "fundamentally disjunctive, sometimes jarring, realistic treatment" which eventuates in multiple interpretations. "Fundamental incongruity" is what he sees "as part of a metadramatic dynamism, at once the origin and object of the play."[7] As perplexed scorekeepers, we are partially off the hook with a Shakespeare who problematizes his sources and qualifies his characters in ways that stretch our computational acumen to the limits – on purpose. In consequence, the play can hardly achieve a communal moral order if spectators and readers can come nowhere near consensus on the fundamental ethical natures of the principal characters.[8] These problematic characters and circumstances are, in turn, reminders that the modern world has also given us moral issues about which we cannot yet make up our minds, such as legalizing prostitution, an issue as alive then as now, not to mention the play's very particular challenges: whether the preservation of a girl's virginity outweighs the value of her brother's life, or whether a man who has unjustly taken life owes his own in repayment – the principal of talion in exact equivalents alluded to in the play's title. These are some of the koans of the ethical life, each a point of contention worthy of considered discussion, but likely to induce inconclusive opinions. We may just resign ourselves to the problematic. But at the same time, we cannot turn off the instinctual search for justice in these stories: we believe in the accuracy of our liking and disliking in relation to expected rewards and punishments; and we continue to look to endings as progress toward the stasis which justice and consensus alone can provide, and on which all readers can agree. Hence, the pursuit for the hermeneutic terms in ethical values (reflecting universal human values) by which endings may be justified, going further and further back into the workings of human judgment.

Given our imperative habits of social evaluation, it seems a small step forward to say that we are a moral scoring species, *tout court*; ethical adjudication is in our natures and it is a large part of how we live and why we read. Clearly the nature/nurture debate will continue long into the future when it comes to reasoning out just how such baseline operations were installed.[9] The problem before us is to investigate the levels at which fairness or fair measure in a reciprocating social economy is performed, not only in the play, but in our minds, as well as in Shakespeare's mind, and in the "minds" of his respective characters, and how generic we should expect these operations to be. If Shakespeare missed the mark in computing justice through the outcome stances of his play, we must work out those particulars in which he tallies justice precisely as we do, in sharing the same computational instrument as we do, or in tallying justice in eccentric and unacceptable ways, using that same computational instrument. (Wittgenstein and Dr Johnson expected, after all, that Shakespeare's justice would accord with theirs.) Explaining Shakespeare in this regard entails the creation of an explanatory tool whereby we can understand the closural values of the play as conforming, or failing to conform, to the values behind our instincts for justice based on ethical adjudication, the desire to punish non-cooperators, to reward reciprocators, and to rebuild the social order that is most conducive to the greatest fairness for the greatest number.

But before characters, justice, and the ending of Shakespeare's play can be addressed, there is controversial work to be done simply in defining the ethical reader. It is perhaps sufficient to treat the human ethical praxis as a self-evident and self-defining universal – a product of reason, experience, and custom. We might begin with the design perimeters of ethical scoring and ask hard questions about what our brains know about the survival advantages of our compulsive invigilation of other behaviours in relation to ourselves and our loyalty groups, how those groups are established, the degree to which they depend on cooperation, and the degree to which cooperation depends on sophisticated social monitoring.[10] In circular fashion, this social monitoring is essential to the practices of reciprocal altruism, which are so critically germane to all forms of cooperation.[11] We have to know who plays fair and who cheats, backed up by our social emotions. With this we may come close to an axiom based on selective design behind the neural platforms, a design that tilts the thinking that conditions the behaviour that expresses the inferential values favoured by the entire system.

In the work of Patricia Churchland, these values derive from the caring instincts that may be extended beyond children to family and loyalty groups on a reciprocal basis.[12] In the work of Joshua Greene, they pertain to the emotional and cognitive networks that make group cooperation possible through deferred gratification, promises, and expectations of loyalty regulated by such emotions as honour, conscience, and shame. In his words, "morality is a set of psychological adaptations that allow otherwise selfish individuals to reap the benefits of cooperation," which has little to do with moral truth, per se, and everything to do with negotiation and expediency in a consequentialist world subject only to the best pragmatic reasoning available at the moment – the world of *Measure for Measure*. The values and necessity of cooperation remain paramount, which is sufficient pressure to have designed the ethical brain.[13]

Both approaches assume the long-term learning through genetic instruction that comes with the heuristics of adaptive selection. Yet both researchers, at the same time, are keen to apply the brakes on fancifully reductionist applications, inferential rules, or material determinism. Not only are they fully mindful of the phylogenetic instruction which comes through the reproductive failures and successes of our ancestors over huge swaths of evolutionary time, but of the successes of our more immediate ancestors in compiling wisdom templates and customary belief-practices apt for transmission from generation to generation, as well as the learning that comes through personal experience. Churchland is especially concerned with avoiding the "if general, then genetic" fallacy, pointing out that certain kinds of problem solving lend themselves to certain kinds of universalizing solutions, such as building boats out of wood the world over, which consilience requires no genetic backstory.[14] Yet both are committed to the features of the designed mind that *do* influence our ethical values, with Churchland leaning toward the hormones and peptides such as oxytocin and argonine vasopressin, both of which lend their scrutinizing support to minds caught in moral dilemmas. Greene, in parallel, stresses the rich dialogic activity between the rational dorsolateral prefrontal cortex and the more emotional ventromedial prefrontal cortex, prefiguring his theory of the "dual-process brain" through which we work out the felt qualities of our ethical computations by defaulting to the emotional brain or by consulting the planning brain.[15] In these scientific matters, I must defer to their engaging experimentation with game theory, stroke patients, magnetic resonance imaging, the ethical lives of babies,[16] and our mammalian ancestors, whose rich social lives in preverbal and pre-rational terms give weight to the ethical brain as a primate inheritance, combined with our counterintuitive yet innate capacity for empathy.[17] Critical to their arguments is precisely what we *can* know, with a high degree of scientific certainty, about the hardwired preoccupations of human thought, and the social values about sociality and survival that hum along under all our cultural beliefs and experiential understanding, girded up by the embedded intelligence of our computational biases and emotions.

The precise boundaries between ancestral design and cultural understanding will continue to waver, and it is a good debate. In the discussion to follow, however, the intent is to re-examine *Measure for Measure* from the perspectives relating to the most fundamental biogenetic values embedded in the ethical brain – that suite of "psychological capacities and dispositions that together promote and stabilize cooperative behavior," namely, the neural properties that produce the basic family and group instincts that define our species. Readers who cannot go along with this premise may still gain value from the argument concerning Vincentio's eccentric solution to the woes of Vienna, but the argument will, I think, lack some of the force in showing how his preoccupations align with the preoccupations by which our design-tilted brains do business with the social world.[18] In that sense, I am suggesting that the ethical challenges of the play, to be better understood, take us all the way back to our ethical instincts, and that we cannot come to much of a consensus about Vincentio's strategic devices without understanding them in such phylogenetic terms – and that understanding depends on the best thinking we have concerning how and why we are such compulsive ethical invigilators. Such an approach will be tantamount to an extended but necessary hermeneutic loop, and I can imagine that some readers are already in resistance mode, which is only natural. But if the play gains in interpretive clarity based on the perspectives of cognitive psychology, then the science must first be made sufficiently plausible to legitimate the reading – a contentious but worthwhile enterprise.[19] If we have the right tools for getting at the universals in human nature through the mental designs that constructed them, we have a new way, simultaneously, of reaching further back for the themes of the race potentially reflected in the closural positions even of these problem comedies. That argument must be built in stages.

If the biogenetic origin of the ethical species is a first axiom, there is a second and less convenient one. Humans compulsively score social behaviour, but we do not all agree on the social and judicial significances of those moral weightings. Social narratives are computationally treated as the motivated lives of self-propelled agents striving after their respective and contingency-adjusted goals in the unfolding contexts of decision and action by which fortunes are altered and stories assume their shapes. Rating such behaviour is vital to us, in art as in reality, and we are called on to explain ourselves when our evaluations differ from those of others. In this ever-changing environment in which new social inflections release recalibrated innuendoes, the computational demands assume a remarkable complexity on an "if this, then that" basis. Insofar as ethical ratings are narrative based and cumulative, entailing constant revision and modification, they must be linked to the working memory through which the mind quite automatically adds new features in temporal sequences, while relying on editorial chunking and filtering to determine how each new bit will, or will not, become part of a porous, emerging, and recalibrated memory sequence coloured by ethical values.[20] All such information systems have their capacity limits, again a factor of biological design, and we know that information reduction necessitates systemic editing and probability schemas, not to mention experiential biases concerning the ethical intentionality of others. Someplace along the way in that data assimilation, readers make up their minds about an Isabella or a Vincentio, the proto-nun and the devious magistrate who, nevertheless, just may unite in marriage and live happily ever after. How do we deal with the vicissitudes of moral vetting and the solipsistic predicament of each reader? We all score as best we can and must if we are to hold our places in social negotiations, yet we do so with differently motivated minds abetted by a powerful assumption of the incorrigibility of our faculties and judgments. Shakespeare delivers his representations to our attention, inviting us to impose our own criteria for fairness and reciprocity. That leaves us in a state of nearly blind subjectivity. Yet the question remains whether the best we can do is still done on the basis of maximizing our interests through social and material cooperation.

Back and forth we go. But always we return to that essence we share as compulsive moral invigilators and to the world view we share through this common architectural platform. The discipline arising from such structures Michael Gazzaniga refers to as "neuroethics," which is the study of moral problems in relation to the cognitive mechanisms that inform our thinking and decision-making: a "brain-based philosophy of life."[21] Societies, after all, function through consensus and thus on the similarities of mental states. Crucial to our purposes is that social interpretation depends on these systems, "content-imparting mechanisms," based on an evolved psychology that imposes constraints and conditions on all cultural innovations.[22] Steven Pinker calls these processor mechanisms "primitive or irreducible," and hence "all higher-level concepts are defined in terms of them." Human nature, for him, is "innately endowed with an understanding of the basic categories of the world," and cultures unfold entirely in compliance with them.[23] That in turn implies homogeneity in the systems that define the species. How may these divergent monitoring mechanisms be reconciled? Can a brain-based philosophy provide a useful hermeneutic perspective for *Measure for Measure* and for literature in general?

The investigation to follow, then, is not only about Shakespeare's potential shortcomings but the default processes by which every reader knows that such plays are about rewards and punishments meted out through imposed alterations to the characters' fortunes, computed according to the binary calibrations that pertain both to law and to status within social communities.[24] To that end, we require an understanding of the social values embedded in the operations of those "primitive and irreducible" processor mechanisms, which in turn entails speculation about the ancestral environments which, step-by-step, inaugurated the genetic adaptations that shaped the designs, that shaped the thoughts, that made for the most efficient replicators, that sheltered the cultures, that shaped the behavioural repertoire, that built the house of homo sapiens. No easy reductionism here. Yet it had to have been by that same sinuous means that we acquired the basic social intelligence which shaped the behavioural displays by which that intelligence was passed along – an intelligence that included the prioritizing of attention toward resources, reproduction, material security, the integrity of the family, fear of competitors, credit within negotiating systems, sexual jealousy, cooperators and cheaters, as surely as it equipped newborns with a desire for mother's milk. The list is long and growing. At the same time, the brain has limited powers of perception and computation as well as interested factors which enhance self-oriented benefits through biased frames of analysis. All out reductionism has been dismissed, and yet design has its latent contents, without which the design is pointless and ineffectual. Hence, mental outcomes may be called reductionist to the degree to which they trope and bend the mind. In the words of Richard Alexander, "evolutionary reduction, when it is successful and accurate, tends to deepen our understanding of all our immediate and primary behaviors, motivations, and emotions because their evolutionary significance and the involved compromises are almost never a part of our conscious knowledge before we pursue them deliberately."[25] Directed actions following unconscious computations and systemic prompts, in a sense, define the human family, and thus are at least partially archetypal and predictable. This may provide our only analytical recourse to juxtaposing Vincentio's eccentric means, bed-tricks and all, with an order of justice we might recognize and acknowledge as our own. (The play has not been forgotten!)

Because this approach entails a hypothetical assessment of the mental values of the working mind, and because that brain was largely engineered at the latest under Pleistocene conditions, we must face the awkward prospect that how we *sometimes* weigh our modern circumstances still carries so-called Pleistocene values. Jonathan Kramnick, although an opponent, nevertheless states clearly a version of this idea: "to be a literary Darwinist is thus to take as a first principle that present-day habits of mind may be explained by selection pressures from an antique environment."[26] Accordingly, the mental values behind evolutionary successes are embedded in promotional design and are structurally bound to evoke those same values when the brain senses, again, the environmental crises to which they correspond, now as well as then.

Yet how plastic the brain can be in making its analogous inclusions of modern social stimuli among those read by an ancestral brain is a point to ponder. Our brains run on that legacy, and we are left with the complex discriminations to be made between social content inclined by the genome and social content inclined by conditioning, religion, personal experience, or ideology – all of which emerge directly or obliquely from the drives and tropisms of our natures. Pleistocene logic genetically installed may seem an inconvenient assault on our liberationist views of the self, but if we evolved as an adapted species to adaptive ends in any historical sense, then so did our brains, the designing of which preoccupies fully 60 percent of the genome. I am arguing that those adaptive ends, through the emergent properties of our neural circuitry, include the innate rudiments of the justice that Vincentio seeks to realize through the highly planned social machinations he imposes on the show society of the play. How can that hypothesis be tested?

Let me illustrate the point about baseline values and open interpretation. Brain design represents a form of slow-motion learning, giving rise to our bedrock social instincts, including the anger we feel toward bounders and blackmailers, not only in social reality but in fictive worlds – anger that is often justified as a critical prompt to protect the self or to punish the offender, even at a cost to ourselves, as part of our loyalty to a group.[27] That we can override with self-congratulating philosophical perspective the little surge of pique that is felt when someone cuts into the queue ahead of us is a genuine possibility. Such an override may arise in Christian charity, an unwillingness to spoil our calm with things relatively inconsequential, or a brush-off pessimism about human nature: that some are more brazen about seeking advantage than others. The point is that the surge of anger itself remains a form of felt philosophical analysis of environmental change, and as such, continues to define us by constituting a first-order hermeneutic response to modern stimuli read as ancestral cues.[28] We have to interpret our anger in relation to each newly registered stimulus, many of which no longer resemble ancestral circumstances, and we have to assume that our emotions, from a first-person perspective, have their own "opinion" about fairness.[29] Our brains continue to fire according to ancestral coding; the environment will have changed, but not the world view.[30] Mental plasticity has allowed for an analogous application at the level of a generic ethical schema. Ergo, our limbic system dislikes unfair advantage takers and tells us so, according to values installed to our benefit through designs relating to ancestral problem solving. The sum of those inferential and attributional biases identifies us still and shapes our evaluations of social circumstances. Where stories are pitched and resolved in accordance with those ancestral values, we should be able to identify them. Intimations are that *Measure for Measure* provides just such opportunities.

At the ground level of this analysis is a brain basic that is critical to our ethical take on the world: our compulsive binary mode of thinking. Just as colours do not exist in the world of nature but only as "ontological" qualia in the brains that interpret light waves as colours, binaries are also a thing of the brain rather than of nature.[31] Our neural systems are responsible for all polarized perceptions of the world.[32] "Contrary to common sense there is no unique 'real world' that pre-exists and is independent of human mental activity and human symbolic language; ... what we call the world is a product of some mind whose symbolic procedures construct the world." This sums up the constructivist position that prefigures Nelson Goodman's "philosophy of understanding."[33] Nature knows nothing of bad or good, any more than it does of its own apparent antinomies and polarities, which we recognize as up and down, light and dark, birth and death, drought and flood. More pertinently, our ethical binaries take instruction from the limbic states of pleasure and pain, which themselves polarize experience through their hedonic properties. Through our simplest computations we come to understand our own existence in terms of gains and losses, hence things beneficial and things detrimental, easily associated with pain and pleasure and thus with good and evil. Such binary properties may then be imposed on the elements in the environment apt for diametrical status, or projected into the realms of the gods and revisited upon us through the clerical inventions whereby creation, history, ritual, and the entire social order are moralized and managed according to the binaries of divine law.[34] Humans have thereby converted themselves into a binary-oriented species at the neural level at which reality is constructed in human terms (terms in turn translated into the value-substrates of culture). It has become one of the most compelling of all the emergent properties of the mind, a category of consciousness, and a schema of the social and legal world orders.[35] Binary logic features among the "strategies by which ordinary people penetrate to the logical structure of the regularities they encounter in a world that they create through the very exercise of mind that they use for exploring it."[36] By imposing these binary interpretations on the phenomenal world, our performances in natural and social environments have been improved, and that is all that survival requires.

Biogenetic or not, the binary interpretation of social behaviour is a brain-based, world-viewing frame through which perceptions of reality are organized. We are then challenged to conceptualize this operator as our basic mode of social scoring, most pertinently in evaluating persons according to their thoughts and deeds in relation to the rewards and punishments we concomitantly wish and expect for them. For Michael Ruse, "Because of our common evolutionary heritage we share an ultimate moral code, and can indeed make judgments of right and wrong, distinguishing them from personal preference."[37] Colin McGinn, specifically assessing Shakespeare's moral sensibilities in the plays, likewise determines that while he is not morally preoccupied as an author beyond getting his characters described with perfect psychological fidelity, the characters themselves "are above all ethical beings," upon whom we impose our "moral evaluations – admiration and disgust, approval and condemnation."[38] As James Q. Wilson sums it up, such judgments "are not arbitrary or unique to some time, place, or culture," but universals which bind us together "by mutual interdependence and a common moral sense."[39] Antonio Damasio deals with the externally directed attention, which includes "a host of tasks that involve judgments of people or situations within a moral framework." His efforts to link these integrative computations point to heightened activity in the "posteromedial cortices," but however that proves out, there is certainty about our compulsory binary computations and the neural competence to perform them.[40] Pascal Boyer grounds his study of "the evolutionary origins of religious thought" on this platform of emergent mental properties: "our evolution as a species of cooperators is sufficient to explain the actual psychology of moral reasoning, the way children and adults represent moral dimensions of action."[41] These and many other studies offer hermeneutic hope that brain systems and their thought productions will coincide with the experiences of literature through an elucidation of the social values they prompt. (If this is an analytical blind alley, well, I'm in terrific company.)

We can now circle back to clarify a few remaining complications in the ethical scoring of the play and its characters. *Measure for Measure*, by dint of its title, suggests a thematic application of tit-for-tat, equivalent punishment for a perceived quantity of injury, yet gravitates toward a justice based on the imposition of rehabilitative constraint. The latter is conceived, not as a wishful golden rule, but as an armed one, despite the showing of mercy: that what others do they should expect to have done unto them. It is the dark side of self-interested reciprocity. Had Shakespeare stuck to the talion formula, he would have remained in more familiar territory when it comes to the containment of anti-social or criminal behaviours. The brain, in vital circumstances, can be rigorous. Survival logic may hold for the justice that brings equivalent harm to those who have harmed us. Or does it?[42] Survival logic may read social environments in even more fundamental ways: if others cooperate, cooperate; if they do not, banish them – the Golden Rule tweaked yet again. In translation: continue to help those who are likely to help you in the future (even if they have not in the past), and expel those who will never reciprocate. There is a subtle difference, and Vincentio may opt to do justice according to the values of reciprocators. Do our brains compute justice in contrasting ways: the procedural justice of Angelo or the utilitarian justice of Vincentio, the one bent on punishment, the other on the rebuilding of communities of the morally like-minded? For adaptive reasons, they undoubtedly can. How then are these contrasting visions to be reconciled: retributive justice and redemptive justice? Is talion a base and unworthy principal while tolerance for reciprocators is divine?[43] Vincentio has been thought to abandon talion for mercy, but I think that is misleading. Colin McGinn's assessment, that "Shakespeare's characters inhabit a harsh world and tenderheartedness is unlikely to survive its rigors" is probably about right.[44] The Duke's consideration for Angelo is not unqualified or disinterested pity. But it may be calculated remission, because forgiveness is a second chance for becoming a cooperator, and can even outperform the rule of non-cooperation with non-cooperators. As Joshua Greene points out, in the scoring game, even forgiveness has "deep biological origins," and with well-chosen candidates can prove fully adaptive.[45] We are now looking for the quid pro quos of the ancestral modes of justice involving cooperators and deceivers.[46] That is our ultimate destination: a brain-designed set of values reflected in the play's ultimate social order.

But hold that quest in abeyance for just a bit longer. What if Shakespeare, the provocative maker, deliberately designs the play so as to baffle our ethical ranking skills? That pertains to the second axiom above: that our faculties become disoriented in conjunction with computational or emotional overload. Will that not frustrate any hope for closural consilience? We do not want to be found solving riddles for which there are no answers. All authors devise ethical characters subject to moral scoring, but induce them into error through inadequate information, or place them in narratives of such nuanced complexity or ambiguity as to surpass their unaided computational capacities, as well as our own.[47] On the one side is the hermeneute's drive to understand the truths of human nature as revealed in the orders of art even as the artist devises social worlds that foil our mental platforms in computing binary values. Or ultimately, the author may baffle us along the way, yet deliver to us a reapportioned final world that chimes with our deepest social drives. On the one hand, our ethical scoring is subject to the "good enough" impediments of adaptive design; we are not truth machines but survival machines, and what was ancestrally sufficient is where further selective design became waste. As David Geary points out, the adapted mind is not a precision instrument in those domains in which what we have is sufficient, including not only our "folk knowledge" or "psychology" but even our "biologically primary abilities" through which we formulate social worlds.[48] In that specific regard, there is no other hand. Our ethical considerations are, all along, potentially compromised, and not only by the playwright's intentional, or not so intentional, showcasing of the disjunctive and equivocal aspects of his story.

Something of those hermeneutic challenges may be glimpsed in Shakespeare's sources and the degree to which their conflation in *Measure* constitutes a surfeit of information leading to divergent moral adjudications. First there is Shakespeare's habit of interiorizing his dramatis personae through their concealed thoughts, conflicting motives, self-justified strategies, and ambiguous desires. We have a lot of scoring and weighting to do, long before we arrive at the play's anomalous closure with its putatively dissonant representation of ethical order. The play has come a fair distance from its schematic Italian sources in which, initially, a *juriste* seeks sex in exchange for the release of a man dear to the woman he propositions. Her options are to refuse and allow a husband or brother to die, or to accept, trusting in the magistrate's promise. It is not an easy choice, but its terms are computationally limited and heroines have gone both ways. When she offers her honour and the promise is not kept, the story grants her recourse to a higher order of justice embodied in a ruler with an exacting sense of retribution – which does not exclude marrying her to the malefactor in anticipation of his execution, whereby his wealth is transferred to his victim – a little shaking of the superflux before the *lex talionis* takes over. The plot type is referred to as the "monstrous ransom," which, in the immediate Italian sources, is troubled by the heroine's own reconsideration of her status as a now married woman and the relenting duty she feels toward her husband. Initially it teases our moral calibrations only by the implications of the bribe, the nature of the husband's crime, and the rigour practised by the adjudicating ruler. But by degrees the story is transformed into a perverse romance in which the union is rescued from the rigours of the law. That sets the model for Shakespeare's plotting transition from deterrent show justice (kill Angelo) to an armed fairy tale (let the good magistrate Duke marry the victim while reorganizing the social order). Shakespeare troubles these matters even more by adding a framing plot concerning the administration of the law, a subplot dealing with plebeian mores, a clever heroine headed for a convent, and the meddling friar as eccentric referee, including all his strategic lies by way of trauma therapy, who then discovers his own emerging romantic interest in the beleaguered heroine. Quiet, ethical score keepers at work!

The story has always turned on the darker elements of human nature whereby gross victimization and plangent innocence enter into contest around questions of retributive justice and where that justice may be found. The "monstrous ransom," by definition, entails cost-benefit calibrations that, as social conditions are altered, can be rendered incrementally more difficult to score. The magistrate who offers to save a life from an excessive application of the law by sexually violating the victim's sister or wife will never fare well in the moral scoring of nearly anyone who hears the story. But even among the earlier versions from St Augustine to Giambattista Giraldi Cinzio there are variations that adjust the nature of the moral experience. Add, by degrees, a romance dimension through which the imperatives of love conflict with the imperatives of justice and the story becomes a laboratory in bewildered reasoning and empathic engagements. Discussion will invariably follow, largely in binary terms of merit and demerit, now according to conflicting scales of punishment and empathy. Scoring overload becomes a factor, leading minds intent on settling matters to indulge in selective evidence, tie-breaking principles idiosyncratic to the reader, or intuitional adjudication. Anxious to fix its claims, the mind can sometimes settle early into a position and thereafter rationalize away the inconvenient data. That too may be an adaptive strategy, but it will lead to interpretational diversity. Lisa Zunshine addresses this problem: "evolution ... did not have a crystal ball: the adaptations that contributed, with statistical reliability, to the survival of the human species for hundreds of thousands of years and thus became part of our permanent cognitive makeup profoundly structure our interaction with the world, but *even when they function properly*, at no point do they guarantee a smooth sailing through concrete complicated situations or the instinctive knowing of the exact origins of every aspect of our personal memories."[49] More on that in due course, insofar as the course of right for the Duke, as for Isabella, never ran smoothly. That makes for a good story.

As early as 1556, in Claude Rouillet's *Philanira* (a tragedy in Latin), the female principal allows sexual congress with the judge in order to save her husband's life. The judge nevertheless mocks her sacrifice by slaying her husband and delivering his body to her. But there is recourse, for when the ruler arrives and hears the story, he compels the judge to marry her and then stand for execution. Rather than finding justice, however, Philanira feels double dismay in losing both husbands, and turns suicidal. Already the calibrations of love cut across the calibrations of justice, as they do in the case of Shakespeare's Mariana, forcing adjustments to our scoring of the sinned against and the sinning. Other factors pertain in this play, including a condemned husband in the place of a brother, a more serious crime, and a tragic rather than a comic ethos. But we recognize the combined marriage and execution motif as the ruler's double tit-for-tat: first resource the widow, then eliminate the scoundrel. In Giovan Battista Cinzio's play *Epitia* (based on Decade 8, Tale 5 of his *Hecatommithi*, 1565), the heroine is a student of philosophy and can therefore advance sophisticated arguments on behalf of her brother's life, turning matters of law and mercy into a formal debate. That too we recognize as an interpretational crux in ethical scoring. Now the monstrous ransom is surrounded by considerations of right and wrong pertaining to the law, with and without equity. Cross-computational economies emerge. An awkward subplot is also incorporated in which an adolescent rapist, driven by the passions and power of love, is granted marriage to the girl as an option to prison. The lop-sided principle on which this solution is based – where are the girl's wishes in all this? – nevertheless gives us pause for reflection, especially regarding how the two plots do and do not share common principles. Epitia, meanwhile, remains coldly philosophical, even upon receiving her brother's dead body, before going to the Emperor. Clearly she has inspired features of Isabella's problematic character.

Cinzio, in this play (ca 1572; published in 1583), asks, as Shakespeare does: what if the brother could be saved by subterfuge and the play turned to a last-minute comedy? The principal invention is the substitution of another criminal in the place of the condemned brother, the work of the head jailer, all of which we recognize in *Measure*. Juriste, the corrupt magistrate, gains in complexity because he loves Epitia and proposes marriage to her, to which she acquiesces, he thinking all along that he might keep her even if he allows her brother to die for rape to satisfy the rigorous Podestà. All he had to do was persuade her of the weight of her brother's crime before the law. So then, should Epitia go on to marry a man she cares for, even if he goes back on his word concerning her brother? The conflict between love and justice is hers to resolve. Meanwhile, to his credit, Juriste wanted to change his mind, but the Podestà acts too quickly. Now Epitia goes to the Emperor for justice against a promise breaker. Angela, Juriste's sister, then pleads for her brother's life, but to no avail. New plotting elements invade the action, providing themselves as free-floating "theatregrams" apt for modification and variation. In this simple and inadequate summary there is already a vocabulary of motifs that Shakespeare reworked to more complex and equivocal ends: Isabella becomes a novice in a nunnery and Angelo makes no effort to recall his order for the execution.

That Shakespeare knew this source is hardly to be doubted, but he also had a reworking of the original novella in a play by George Whetstone entitled *Promos and Cassandra* (1578). Somehow, Whetstone managed to come up with the same solution to saving the brother: the substitution of the body of another criminal. Promos, the newly appointed judge, takes up an old law against adultery and, as in *Measure*, levels it against a young man and woman destined for marriage, but who have anticipated its legal consummation. The judge demands death and sticks to it. When he meets Cassandra, he falls for her and pledges marriage, but lust drives him to arm his offer with the promise of ransom. She grants, whereupon he proceeds with the beheading, feeling no remorse. More ideas for Shakespeare. Meanwhile, low-lifers are escaping the ruler's notice, for which he is reproved. A subplot is born. Promos remains truly hard, corrupt, and rather like a villain. Yet, when Cassandra turns him in, he realizes he is caught, and like Angelo acknowledges the justice in his own death. More plot grist. Cassandra is sentimental rather than stoic, and by those promptings imagines suicide on two occasions. In the end, however, she marries Promos to recover her honour and talks about being a faithful wife, as in the original novella. Paradoxical to say the least. Meanwhile, the jailer takes pity on the brother and rescues him, completing the tragicomedy. Andrugio, the rescued brother, in the interim, flees into exile before returning in disguise. Turn him into a displaced duke and the trickster magistrate hoves into view.

Shakespeare stirred this compound broth and came up with *Measure*. It would be entirely supererogatory, here, to enter into a comparative exercise; each work offers us its own unique configuration of personalities and events in its unfolding of plots for audiences to adjudicate. The point is that each also poses its unique level of difficulty to our computational appraisals. There is enough in these summaries to show how Shakespeare benefited from a cumulative story tradition – one that supplied not only the plotting materials but also the implicit moral challenges that he molded into *Measure for Measure*. Without conscious instruction, our brains go about the business of interpreting each inflection, each meme, in relation to moralized goals and outcomes on the part of each character. The play as readers have it is the one they must score.

We can now return to the question of the ethical themes of *Measure for Measure* – those pertaining to our problem-solving capacities as enculturated and experienced subjects, as well as creatures imbued with the design platforms which make ethical scoring the default system whereby we search out meaning in social representations.[50] Cooperators though they may be, humans are, nevertheless, shockingly inclined to "give a bit less than they get." They can distort their record keeping to their own advantages and rationalize their own truths and justifications. They remain individuals with a capacity to betray if advantage seems assured. Welcome to the kingdom of humankind, where a puritanical Angelo, once exceptionally empowered, is sadly tempted by corruption the moment he considers himself above the law and social reproach.[51] And welcome to the realm of Duke Vincentio who, in the presence of deceivers and defectors, must turn over his options for restoring a maximum level of inter-cooperation in tribal terms beyond the simple rule of law – the challenge posed by the leader of a would-be cooperators' society. Sanctimoniousness is yet another means to undeserved rewards through the crafted construction of self-serving values treated as beyond all reproach. As Richard Alexander confirms, "because of histories of genetic difference ... nearly all communicative signals, human or otherwise, should be expected to involve significant deceit."[52] These are the instincts of selfish replicators incited to push the deception detectors of others to their limits, as in the case of Angelo. But they are also specialists in identifying sham loyalty, emotions, and displays, as in the case of Vincentio. Both computational capacities are themselves the result of ramped up acumen over evolutionary time, and may account largely for the big-brained creatures we have become. Cooperation and defection pose a complex environment with survival implications. Children learn deception at an early age and hardly consider it dishonest. Everyone cheats to get parental resources. Mothers have a limited reserve of them and the survival of the fittest begins with the interested pursuit of her energy and supplies, including the child who resists weaning. In their physical weakness, infants readily resort to blackmail, cuteness, or screaming, and so the artisans of deception serve their apprenticeships.[53] When characters, alone or in collusion, fashion themselves as volatile mixtures of both strategies for the advancement of the self, ethical assessment easily loses its way, clings to convenient explanations or biased reasoning, or is entirely corrupted by hypocrisy. Cheaters and cooperators rolled into one – what a species – yet groups have means to readjust the ratios through detection, punishment, or exclusion.

With that, we should have almost enough to proceed. Arguably, this is the world of *Measure for Measure*, wherein Vincentio becomes the last computational resource for comprehending and outmanoeuvring deception – the agent who, without benefit of legal instruments, reconstitutes a society of cooperators.

The Duke is in a bind as the play opens because he has destabilized his state through an excess of leniency, providing opportunities for exploitative opportunists, including the stern and principled protege with whom he has replaced himself. Angelo's hypocrisy and abuse of power challenges moral detection; even those made party to his wickedness are nearly silenced by his overbearing authority. In this manner he seeks to intimidate Isabella. The Duke, initially, is unable to decide whether licence is self-regulating, whether states must intervene through the law, and whether overregulated states will be deemed tyrannical in outlawing harmless indulgences and all too human pleasures. This is a collective confrontation with the cooperator's challenge, the challenge posed by unregulated societies seeking to exploit limited resources to maximum self-advantage through individual self-monitoring. Bullying, hoarding resources, or deceitful appropriation constantly imperil social systems. In response, cooperators will build coalitions of the victimized to regain civic order. Under Vincentio's tutelage, his state makes an emblematic step toward social homeostasis by enforcing compliance on two diversely unruly members. This he achieves as a duke-in-disguise who plays the amoral agent in relation to self-justifying ends: "craft against vice I must applie" (III.ii.299).[54] This theatrical device Shakespeare alone brings to the interpretation of the seminal story. In his reading, he no longer resorts to a representation of the good magistrate pronouncing his verdict as magistrate. Vincentio replaces the talion justice of a legal proceeding with the improvisatory and utilitarian justice of a trickster whose pranks alone constitute social judgment; it is a theatrical solution to a theatrical representation of a social crux.[55] In this he substitutes for the old law of talion justice an even more ancient survival code pertaining to self-regulating societies: help those who are likely to help you, marginalize those who never will, and forgive potential new cooperators for their past infractions. In his mental economy, there are those in his debt, those in his sights, and those reformed by shock experience. The Duke is scoring closely, reading intentional states as never before, and improvising the equivalent of retaliation through enforced domestic arrangements. Paradoxically, through the imperatives of romance values, he regulates a disorderly state. But can state justice be achieved in this manner, particularly after framing the play in legal terms? And will tactics deemed devious in themselves ever find universal approbation as a means for social regulation? This is a stumbling block for many, despite the fair outcome.

Along the way, Vincentio, in order to reimagine justice, requires privileged information, and much of the ordering of the play is in creating those opportunities. This feature of the work may also be interpreted as a thematic *idée force*. Information procurement is vital, and arguably throughout the era of "the talking ape," the principal means has been gossip. "According to anthropologist Robin Dunbar humans spend about 65 percent of their conversation time talking about the good and bad deeds of other humans – that is *gossiping*."[56] It has been the medium of information exchange of choice concerning deviancy, outsiders, trustworthiness, and altruism throughout recent social history. It is not, in itself, an adaptive trait – it is too recent – but by all means it was applied to a furtherance of the values promoted by the hardwiring that established the computational capacities for cooperation as a marker of the species. Once talking became our forte, gossip became the most efficient means for legislating the economies of reciprocity based on honour, conscience, and shame – the social emotions whereby collectivities control the ethos of exchange. Language merely extended and facilitated our instincts. We need to know who has power, who sleeps with whom, who is loyal, who tells the truth. Moreover, "trading gossip is one of the main things friends do, and it may be one of the main reasons friendships exist."[57] We make alliances among the trusted by sharing information, cheap for us to deliver, but of tremendous value to others. In the words of Jerome Barkow, we are endowed with "psychological mechanisms that evolved in response to selection for the acquisition of social information."[58] It is our capacity to process such social information to our benefits that is adaptive.

Theatrical gossip incites us to invest our attention and feelings in the lives of imaginary characters and to employ our psychological modes in the understanding of other minds.[59] Through information exchange, as dramatized, we participate directly in the panorama of instability produced by secretive advantage-takers, ultimately contained through machinations made possible by exposure through exchanges of information among cooperators. A new society begins to emerge as damage is controlled, favours are shared among the victims, IOUs are called on the basis of guilt, and the group comes to a collective ritual, staged by Vincentio, whereby in-groups are restored and out-groups (or individuals) are marginalized, based on the overall fitness of the group.[60] The principle matters, because through his disguise, Vincentio is an information collector and thereby gains privileged insight into the deceivers and malingerers who destabilize the social machine. The entire play is designed around the movement of this information. But more pervasively, in discussing plays in these seemingly remote terms, there is a new vocabulary for defining the archetypal shapes which emerge not only from Vincentio's quirkish inventions, or the social values of a culturally unified population, but from the arguably innate predisposition of the species to score justice and fairness as a precondition to the cooperation which alone has permitted its spectacular success.

Let us suppose that the play becomes an experiment in social justice meted out through a crafted combination of trickster planning, exposure, pardon, and constraint. The most difficult equivalence or "measure" to be drawn is between the implementation of the law of talion for crimes committed and the improvisatory settling of accounts through comic justice. It is a computational exercise as messy as the world Vincentio had to deal with, a world of corruption, yet peopled by those, for the most part, with better potential. Rules have not served well for stabilizing the state, whether strict or lax. In turning trickster, Vincentio turns consequentialist by reverting to best possible scenarios achieved through a reading of psyches and risks in the cooperators' game. Trickery becomes his way of outing and assessing those psyches by putting them through behaviour tests of a kind he is able to devise only by the aid of his disguise. The argument in hand is that his trickster justice is linked to long-standing social benefits only obliquely related to the law and far more directly related to the biogenetic wiring and control mechanisms of cooperation. As a secret agent, he opts to rebuild a functional survival community according to the arguably gene-assisted values of reciprocal altruism. Moreover, his moral economy is reproductively centred (as is Isabella's for that matter, not to mention Claudio and

Julia's), and by implicating the entire cast in this transposed economy, his machine becomes collective and inclusive as well as mirthful for most. Some readers are uncertain that the arbitration of a social world by an enlightened but potentially interested trickster-magistrate, tantamount to an armed prophet, is an adequate vehicle for the performing of justice. But that is less and less the issue. What we witness is an astute puppet master who designs the redemptory activities whereby the peccant are tricked into becoming future cooperators.

Through enforced marriages – and this is key – Vincentio performs utilitarian justice according to the material rights of brides as mothers or prospective mothers by concentrating on their benefits and the legitimation of their children, while, at the same time, curbing male sexual freedom. After all, this is the domain in which nearly all the infractions are committed which have turned the state upside down in the first place. Freud would have called Vincentio's values the necessary entente that defines the discontents of civilization; monogamous unions which furnish sex for security in the interests of the offspring.[61] Why, we may ask, should this issue take over a play that is about crime and punishment? Can the evolutionary backstory to our ethical brains supply a rationale for Vincentio's prioritizing of the conditions for reproductive success, or is this an argument of necessity to cover Shakespeare's eleventh-hour drift into romance? Can we build an analytical bridge between talion justice for attempted murder and the emotional values that stabilize the family?

Claudio and Julia as loving replicators are precious to a self-preserving society, yet they are the near victims of an overzealous interpretation of a law paradoxically designed to stabilize family relations. Such victimization might very well be deemed an offence not only against sympathetic innocents but against the genetic future they represent – our brains are wired in relation to these values, as Richard Dawkins has argued so insistently in *The Selfish Gene*. The law, read to the letter, has become an ethical crux, disconnected from the imperatives of fertility and the family-oriented society. Hence, we feel pity for the engaged couple, despite their infraction of the code. What then if Vincentio adjudicates according to the self-preservation side of social thinking, bringing temporary order to his state by imposing marriage, not death, on those who have enjoyed its fruits, while compensating Isabella by raising her up to a duchess? Do we not have a kind of justice in equal measure based on a hopeful perpetuation of reciprocal altruism through collective cooperation? Once again, Vincentio was not a mental robot acting according to some automatic dictate of his brain. Quite the contrary, he was a most adept and rational problem solver as evinced by the complex scenarios into which he draws, unwittingly, an entire social set in order to generate the experiences which will segregate the actual and potential cooperators from the defectors. Tricksters can be mentalists who engineer redemptive experience. At the same time, those same stratagems are aligned with solutions that chime with the socializing platforms of a designed brain pertaining to social and reproductive success. Something here may be reaching all the way back to the level of our species that emits the emergent thoughts favouring the very outcomes he pursues through all the levels, constraints, and conditions outlined above – the emergent thoughts which in turn make sense of the tropisms embedded in so much of our social and cultural behaviours.

Ben Jonson was ready to cast us into a world of gulls and knaves, of victims and deceivers, by deleting the psychology of cooperators from his theatrical vocabulary, which may point precisely to the principal difference between the two authors. Jonson's mode is satire, a form of metonymy in which the unfavourable traits of human nature constitute the whole range of human potential. Containment of excess and deceit is his only concern, corruption which comes about largely through the efficiency of cooperation among hucksters, debauchees, or confidence men. Vincentio, by contrast, reconstitutes his world of reciprocal altruists by including those who can learn to become cooperators. Law may demand its due in terms of talion, but the instincts behind the cooperators' economy have more elasticity, for the controlled pardon of defectors is a great inducement to future cooperation among those capable of calculating long-term benefits. Those of duller spirit may be incorrigible, but Angelo performs the rites of potential re-inclusion through an emotional display, not only of fault and confession but of an honest acceptance of his deserving according to the law. Forgiveness, in the reciprocator's economy, is not weakness; it is a calculated risk toward realizing future benefits both individually and for the group. The transaction relies on the refinements of computation through which we score fairness and on the subtle computations through which we compute advantage. The beauty of this innate platform is that cooperation emerges out of self-interest on the part of all participants. The distinction is vital; it is bedrock. Apology is a pledge of intended future cooperation.[62]

Hence, the Duke, through privileged observation, makes decisions pertaining to social rescue and social exclusion based on the default computations of a primate brain. Alexander calls this the "assistance and ostracism" principle, which is the basis for enforcing the moral systems attached to reciprocally based societies: make alliances or evict and exile.[63] Comic justice replaces the laws pertaining to crime and punishment with the pre-legal principles of exclusion and inclusion and in that sense reflects the mechanisms of ethical scoring integral to ancestral social regulation. Vincentio is perfectly mindful of merited punishments according to the full rigour of the law, but he takes calculated risks based on human ethical potential. That is his job in the play, in full contradistinction to the scoring by which we might otherwise consign Angelo, possibly in keeping with the instincts of Wittgenstein and Dr Johnson, to a talion-based punishment. This new economy emerges in the words of Mariana: "They say best men are moulded out of faults / And, for the most, become much more the better / For being a little bad" (V.i.437–9). Self-interested rationalization, we presume. But what if she is right, in Angelo's case? May not the measures apt to control folly, such as public apology or regret, be shammed and thus fall short of the requirements for containing vice? To that question, all that the Duke had devised by way of entrapment and confrontation constitutes the rites deemed sufficient to reprove and reform a villain. Moreover, it was not an arrangement without teeth, for the expected cooperation is heavily reinforced with moral and judicial leverage.[64] To quote Joshua Greene once more, "enforced cooperation is surely one of the driving forces of history. Chiefs and kings and emperors have used their increasingly large carrots and sticks to enforce productive cooperation."[65]

On another front, however, the Duke scores matters entirely differently. He shows his disapprobation of Lucio and backs it up with indignation – the body language which states that no cost is too high in exacting punishment – the stance of the altruistic punisher to whom fellow cooperators look as the instrument best equipped to pay whatever is necessary to protect them all from the incorrigible. This is an instance in which we might observe just how our "capacity to track little, nameless, but remembered acts of strong reciprocity and altruistic punishment might have been a central evolutionary achievement."[66] The possession of incriminating information fashions Vicentio directly into the role of the altruist with the power to confine and punish. Not surprisingly, when the Duke reveals his identity, Lucio realizes that his situation might be worse than hanging (V.1.358). Angelo, it will be objected, had played the irreproachable card in conjunction with his Puritanical facade, making deception the principal vice in this play, in keeping with McGinn's assertion that it is the primary vice in all of Shakespeare's plays.[67] When he is discovered, where, then, is Vincentio's indignation? Darwin himself once wrote that if a wicked man is "incorrigibly bad nothing will cure him."[68]

But the Duke is satisfied that Angelo is not incorrigible and that he has been shamed into remorse – unlike anything he could manage with Lucio. The latter is a parasite, a man who manipulates the reputations of others unconscionably, yet remains unmoved regarding his own honour or personal credit. Vincentio's emotional instruction is justification enough for his severity. A minor menace to the state, yet Lucio has touched a chord as an irremediably brazen slanderer, which disqualifies him as a future co-operator – a reprobate mind, beyond correction for his *lese majesté*, not to mention his abject insensitivity to his own child and its mother. Lucio is to this play what Caliban is in the school of Prospero's island: one "on whose nature nurture can never stick."[69] For the moral animal, "reputation is the object of the game."[70] Without it, an individual's reciprocity capital is nil, and to disregard its importance is to excuse oneself from the cooperator's union as an unwitting defector. But worse, Lucio violates the innate instructions for paternal caring and its role in the social order. This play is neither about universal brotherhood, nor about a religiously framed ideology of unconditional pardon. Lucio's handling is critical in that regard. His exile to the anonymous suburbs with his sub-valued wife is singular and necessary. Lucio may appear arbitrarily picked-on and scapegoated, but as a subset of deceit in action, the Duke's scoring of him tells an epic in the monitoring of ancestral values and the regulation of society. Good magistrates select cooperators.

More computational challenges arise with the romance motif. Just when did Vincentio find himself attracted to Isabella, and from what moment were his actions recalculated to accommodate that fact? Concomitantly, when did Isabella detect his interest and begin to calibrate her advantages through cooperation? In Act V (scene i, 385–97), the Duke begins to excuse himself to Isabella for letting her brother die. Why would he do this, knowing that Claudio's life has been secretly spared? Why would he not confide and bring her release from her grief at the earliest moment? Readers may be quick to blame. Yet, timing is everything for tricksters because the actualization of planned situations is the critical means to the production of redemptive experiences for others. The bed trick is a case in point. In such matters, Vincentio calculates long-term benefits over present means; what trick does not? Included among those long-term benefits, however, is a marriage proposal, lightly offered. Timing here, too, is of the essence. Claudio's deliverance, too soon announced, would have diminished the force of Angelo's purgatory and redemption. But the effect of that delay is that Vincentio is simultaneously testing Isabella for signs of empathic kindness, given her hardly reassuring attitude toward her brother. Virtue, after all, has placed her in a terrible self-representational bind insofar as it is based on ethical absolutes, which make actions good or evil in themselves without consideration of collateral or contingent causes. This has brought her much censure over the critical years, and yet in matters of virginity, it is also a physical absolute and a heavy signifier. Most critically, she is in a bind precisely because those with the power to choose their partners demand virgins, even though they will also prefer them to be sociable, forgiving, and kind. Or perhaps Vincentio is avoiding the announcement concerning Claudio before broaching his affection in order not to coerce her decision through the blackmail of gratitude. Readers are scoring, because the marriage proposal, requited or not, is hugely significant in bringing an ethos of romance to this story of crime and punishment.[71] The paucity of hard evidence about Vincentio's mind will not deter the moral invigilators, because brains are brilliant, if at times unreliable, at working with the nuances of intentionality.[72]

We can try our own. Here is the basic issue. If Vincentio was suddenly smitten or imperceptibly moved by love, he must, at some point, have been impressed by Isabella's beauty, intelligence, and virginity. We are given a hint that he too may have been taken by surprise. He eavesdrops on her conversation with Claudio in which she tells him that his sin is not accidental, that mercy on her part would prove a bawd in assisting his lechery, and that imminent death is his best course. Following hard upon this rebuke by a sibling, the Duke introduces himself: "The hand that hath made you fair hath made you good," a goodness that is not cheap either, but a grace in her soul that will keep her ever fair. It is a little received Renaissance truism that inner virtue chimes with outer beauty. To this he adds that he might thank Angelo for his assault upon her in bringing her so fortunately to his attention (III.i.179–84). That is hardly a declaration of love, but it is a declaration of appreciation and interest in the perceived traits so often associated with liking. Otherwise, his aspirations remain curiously hidden until the final scene when we are invited to retro-sequence the thoughts that led from seeing to loving. It is one of the most minimal courtships on record, yet presumes a plenary process. Is it because, even as the lord of a realm, he recognizes the limits of his prerogative over the beloved's right to choose? Is it that he may command Angelo and Lucio to love those whom they have injured, but not presume upon an autonomous woman's rights – rights which trump the obligations implicit in favours received, including the protection of her virginity and a brother's life? Then time passes, contingent issues are resolved, and there is nothing left except an ultimate offer of bi-lateral reciprocity: he offers to her all he has in exchange for all she has, the material imbalance presumably offset by her virtues (V.i.534). Even marriage offers are based on reciprocity in deeming that bargain best in which both perceive the benefits of the union to be the greater for themselves.

Readers at such junctures instinctively begin to score the future of the couple based on the characteristics they have witnessed and move on to all they know about the exchanges of married life concerning divisions of labour, investing in the offspring, domestic tranquillity, attentive support, and so much more. Seminar students insist on crystal balling the domestic prospects of Isabella and Vincentio according to indices hinted and inferred. Accordingly, there is a temptation to reinterpret the forgiveness scene in which Isabella is invoked to pardon the man who had offered to abuse her, quite pointedly against the Duke's advice. For in resisting Vincentio's putative dissuasion, Isabella provides the only instance in the play of her empathic generosity. Could it all have been a test, or the probative exploitation of a moment? Isabella has presented herself as incredibly hard to get, and the Duke does not have a lot of time to find out whether she has a more tender and sociable side. This moment of selfless consideration for Mariana is her only chance to intimate another side of her character. If reproductive viability is the order of the social economy Vincentio seeks to restore, then family values also set the terms of cooperation. Binary minds and scoring readers, perplexed by evidential paucity, nevertheless cling to outcomes that align themselves with the order of thought that pertains to the most intimate reciprocity of all, bound by emotional exclusivity and mutuality.

At this juncture we are staring at a fairy tale and hence cannot rule out that the heroine is very sensibly inclined to mate with a male who holds territory and who has "high status in the dominance hierarchy." Fairy tale heroines, high or low, always marry princes or kings, the low born having offset their poverty by their virtue. If there is an archetype in those tales connected to emergent impulses, it is in the exchange of virginity not only for love, but status.[73] Of course, in *Measure*, the deal is not yet done. Isabella may also stay by her first conviction that her chastity is simply too precious for mortals of any stripe and remove herself from the gene pool. It happens.[74] Her mind remains a mystery even at the end, but readers will write their own endings from the scattered innuendoes, and in the process will self-compose the play's genre as they like it. Possibility and predisposition may be enough for conviction, much as we supply or deny fairness to the arrangements at the play's end.

Isabella, in fact, provides the central crux of the story tradition, touching as it must on her personality and potential as a future spouse, namely the choice she must make between kin loyalty to her brother and her obligations to her own unborn children.[75] To a degree, this choice, too, is between group cooperation and the selfish genes. I'll try to explain. The biological rationale for virginity is made equivocal by Isabella's religious vocation, but in bioethical terms, demonstrable chastity is of value only as assurance to a prospective mate that he will be the true father of his children. Isabella need not even be conscious of why she is so ferocious in its defence, but something from within has powerfully spoken. Whatever virginity means to her, the Duke has his own biogenetic drives for favouring it. If, as a man of status, he is interested in her romantically, he must reason in dynastic and jealous terms. The stakes are high and support the view that "coyness can actually pay a female's selfish genes."[76] Even Angelo thought so: "Can it be that modesty may more betray our sense than women's lightness?" (II.ii.169–70) What he sees in her moral character has made him love her, despite his malignant intentions.

Yet social scorekeepers may assume that the Duke is simultaneously troubled by the principled severity with which she refuses her brother. Even though her honour in anticipation of a Cinderella tale is not a conscious feature of her thinking, her emergent thoughts may still have enforced the logic of her genes. That would seem hypothetical were it not so prominent a feature of sexual negotiations across cultures, and had Isabella not said so herself. When the Duke asks her how she will respond to deputy Angelo, she does not say that her chastity devoted to the service of God trumps all considerations and makes her fearless of death, but that "I had rather my brother die by the law, than my son should be unlawfully born" (III.i.188–90). Discount it however one may, Isabella justifies her priorities from the perspective of her unborn children. Shakespeare hands us that one. Rightly or wrongly, a woman's honour is so defined when it comes to marital negotiations. Isabella goes so far as to construct the survival of her brother at the cost of her virginity as a form of incest (III.i.138–9) – a taboo which makes sense only in relation to the reproductive fitness gained through exogamy. Her calculations, however she meant them, have to do with mate selection. She has our attention now because she has reverted from religious to reproductive values in the defence of her honour. Who did she have in mind all the way back in Act III?

Duke Vincentio is decidedly the most difficult character to evaluate, for by implicit authorial design, the play's closing moral configuration results from his ethical arbitration of a society through probative trickery. To be sure, for Wittgenstein and Dr Johnson, his results do not constitute a moral order worth mention. Even so, Vincentio's achievement is the product of his own implicit exercise in moral scoring in binary terms on the assumption that if he is mimetically human, his brain works just like Wittgenstein's. (We're both smiling.) But in some sense, he arranges the prospects of those under his spell of trickery according to ethical terms, and our interpretive challenge is to establish the working social basis of those terms, whether or not they constitute a thematic vision of moral order rising to the challenges posed by the condition of the realm.

The solution posed here has suggested itself through the recent investigations into the biogenetic origins of the computational cortex in relation to generic conditions pertaining to survival, which depend on the rudimentary capacities of ethical computation. Something has built up the wiring that incessantly produces emergent computations of fairness in conjunction with the emotions that make the most basic forms of cooperation possible. Such mental orientations, in their way, are as basic as all the more mechanical designs that have contributed to the viability of our species. These kinds of arguments do, indeed, go a long way back to make a point about the values implicit in the social engineering of a theatrical trickster. But if the principle stands, saying what it does about the substrates of our behaviour, we need only ask, hypothetically, what it offers to explain about the social templates actualized according to Vincentio's plans. Are his cooperators' values tantamount to an expression of a mental grammar? Biogenetic backstories may be carelessly opportunistic in being far too specific about habits which arose too late to have benefited from the selection mechanisms of evolutionary time, but the moral properties of a binary mind which sustain our fundamental capacities to care and cooperate are ancient and determinant. The present argument is either comprehensively right or wrong, and if it is fundamentally right, it will not crumble under the fine tunings which still elicit debate among bioethicists and cognitive scientists. There is no niggling this phenomenon into oblivion by hunting down the differences in analytical vocabulary among specialists. In keeping with these values, I am arguing that Duke Vincentio abandons the law of strict talion for the law of exchange ethics by using the imperatives of romance to stand in for a combination of reward and punishment, bringing comic justice into a common circle with the ethical brain. New to these problem comedies is that the ethos of romance is extracted out of conditions pertaining to legally bound societies in which the efficacy of law itself is under investigation. Shakespeare, in his storytelling based on legal cruxes, sets out in search of the spirit behind the law and finds it in the scoring of deceivers and cooperators, which runs deep in the ethically tinted emanations of our primal thinking. Inferences must, nevertheless, be drawn with infinite caution.

When the Duke opts for the bed trick, he opts for a device in vogue in the literary culture of his age. For its efficacy as a moral instrument, he had the authority of the historical accounts of Amasis, king of Egypt, Theodoric of France, and Pedro II of Aragon, who were impotent with their wives but not with their concubines, yet who, through bed tricks organized by their courts, their spouses, or their physicians, took full enjoyment from their scorned wives, thereby begetting legitimate heirs for their respective realms. Equally telling, the trick served, in the case of Pedro II, to reform his ways and rekindle affection for the queen.[77] The impulse behind such arrangements in history was the restoration of social order through legal sexual union and the begetting of offspring. How should such transactions be scored, and how do we arbitrate among the social benefits, the prerogatives of tricksters to arrange them, and the rights of the participants not to be deceived? The device is cultural, the adaptation of it elective on the Duke's reasoned part, but the fairness it produces for both Isabella and Mariana belongs to generic calibrations both cognitive and emotional concerning care for mothers and their offspring.

All that I wish to test here by looking at the brain's own grounding in binaries and reciprocity is whether, through these innate values, we may understand Vincentio's choices in relation to priorities associated with human nature itself. Vincentio is confronted with hypocrisy, legal bullying, abusive sexuality, judicial murder, slander, and loose morals in the suburbs, all of them subject to legal management. He can score these matters as well as another. But his option is to restore order in terms of family law pertaining to bonds and promises and the rights of married women to protection and resources. The immediately pressing matters of the play are that Mariana is betrothed without a husband or a child, Lucio's bed-partner has a child but is unprotected, while Juliet is soon to have a child yet destined to lose both husband and father. Trivers did the mathematics for abandoned women left to care for children on their own.[78] Lucio has implicitly said to himself, I can leave as soon as I'm assured that the investment of the other parent is so great that my half will be taken over by the nurturing instincts of the mother, leaving me free to carry on with my philandering. His ethical values in the matter are revolting: "Marrying a punk, my lord, is pressing to death, / Whipping, and hanging" (V.i.520–1) – his public image coming before the needs of his own children. Vincentio is surely scoring this crux in a double capacity: as a social engineer in sympathy with abandoned wives, and as a ruler concerned for the future of the state through the stability of the family based on bonds and obligations. In coercing the formality of marriage, relationships fall under legal control, thereby obligating the sexual participants to assume their material duties. Simultaneously, Vincentio has conducted a benefits-to-costs analysis between the interests of the family and the aggression of the law against criminals. The play, after all, encourages spectators and readers to score the entitlements and losses of all three women in terms of fairness. By intimidating Lucio, arranging for a deceitful coupling through the bed trick, and saving the life of a devoted and well-intentioned fornicator, Lucio's partner, Mariana, and Juliet will enjoy essential benefits: the protection and resourcing of offspring. These are ancestral values, the most adaptive of them all and vital to social order.

This brings us to the pattern of the play in general, namely a highly particularized form of hero-centred, potentially romantic, quasi-tragicomedy.[79] The Duke's departure and negotiated return frame the action. In this heroic mode, he is the equivalent to the duke-in-exile whose kingdom has been usurped (as in Marston's *The Malcontent*), now merged with the ruler on holiday fleeing the burden of office (as in Marston's *The Fawn*). The substitute he leaves in office becomes the potential tyrant and usurper destined to re-enact the monstrous ransom motif. The general political ramifications pertain to internal disorder and the royal prerogative in managing social riot and excess, even in ostensible absentia. Without sharing his game plan in advance, the absent-yet-omnipresent Duke, through his own moral instruction, hits upon several efficient stratagems through which to intercede in current affairs, limit crimes, and set up his own return. Justice now must fall outside the courtroom, abused as it was in Act II, and proceed along lines of rehabilitation, shame, the protection of reproductive prosperity through family renewal, and the exclusion of reprobates. These are values for the nonce, if not the future measures for escaping the disorder of a failing state. This archetypal action of the epic hero is accompanied by the emotions associated with high-minded deeds through which the protagonist assures his reputation and contributes to his material and, potentially, romantic survival. His deeds all along have been evaluated according to the binary scoring templates or prototypes through which we understand the ethics of action.[80] If the moral values implicit in those deeds align with our own, the play completes its experiential argument, forming a community of understanding and cathartic joy in deliverance. That requires no further elaboration here, given all that has been said above.

But the problem of means and the problem of the disorder that remain latent in the society – two very distinct challenges – continue to pester the minds of readers. The festive return and proto-romance are darkened by residual menace, as well as by the diversity of our views concerning the Duke's methods. The problem of means arises because the deception scenarios he creates are stories-within-stories, wit-crafted trickster narratives fulfilled unwittingly by the actors he enlists. Tricks have an ethos all their own, which may benefit society yet not allow their perpetrators to escape opprobrium. There is no right answer to this crux. Even social benefactors who operate by subterfuge are often deemed to merit exclusion. These concerns, for many correct-minded students, continue to overshadow the mythical order of oppressed goodness and legitimacy seeking restoration in a world of calloused corruption.

However we rate Vincentio as a trickster, he remains a man on a mission based on ethical weightings and a responsibility to enact justice. These are profoundly tilted cognitions in relation to all the emotions that guide us as creatures of guilt, negotiation, anger over injustice, and cooperation – cognitions which, paradoxically, include our capacity for deceit when its advantages outweigh the dangers of detection. These are the values of survival – those to which our brains are most inclined to give attention, particularly in terms of mate selection and exclusivity, distribution of material resources, and defence against outsiders and cheaters. Principles of reciprocal altruism form a large part of those ethical calibrations. Granting the trickster his rightful agency according to conventions, wit, and the legitimacy of his ends is surely a necessary pre-condition to the validation of his means – a concession more readily forthcoming in the Renaissance than in our own age of over-sensitized political rectitude. Granting to Vincentio the right to replace talion with the constraints and rewards arising from his sense of social justice, however playfully, is the principal convention of the play's imaginary world. That comic justice replaces legal justice with arguably better, more constructive social results necessitates a paradigm shift in design which some readers resist. Vincentio's justice, grounded on the conditions of the family rather than retaliatory law, continues to baffle, but perhaps less so when those themes are aligned with bioethical values. This is not meant to browbeat the reader with science, but to re-establish the play on some of the most apparently forceful orientations of our natures. For James Q. Wilson, science may be taken for the enemy to morality, but in fact, it is our best recourse to what morality is, why it emerged, and why it is so powerful.[81]

As a tenet of our cardinal natures, the binary moral invigilator system, with its embedded meanings, becomes not only an instrument for social evaluation but the system producing the emergent meanings we most expect to find underlying the behaviour of all humans. For Robert Wright, what is emerging is not a denial of culture, or an endorsement of social Darwinism, but a new way of scanning the world's peoples by focusing "less on surface differences among cultures than on deep unities," as I have sought to do for this play. Those deep universals would appear to include guilt, notions of justice, feelings of empathy, the protection of children, and sexual jealousy. The list of such architectural abilities is constantly under revision, but the generic nature of these traits and emotional values continually suggests a means of preservation well beyond the more fitful capacities of cultural transmission.[82] With this explanation, we may be a tad closer to imagining archetypal themes behind Shakespeare's comic order, to the degree they coincide with the values that produced the brain platforms that give rise to our ethical instincts. The norms are in the brain's own way for shaping emergent thoughts, modified by experience, culture, and circumstances; they have nowhere else to exit. What matters to that architecture need not dictate what must matter in the assessment of social art, but in defining our natures, it is never a bad place to start.

CHAPTER THREE

On the Emotional Intentionality of Criminal Protagonists

The Yorkshire Tragedy

The causal relationship between the emotionality of literary characters under stress and their consequential acts may never be entirely resolved, largely because their motivations need be no more cogent than those of persons in everyday life, whose motives do not always conform to the templates of reason and expectation. That question becomes particularly acute in the case of certain tragic protagonists who find release from their unbearable circumstances in counterintuitive acts of violence most easily accounted for in terms of temporary madness, yet a madness still bearing hints of meaningful intentionality.

Can and should we seek to know more? In order to plumb these depths, we can turn only to the mind with all its systemic mechanisms for shaping desire and will. Crimes of passion are among the most mysterious, and novelists are often inclined to leave that mystery in place, even though we, as well as characters in the story, feel the compulsion to provide explanations of the twisted logic that carries the crazed mind into action. In Dostoevsky's *The Idiot*, as Lubomír Doležel rightly points out, the murder of Nastasja Filipovna remains the work's "deepest mystery. Myškin never asks, and Rogožin never tells, why he committed the crime." The lawyer appears with an explanation, that it was brain fever, and the jury accepts this explanation, but it explains nothing; "it just disguises dark passion as a pathological nature event." Doležel realizes the problem but is not, himself, prepared to explore further, concluding that "the tragedy of Dostoevsky's heroes is a tragedy of those who lose control over their acting and, consequently, over their destiny."[1] With that we too may be content, or we may wish to press on with the questions posed by such works and what they tell us about human nature in its darkest and most enigmatic manifestations. Irrational deeds cannot be entirely so to that part of the brain that set them in motion. Hence

Dostoevsky challenges us to know, for such crimes tease our explanatory science of mind and the ostensible infallibility we grant to our analytical capacities concerning human motivation. If authors imitate what is or may be true, we may ask how those mimetic events originated, which often lead to matters of the psyche and to theories of mind. Such an inquiry is doubly legitimated when the criminal acts forming the literary *récit* originated in historical events.

The work chosen to exemplify the challenge is the anonymous, early seventeenth-century play *The Yorkshire Tragedy*; it is a compact dramatized story that investigates a sudden cluster of criminal acts. In its design, it offers a narrative of the interpersonal encounters of a protagonist identified to us only as "The Husband," who, in a moment of desperation, executes the most counterintuitive deed a parent can commit: the slaughter of his own progeny. Characters enact their parts in this unfolding drama in ways we assume to be explanatory; the play reveals by degrees the exacerbations that precede the sudden violence – events which are clear in themselves, yet enigmatic in their effects on the protagonist. Together, they form what we have of the story of a man plunged into a mental state that induces him to commit murder. Through these episodes involving the protagonist and his wife, then gentlemen from the community, and finally a Master of the College, we come to understand that The Husband is a victim, by fate or his own folly, of a devastating change in his social and material environment. As an addicted rioter and gambler, he had squandered his estates in dicing, which had brought him to a full acknowledgment of hopeless and irrevocable destitution, for he had nothing left to wager and little left to live for. Devoid of imagination and obsessive in his strategies, he could only imagine that by bullying and abusing his wife, he might come into possession of her jewels in a last ditch effort to turn fickle fortune in his favour. When his wife returns from a visit to her uncle with a counter offer to gain a new livelihood through service at court, thanks to her uncle's preferment, the proud husband not only disdains the offer as beneath the consideration of a man of leisure and pleasure but enters into a state of despair over the loss of a final bid to recover his credit among the rioters who had bankrupted him. These are circumstantial causes leading to his delusions and to the ultimate expression of his personal collapse in the destruction of other lives, determined upon by a perverse but undisclosed chain of mental events, and administered in a state of cool and mechanical distraction.

The initial challenge posed by this play, as by Dostoevsky's work, pertains to the nature of temporary madness, runaway emotions, or the operations of what Antonio Damasio refers to as the "cognitive unconscious."[2] In all of these potentially explanatory conditions, a significant degree of conscious control is lost to a mind still capable of formulating decisions and actions. The question is whether such actions can include pathological, subconscious, or systemically induced murder in relation to despair, anger, abjection, or confusion. One may demur that regarding tragic protagonists, we need not ask precisely how, in cognitive and affective terms, such deeds are generated. We may simply fall back on the words of George Steiner "that there are in nature and in the psyche occult, uncontrollable forces able to madden or destroy the mind."[3] Where the moral order is concerned, all that matters is a demonstration of circumstantial provocation in relation to the knowledge of right and wrong. But establishing even that finishing line by which the two are distinguished takes us through an implicit cycle of considerations concerning a theory of mind, the origins of motivation and moral standards, and the levels of cognizance in volition. In that regard, every word and gesture, interpreted by our best estimative psychology, becomes evidence, all of it dependent upon characters in action in relation to thoughts, in relation to circumstances, in relation to the components of brains and their emergent mental events.

At such junctures, we are compelled to fall back on our default notions of social cause and effect, or we must turn to the psychological theorists writing during the era of the play to account for the world views most likely pertaining to the author's sense of human psychology, or, finally, we must turn to the present state of research in the cognitive sciences concerning psychological causation, intentionality, and the ways in which the brain makes up its mind through its stratified modes for evaluating the environment. All such approaches acknowledge that we need a theory of action to resolve the motivational enigmas of the play in the interests of assigning moral responsibility for the acts committed.

But how do we produce a theory of action? Curiously, scholars do not hesitate to look into seventeenth-century theories of psychological motivation, yet many are reticent to turn to the very latest forms of analysis – theories unknown to Renaissance authors – to explain early modern behaviour.[4] But if cognitive explanations of human nature are cogent in themselves, they should, in qualified ways, be valuable as explanatory tools for the assessment of character, literary or real, in all ages. This is because, in phylogenetic terms, human nature has not changed in measurable ways since the Pleistocene, and hence, in subsequent years, the genome that defines the species has remained relatively constant.[5] That biological endowment has generated and met many different forms of cultural expression and historical contexts, but the design of the brain in terms of its fundamental apparatuses for the production of thinking and feeling states is common to all members of the species. That humanists may be disinclined to subject fictive constructions to the dictates of an evolutionary theory of human nature is understandable on the surface, but only at the risk of bankrupting all artistic representation of the human condition as alien to what the species is and does in the phenomenal world. To them I can only reiterate the words of Ian McEwan: "that which binds us, our common nature, is what literature has always, knowingly and helplessly, given voice to. And it is this universality which the biological sciences, now entering another exhilarating phase, are set to explore further."[6] This remains a delicate business, for if the characters in Renaissance plays are generically human, then Renaissance humoural psychology has no privileged insight into the workings of the Renaissance brain; inversely, if they are not generically human, there is no reason we should have any interest in them except as historical curiosities.[7]

Even in granting this point, however, problems remain. The study to follow deals principally with emotions, and with the axiom that "behind the notion of a commonly held stock of emotions lies that of a universal human nature."[8] To subtract an implied psyche from the depiction of character by limiting considerations to aesthetics or to the functions of agents within plots remains one means of analytical escape from the plenary representation of personhood. This question will reemerge as the study unfolds. Meanwhile, I can proceed only on the assumption that literary characters act in accordance with what authors have perceived the species to be like; that they laugh, cry, feel anger and frustration, love and desire, or commit murder in the presence of environmental stimuli that we acknowledge and recognize as appropriate and true; and that, hence, what we can know about these states in daily life may appropriately be assigned to characters. Without that prospect our literary enterprise is gravely impoverished and irrelevant. That frame of analysis, which cognitive scientists call folk psychology, however, is only a beginning.

The greatest challenge to a confident reading of *The Yorkshire Tragedy* with regard to the volitional phases of an impulsive and darkly rational crime derives from unresolved questions about the competing modules of the brain capable of formulating action plans backed by executive functions. Our brains are ours, they generate our decisions to act, but they are ours only by fluctuating degrees of knowledge and control in accordance with the brain's own genetic design in the production of emergent mental states. Criminal intent is one of the liminal areas of volition through which we may seek to establish borderline elements of the human. What is possible for one is possible for all; of what it consists, we need to know. The challenge is thus unavoidable, but the work is frustrating in the extreme.

On the matter of the protagonist's emotions, the play is anything but expansive, and yet it is inconceivable that such painful encounters, such devastating loss and acute despair, did not entail powerful and driving passions. The emplotted design of the play is a waxing and waning of intense feeling, a rising firestorm of some kind that coincides with, or brings about, the loss of all inhibitions to murder. More generally, any consideration of hope or despair entails hedonic states, what they are, and how they impinge on the shaping of thoughts and goals. As imitations of the human, all literary characters throughout time must embody this principle. In the words of Anthony Paul Kerby, "all such states make sense only against a background emplotment, against a drama one is cognizant of."[9] He is speaking about how emotional states are explained to the self through the order of events to which they are attached, and how they may serve as prompts to redress the circumstantial changes that have provoked the alarm through conscious decision-making. We know ourselves through narrative, including those that rehearse choices.[10] That emplotted design is all that we have to signal the trajectory of mental events; drama is like that.

In brief, playwrights do not describe emotions; their characters enact them. Hence, can we "read plays and novels as the closest thing to a controlled experiment involving high-stakes human emotions?"[11] Apparently we can, just as we read the realities of our social lives. For our own well-being, we need to sense what others are thinking and feeling, and, arguably, stories provide trial contexts through which these same skills can be exercised and tested. The author of *The Yorkshire Tragedy* leaves us with our work to do because he too is an outsider to the historical mind that performed these very deeds. Yet in telling that story, he provides us with the actions and events that correspond to the inner emotional and volitional life of the protagonist, which we are seeking to interpret in motivational terms. Our folk psychological habits may be lacking in explanatory sophistication, but they are powerful witnesses to our cognitive instincts in confronting social worlds; we are obsessed with the minds of those around us, in reality and in art, and we never cease to speculate both privately and collectively on the rationales of others. Such obsessions decide for us what is important about the play.

In such states as those represented by the protagonist, emotions are no longer something which individuals possess, but something which they have become: "something we *are* and not merely something we *have*," in Kerby's words, revealing "value directions in our lives."[12] Or to follow the thoughts of Charles Taylor, all-consuming emotions represent natures, major goal-directed orientations, and thus dictate "what matters to us."[13] In establishing priorities, they also define our moral beings, for the pleasant and unpleasant feelings of the limbic system erect their own realm of goods and evils, of benefits and dangers. They place their evaluations upon the world and prompt the organism to strategize, not only to control the world in its favour but to advance and reinforce the paths to positive or rewarding feelings and to diminish unpleasant or negative feelings.[14] For Taylor, human emotions are always seeking an "adequate form," because emotions demand cognitive interpretation and interpretation generates actions, completing the circle whereby feelings shape engagement with the world to reduce suffering or to enhance pleasure.[15]

It would thus appear that under emotional tutelage, the individual becomes a self-actualizing entity, a self-defining agent by constantly denying old identities and embracing new ones. Such changes in The Husband are adumbrated by the servant who remarks in passing that he was "a man before of easy constitution" (vii. 592). What sustains the self is not some fixed quality of being, but a working consistency of the self-narrative that gives the illusion of an integral, harmonious, and essential personality.[16] When that illusion is radically disturbed by folly, fate, circumstance, or suffering, it becomes hostage to its own emotional defence system – a system designed to protect the cogency of the narrative and referential self by fighting back, not always at the highest level of considered options and ethical duty, but according to the brutal logic of an animal trapped as though still in an ancestral milieu.[17] Just what constitutes the values of that ancestral reasoning, elicited by a brain designed to maximize its self-interests and those of its genes in a competitive world, and the contribution made to those endeavours by the limbic system, remains seminal to the investigation.

That is the paradox of the emotions and a presiding critical concern of this play, for we are surely meant to understand that emotions are an integral part of the mental economy that leads to the murders. Thus we ask, through what form of emotional prompting is a man brought to slaughter his own innocent offspring? The answer would seem to be emerging in this simple framing of the emotions as the brain's crisis makers and managers, the agents of lost control, the temporary winners in the psychomachia between reason and passion. But then, how do the emotions make us what we are, decide what matters to us, create their own "moral" priorities, and condition our actions in the world? How autonomous can the emotions sometimes become? The play demands an answer so that we can resolve our craving for understanding, even though certainty may evade our grasp as we vacillate between equivocal explanatory options.

John Stow described the act which defines this work as a "strange crueltie" in the modest entry in his *Annales or Generall Chronicle of England* concerning the execution by pressing of one William Calverley at York Castle on 5 August 1605 for the fatal stabbing of two of his three children. Beyond that fact, he mentions only that Calverley left his wounded wife for dead before riding toward the town in an attempt to dispatch a third child who had been sent for nursing, and that he remained mute throughout the ensuing trial. From this lean and spare account, together with a series of crime pamphlets on this bizarre event,

The Yorkshire Tragedy was devised and put on the boards by The King's Men in 1608; it was printed that same year, its authorship attributed to William Shakespeare, who might in fact have touched up a scene or two.

Of singular importance is that the anecdote bears the authority of history, making what has happened, by its very definition, possible and in need of explanation. Those happenings entail antecedent events and their attendant motivations, and by inference the minimal mind states necessary to bring goals to action. The dramatist, as historian, supplies the social correlatives to the formation of murderous acts. At the same time, however, in framing that dramatic relation the dramatist believes there are grounds for situating that rising and falling crisis within a tragic template, conditioning the audience response by suggesting some form of extenuation of the crime through remorse. His best explanatory model was that of demon possession, still considered plausible in certain domains of early seventeenth-century thought and legal practice. Even figuratively employed, however, it bespeaks the dramatist's own conviction that the protagonist was under alien control, that his actions were, for a time, not his own, or that as his own, they represented the hostile takeover by some "evil" part of the brain, an emergent property of his own psyche (which is the case even when actual demon possession is involved). Moreover, there is something it is like for a man to find himself in such a state, including his beliefs about the world, which to him are everything. Equally, there is something it is like to sense the departure of those "demons," leading to a recovered sense of identity, remorse, and acquiescence to justice. The playwright recognized in such a form of temporary alienation the substance of the tragic, thereby delivering the protagonist to us as the object of pity and fear for our own vulnerability. The demonic rationale, whether metaphorical or real, is a reading of the psychological riddle, blaming on a wicked agent the evil that otherwise challenges comprehension. It is also a form of aesthetic packaging. But genre and criminal causality are overlapping matters and must be taken up one at a time to avoid confusion.

Critical to any reading is a second source of aggravation conceived in The Husband's mind, without which his desperation would have found no outlet for violence. For phantom reasons he upbraids his wife, scorning her as a "harlot, / Whome though for fashion sake I married, I never could abide" (ii. 165–6). All the while, he demands of her "Mony, mony,

mony, and thou must supply me" (ii. 149) in reference to the jewels remaining to her in her uncle's custody, while warning her, meanwhile, that any attempts to reform his riotous living would be futile. The playwright places the gambling losses and spousal abuse in narrow juxtaposition and hence sets up the projection whereby The Husband is able to blame his wife for his own shortcomings, while she, desperate for affection for herself and protection for her children (ii. 369), maintains the ill-advised proximity that makes her the ready target for material exploitation and domestic violence. We see the turmoil of an afflicted imagination in such ejaculations as, "Puh Bastards, bastards, bastards, begot in tricks, begot in tricks" in reference to his children. At such a juncture, the truth is no longer at issue, but only jealous imaginings, contributing to his mental commotion. He has perversely fashioned a scapegoat for his own failings, replacing his self-loathing with sanctimonious jealousy and a motive for revenge. Such a transfer is an explanatory add on, or it may alone suffice as the motor to the central calamity of the play. There are mental "values" which, in their ancient terms, may be employed to explain this sudden violence against the offspring of a hated wife, but that anon, after looking at the transfer mechanisms whereby self-blame finds release through blaming another.

Jon Elster examines such "transmutations" as when "an initial emotion induces a belief that justifies and even strengthens it, generating an 'emotional wildfire'" as a manifestation of the mysterious "alchemies" of the mind, for "to avoid pain, the actor has an incentive to *transform* this motivation from a less acceptable to a more acceptable one."[18] Keith Oatley and Jennifer Jenkins, in describing contempt, offer an apt profile of The Husband, possessed by "an emotion of complete rejection, of unmodulated power, treating the other as a nonperson."[19] Contempt may have its adaptive value in certain circumstances, but within marriage it is fatal to the couple, for, as an emotional stance demanding decisive action, only abuse or separation can follow. Yet The Husband must tolerate his wife for her possessions even as she clings to him, offering more than compliance to his every wish. That he is challenged to a duel by a gentleman championing his wife's innocence and reputation can only have contributed to his bruised ego. He is wounded in the melee and soundly defeated, which is cause alone for him to discount the worth of other causes in the world. Thus, by dramatic degrees, the portrait of a dissolute life becomes a domestic tragedy, a score to be settled against the despised wife by destroying what she deems most precious – the children he had come to imagine as bastards and thus tantamount to nonpersons. Such an emotional transmutation becomes explanatory when our working psychology perceives how emotions corrupt beliefs and those beliefs, in turn, incite secondary emotions that take The Husband by surprise when an occasion to act presents itself.

Perhaps no further analysis of his deeds is required. After all, behind all that presents itself in the guise of love is the fact that husbands and wives are not genetically related except through the genetic contributions they make to their children, in whose well-being they should share a mutual interest. But if Richard Dawkins is correct in his assessment of the drives that emerge through the mental wiring that promotes individuals rather than societies, even societies as small as the nuclear family, then it is as natural for spouses to exploit each other in the care of the offspring in order to pursue other adaptive goals as it is to cooperate. Behind this kind of behaviour there is a robust evolutionary backstory that will not be to the tastes of the romantically minded. The drives that promote philandering husbands and nagging wives, or unfaithful wives and jealous husbands, are genetically grounded, going all the way back to the disequilibrium between the male's millions of non-nourishing gametes fit only to reproduce themselves and the female's precious eggs which require nourishing and long-term care. Nature has tricked mothers into remaining the more biologically committed nurturers, who in turn must seek for greater male investment in the offspring. The dynamics of that imbalance can become the grounds not only for cooperation but also for opting out, brutal negotiations, or cultural tyranny. Slowly, we may build toward the hardwired credibility of a war of investments which, through jealousy, may produce one of the most perverse of cost-benefit decisions imaginable. Because fathers have less investment in children, if the mother goes off with another male, the offended husband may prefer to punish the mother by wasting her much greater investment through killing the children, while sacrificing his own far more modest contribution.[20] Mere anecdotal evidence can confirm the psychological devastation done to a mother by such an act; it can scar and empty out a lifetime. The intelligence of the emotions, of necessity, includes the hedonics of such calculations. Whether such jealousy is based on fact or fantasy makes little difference, as in this play or in *The Winter's Tale*. In both, the selfish genes are talking out of the depths of the human psyche. Hardwiring makes the counterintuitive not only possible but downright probable under the right configuration of circumstances – before more reasoned cost-benefit analyses kick in, instructed by guilt and regret. Why do males react so violently, and why will females sometimes take the risk? The best answers are written in the genes. Wilson and Daly include in their analysis a "morbid jealousy" which is "an obsessive concern about suspected infidelity and a tendency to make bizarre 'evidence' in support of the suspicion." Is this a form of insanity, given the innate logic of exclusive possession and anxiety over paternity? It is not treated as such by Anglo-American common law, which is echoed by European, Oriental, Native American, African, and Melanesian legal traditions.[21] What appears here as a clever bit of material determinism that belittles the dignity of our natures requires for its full justification the layered arguments for the emergence of that nature over vast tracts of evolutionary time – arguments which form the substance of Dawkin's study.

But the considerations to follow cannot await such a detour. Ultimately, an explanation is called for concerning how such murderous deeds, arising from a brain in some sense biologically determined, pass from mental reflection of some kind into a state of execution, and how such actions relate to pity and justice. If tragedy is the result of a deliberative action, to what extent is the constitution of the brain in defining our mental capacities an extenuating and explanatory component of the tragic sense of life? To my mind, the question is huge, and the definition of a genre lies in the balance.

There are further features of the narrative design of the play that provide explanatory hints to our folk psychological analysis of the protagonist's mind states and the passions that lead to action. Blaming his wife ensued from a corruption of reason supported by strong emotions, but the algorithm of accelerated feelings was simultaneously abetted by a series of social encounters with gentlemen who admonished or threatened in ways the protagonist could not deny. The reproach and vilification from fellow citizens is now added to the torment of a clinging but despised wife and the loss of a contest of honour. If the intent of these gentlemen was moral belittlement, they succeeded admirably, for they accused without consolation and suggested no strategic alternatives, no sweetening of his disposition, not even the heroics of Stoic resignation. The compounding effect of each confrontation is only to be imagined, but we keep a running score of their emotionalizing potential. All these elements are part of a shrinking psychological world, calling for alleviation through inward or outward violence. It is the direction, values, and modality of that violence that bring us continually back to the seat of the genesis of intention.

The fascinating case of Paolo Barbieri of Bologna might be called up in parallel, for, before fleeing the country, he unaccountably killed his fifteen-year-old-wife Isabella while she was sleeping in an adjoining room. The trial concerned those whom he visited, including his brother, before making his escape. The question was whether they were accomplices to his crime and thus punishable by law. Their fates, in turn, depended on whether the deed was conscious criminality or innocent insanity, thereby compelling the judge to rule on the mental state of the perpetrator at the time of the crime. The initial decision leading to the incarceration of several friends and relatives was overturned only through the extensive testimony of no less an authority than Girolamo Mercuriale, whose declarations are preserved in the Archivio di Stato di Bologna. Fortunately for the friends and relations, Paolo had an established history of mental instability, delusions tending toward violence, and wild diabolical images and thoughts, followed by periods of mental lucidity. Among his delusions was that his wife deserved punishment even by death, all of which was ultimately deemed to be clear signs of *pazzia* (madness) and *furore* (frenzy).[22]

Such a construction might have been placed on the present play, if only as an interpretive position, but the story as told contains no such intimations. There is no prehistory of madness or violence; a servant tells us that prior to his losses The Husband was a man of sanguine temperament. There is no mad fury as the act is carried out, and immediately after the fact, The Husband collects himself sufficiently to convince the Master of the College, standing by for his answer, that he was still in search of financial relief for his brother, imprisoned for debt on his behalf. There is a kind of reasoning as he wielded his knife that it was an act of charity to brain a "lack-land" (dispossessed) child. Full self-awareness in the case is sealed by a later confession to the deeds in an act of contrition justifying his own execution. The play events allow for no other interpretation of the emotions than as being of the most garden-variety kind – contextually justified anger, jealousy, despondency, desperation, self-hatred, revenge (however invalid the grounds) – and their intrusions into the life of the constructed self, conceived in memory and active in a negotiated world. If there is tragic potential in such a portrait, it is of a most particular kind pertaining to the mind under extreme emotional duress, which found its outlet in a parallel susceptibility to conjugal jealousy, thereby establishing a cognitive evaluation of circumstances, no matter how false, together with the rationale for seeking profligate goals. It would seem that we must rule out insanity from the outset. And yet, as a man devoid of all hope, his criminal deeds are performed only because their consequences in the future no longer matter to him. The act made possible by despair, a condition extended to include his offspring, may be the stance of a calm berserker, to coin a paradox, a controlled execution of an emotional act calculated without a sense of future cost or effects.[23] Paradoxically, raw emotions may not be pure insanity, but pure logic of a kind pertaining to their own systemic design. By such reasoning, we are inevitably brought back to the level of cognition in emotional states, to periods of lapsed judgment, and the degree to which such states engender actions meriting either compassion or condemnation, or both.

The moment of the sudden first fatal act is accompanied by a soliloquy, an address to the self as well as to the eavesdropper audience, which ostensibly reflects the passage of thoughts in a conscious mind in the very process of committing a violent crime. In fact, the protagonist pursues a litany of his personal failings as his child enters, is swept from the floor, and mechanically stabbed. Immediately prior to this, The Husband had been damning himself, holding forth on the inability of the mind to resist temptation, going over in his memory the final game in which, by a palsied shake of the hand, he tossed away his ancestry, beggared his children, and caused his brother to be imprisoned. He then tore at his own hair in response to the emotions commensurate with his thoughts, commensurate with what he had, at that moment, become. As he thinks of poison, devils, hell, misery, and the pawning of his own soul for deliverance – a flight of emblematic and metaphorical associations pertaining to self-loathing – his little boy comes in playing with his top and prattling comfort and concern for his father as he is taken up with one hand and a dagger is drawn with the other.

The playwright seemingly provides his conventional best for exteriorizing a mind in disarray, but the mechanical murder proceeds without benefit of a spoken confirmation of motivated anger, retaliation, contempt for the child, or spite of the world. If such a soliloquy, with its unlimited access to interiority, fails to enlighten at the very moment of the deed, we must allow for the total inadequacy of the playwright, or his total honesty in reflecting the motivational ambiguities and vagaries of a surcharged brain. Perhaps he was not up to the task, or perhaps he had carefully considered that mind states resembling demonic possession simply elude complete analysis both then and now. That may represent the limit of our capacities as folk psychologists.

But can recent research in the cognitive sciences bring further insights into the inner workings of the deliberative, emotional, and volitional mind in a criminal/tragic mode? The question of divided motivation arising in a multi-architectural and partially subconscious mind is by no means new, but it is always troubling to a theory of action.[24] How many brains direct us and can a true sense of tragedy be born of their uncoordinated moments? Are emotions able to set their own goals, or do we conclude simply that "emotions, cognitions, goals, action, context, and so on ... all flow into one another until, as Pinker and Damasio argue, the distinctions are difficult to maintain"?[25] We will return to this crux in several keys, for while cognition and emotion are thoroughly integrated in daily practice, they arise from parallel systems and have their distinct emergent states of thought and feeling.

Here, the magnitude of what potentially lies ahead brings us to a reflective pause. We might enter into major speculations about mind, modularity, and volition, pointing out the multiple features of the brain that contribute to decision-making almost on a competitive basis. All such models might apply potentially to the mind of the protagonist of this play, but none can be clinically imposed. That is as true in life situations as it is with literary characters. Fiction is not the problem. Cognitivists themselves have not reached consensus about the interface between emotions and consciousness, about the wisdom and autonomy of the limbic brain versus the restraining powers of the cerebral cortex where, if there is not control, at least there is awareness. Yet insofar as the cognitive sciences are concerned with the competing emotional and cognitive readings of the environment, where those same readings in conflict find representation in literature, cognitive perspectives, at the least, offer to help define the problem.

We could, at this juncture, relegate our protagonist to a state of pure emotions, but his cognitive mind is fully engaged, and it is unclear how emotions with their "action tendencies" actually formulate so directed a deed as murder without first conceptualizing it. Yet the act is not overtly conceptualized in his mind. Such philosophical fine tuning would seem supererogatory were it not so critical to the levelling of moral responsibility necessary to our appraisal of the man both criminally and tragically. To feel guilt for the "acts" of the emotions is to subscribe still to the belief that there was a time when a different frame of assessment might have been imposed, leading to more rational values and a different outcome. Uncertain about the imperatives of the limbic system, we remain unwilling to cede our volitional autonomy to their subcortical logic. Tragedy, meanwhile, takes us to the interface between interpretational options: crime is crime, but it may sometimes be committed by good persons circumstantially compromised and made vulnerable to the latent potential of their own modular brains. Tragedy would thus become a cognitive affair insofar as aesthetic meaning relies on ethical extenuation, which relies on a condition of mind. Insanity and moral accountability, then and now, are points of entry into this troubling debate.

Insofar as this is a play about the emotions, let me add a bit of background on their nature and origins as adaptive mechanisms in ancestral environments. The logic of evolutionary design has ensured that our brains are efficient survival machines, meaning that the emotional components are likewise adaptive. Ratcheted up by selection in relation to primitive circumstances, the hedonic brain began pumping out its feelings of reward or avoidance concerning every aspect of existence. For males those feelings are involved in the search for commodious habitats and nubile mates as well as in avoiding rivals and protecting hunting territories. Such sensations often serve as tiebreakers in moments of indecision.[26] Emotions are acts of judgment calibrated by the limited array of positive and negative factors in ancestral settings, but they are sufficiently plastic to adapt themselves to the complexities of modern social environments, including those represented to us through the imagined worlds mediated by words. Arguably, no matter what the stressor may be, it can be represented to the emotional brain only through the felt sensations it generates as determined by the evaluative thresholds of the limbic system. Those sensations are, in turn, arranged in a sequenced binary based on degrees of positive and negative. These messages can, in themselves, be chronic beyond tolerance in relation to prolonged adversity, catapulting the organism into actions of desperation to escape the aura of its own brain – the feelings that can drive one mad.[27] The play would appear to make complete sense in these terms, and in a philosophical vein may set us thinking about the relationship between the polarities of positive and negative feelings and the polarities of ethical systems, as well as the ethos of comedy, which showcases the emotions that inhere in escape, rising fortunes, and enhanced chances of survival, and the ethos of tragedy, which foregrounds the emotions of undeserved loss and failure in the struggle to survive. The binary oriented design of the brain that produces the emergent properties through which we feel the world are intricately related to moral systems and aesthetic genres.

This basic system of hedonic messaging within the brain's overall economy pertains to the human race generally, and hence to all representations of persons, for entities conceived without emotions cannot, by definition, participate in what it is to be human. Thus, all stressors deemed to be adequate emotion-producing stimuli in the environment correspond to expectations of the autonomic readings that constitute a profoundly omnipresent component of human experience. Characters must feel precisely as humans feel in accordance with the information they confront in their surroundings. If emotions, moreover, are pegged to the interests of our genes, then negative readings grow incrementally as the conditions for future reproduction diminish. Moreover, if social organization during ancestral periods relied for its success on deeply embedded emotional values and emotionalized computations of fairness, trust, and punishment, those circumstances inciting positive and negative cues must also be taken into consideration in reckoning up the human emotional repertory and its base lines. It is in accordance with just such design biases that the killing of offspring is so irrational and unnatural. Yet the emotions in turmoil have been enlisted in the causes of strange overriding beliefs, and they have been perplexed by deeply embedded prompts to lash out at perceived wrongs according to systemic values built into the way our brains can sometimes do business with extraordinary social configurations. That is the abiding paradox of the play: a counterintuitive act performed by an enigmatically deliberate mind, arguably driven by an advanced state of emotional aggravation.

But to make a judgment of the play in both criminal and aesthetic terms, we must settle the question of that brain's freedom in relation to its own design, for when the emotions enter the formula, in keeping with the logic of evolutionary psychology, genetic determinism appears in ways not all interpreters are willing to accept. How should that critical question be settled through our best information about the human nature that arises through biological design? Peter Carruthers argues his way out by acknowledging "a set of innate belief and desire generating mechanisms. [But] how those beliefs and desires then issue in behaviour (if at all) is a matter of the agent's practical reasoning, or practical judgment, in the circumstances. And this can be just as flexible, context-sensitive, and unpredictable-in-detail as you like."[28] Such is his reassurance that design does not interfere with choice, options, and moral reflection, although the emotions can impose their insistent demands in an imperious fashion. The question, with regard to the present story, is whether the emotions can formulate their own goals, and whether the dire deeds formulated by the biases of design constitute grounds for a tragic perception of the human condition. Our dilemma, broached several times, is whether we should allow subcortically motivated actions to assume a status "suddenly felt to hold a metaphysical promise not appreciated from the limited contexts of law and morality in which they had traditionally been examined."[29] That question in turn pertains to a theory of action, and whether runaway emotions constitute the demons of the mind with a power of agency all their own. It is an awkward but perhaps necessary argument to make because it flies in the face of what G.E.M. Anscombe has referred to as "an incorrigibly contemplative conception of knowledge."[30] It is a long-held axiom that emotions cannot act without the formulation of conscious and hence semi-evaluative belief states.

The play's allusion to departing devils conveys multiple meanings. The Husband is possessed, or more plausibly, he acted as though he were possessed. As stated above, that may constitute criminal insanity or temporary emotional insanity relating to shame, loss, jealousy, hopelessness, and despair. Arguably, moral reasoning had become meaningless for him because he was no longer a stakeholder in society. We have the facts; we want to know what they mean. Once more we come to the mind of the protagonist and what it was like to be that person at the moment of the crime, caught up in uttered thought obliquely related to an intentional state crossing its own finishing line that is formulated at the edges of consciousness. Research into precisely these matters is expanding, yet we are uncertain as literary interpreters whether we should be looking in that direction. What if "information that is seemingly meaningful, interpreted, and aspectual can be held in mind without entering conscious awareness"? Perhaps our unconscious mind is as complex as our conscious mind, but simply harder to inventory through introspection.[31] Perhaps there are independent centres of reasoning that can prime consciousness according to their own grounds for processing data, thereby rendering the self a far more complex phenomenon of emergent properties arising from a designed brain.[32] Philip Merikle concludes, despite the imperfect testing, that "performance differs qualitatively across the aware and nonaware conditions," even though, for now, "how sophisticated unconscious perceptual processes may be is unknown."[33] Do we have yet another psychomachia in the making as dissonance arises among differing aspectual narratives in the mind? Such critical moments clearly draw on cognition, emotions, memory, and experience, as well as cultural conditioning, temperamental proclivities, and rationalizing strategies, combined with notions of justice and entitlement, abetted by stress or despair. Defining the self and its powers of agency is an ever-moving target. There are also biological and chemical considerations, perhaps even less amenable to literary analysis.

In dealing with the dramatization of an actual historical event, however, we may well ask whether there are causal explanations that cannot pertain to the literary. Adrian Raine and associates have pointed out that those who go on killing sprees without previous records of violence are often found to have lower glucose metabolism and thus lower activity in the prefrontal region and a higher rate of glucose transformation in the midbrain and a drop in serotonin production, leading to decreased inhibition and greater anxiety.[34] This would seem to be a dead end for literary critics, but a position difficult to ignore for forensic investigators and legal counsel. Where does the self and personhood of literary characters begin and end, if they are creatures acting in a moralized world through mind states arising in a standard issue brain? Arguably, this is *the* question when criminality and tragedy coalesce. We live in a violent society, says Valerie Hardcastle, and "the story is a simple one. Impulses to violence originate in our limbic system, the deep and primitive brain centers for emotions. Our prefrontal cortex, the executive seat for rational planning, then decides whether to act upon the impulses."[35] But rational thinking and its overriding capacities can sometimes be overcome when costs no longer matter. It all comes down to that little fraction of brain time and potential: all too human and tragic; all too human but morally reprehensible and demanding the full extent of the law. That same fraction of brain time brings us back to the troubling interface between the midbrain and the cerebral cortex, the feeling brain and the thinking brain, evaluating environmental changes in parallel and negotiating in tandem the formation of intentions.

Our crux may be summarized, perhaps, in the debate between the appraisal theorists (no emotions without cognitions) and the evolutionary psychologists (emotions appeared long before the higher cognitive processes). For the latter, the cerebral cortex, as an evolutionary latecomer, has not yet achieved total dominion over the midbrain, and thus the eruptions of potentially devastating passions. Nevertheless, any theory of action predicated on the limbic system's ability to autonomously judge the world will meet with resistance from the appraisal theorists, who emphasize the essential role of consciousness in interpreting our feelings as having at least an option to act on them. Extensive arguments may be taken up on both sides. It is a matter of the integration of brain modules by evolutionary consolidation versus the asymmetry of an imperfectly superimposed design with its capacity for lapses into subliminally defined action.

Steven Pinker argues for a harmonized brain in which the reflexive penetration of conscious awareness is sufficiently capacious to entertain all salient hedonic states: "the systems work in tandem."[36] The emotional self is therefore a conscious self.[37] It was the simple magnitude of massively redundant synaptic potential, it would seem, that allowed for incrementally deeper penetration and saturation of information, ever more richly "considered" and "perceived," that eventuated in the self-awareness of an autonomous agent able to hold beliefs, make provisional propositions, anticipate the future, think in binary oppositions, and explore consciously its own salient hedonic sensations. Those most frequently encountered are, in turn, stereotyped and given names such as anger, fear, and joy. This "new" brain could even consider a top-down imposed management of the emotions in relation to community needs and long-term personal gain, even though no form of mind control could ever bring the emotional brain to a halt.[38] But this is the species at its harmonious best, discounting moments of extreme stress.

Martha Nussbaum takes up this difficult question in her observation that "some of the animals we have discussed have emotions without ever having self-consciousness. We have self-consciousness, but do not always exercise it; and we can ourselves discriminate threat from nonthreat, the loved from the nonloved, without explicitly formulating this to ourselves in every case, or reflexively scrutinizing our own ascriptions."[39] We have a capacity for feeling our way into kinds of knowledge or discriminations without the full participation of cognition. This opens the entire parallel enquiry into all the subtle ways the brain is designed to relegate even significant decisions to subconscious processes, although a clear distinction must be made between systems which have been trained to take over through practice and learning in relation to cherished values and beliefs and systems still geared to ancestral needs and the logic of brute survival. Hovering between these two is the reactive knowledge that rules over us from the heritage of what Damasio calls "the genomic unconscious." It is a controversial construction of mind to intimate things known without cognition, and yet there are many levels of knowledge and interpretive understanding characterizing the global economy of the brain that do not participate in the narrow sphere of reflexive awareness. Are there volitional centres capable of hijacking the consciously deliberative mind? The question always comes full circle to the emotions and the executive stances they are capable of forming.

Antonio Damasio defines the "genomic unconscious" as those elements of instinct, automatic behaviours, drives, and motivations that originate in the genetic coding that creates the "distinctive features of our phenotype."[40] He describes such dispositions in terms of "thematic scope," which is often pervasive. Many dispositions prefigure our cultural and social lives, for example, the jealousy of Othello, the harsh punishment meted out to Anna Karenina, the incest taboo that haunts Oedipus, and the cognitive asymmetry distinguishing men from women in general. The genome, in these regards, accounts not only for differences but for much of "the sameness that hallmarks the repertoire of human behavior."[41] These are the phylogenetic traits of the species. Depending on the situation or even the time of day, we are all susceptible to these biological biases, appetites, and desires.

Damasio then offers a challenging argument through which he concludes that even if we are operating in this unconscious register, we maintain the illusion that even the most impulsive actions are carried out under full conscious control. Only with that feeling of self-authorship can we even assume a self or moral responsibility for the actions we take. He believes that the cognitive unconscious is educated, yet maintains the option for impulsive action under illusory control. His argument stresses the diverging ways in which we construct the "neural operations of decision and action." Again he finds himself saying that "we *think* we are in control, but we often are not," and that our biological makeup induces us, at times, to make unhealthy decisions. When he comes to the matter of law and justice, true to form, he sees law as a means for educating instincts in accordance with the collective will. Yet even in the cases of wrongdoers with brain damage sufficient to interrupt the control of their impulsivity, he concludes that they must be seen both as criminals and as neurological patients because their disease should "in no way pardon their actions."[42] His concept intrigues because it creates, yet again, an intermediary phase of volition that wells up from constituent nature, impulses, and passions. We may assume the "feel" (he too uses the scare quotations) of self-determination and ethical responsibility, but genomic motivation represents a motor for will and action that pertains to deep and systemic biological knowledge and calibrations, which have bearings on the shaping of irrationality.[43]

According to this analysis, we could style The Husband a manifestation of the genomic unconscious, who knew, yet subordinated thought to unconsciously determined action. We are still in the realm of the double brain, the asymmetry of responsibility, the ambiguity of criminal intent, and the degree of self-victimization that rescues the protagonist for tragic consideration – the art form of excusable dysfunctionality. Valerie Hardcastle concurs that "whatever the truth about human violence, it is going to be much more complicated than merely a hole in the prefrontal cortex or an overactive limbic system," yet she concedes that "we really don't know whether any of us could decide to change whatever impulse to violence we have if we wanted to." We are honing the problem, but no one is able to seal the packages: emotions, subliminal knowledge, incomplete executive control, tragic limitations, brain design, and no escape from moral and legal responsibility.[44] If, as Damasio argues, our genomic unconscious in so many ways formulates the biases of our behaviour at preconscious and pre-ethical levels, may not the extreme manifestations of those systemic resources account for the conduct that turns us into the unwitting victims of our own endowments? Can it not allow that some tragic dilemmas begin with neurobiology?

The play under consideration calls itself a tragedy, but, as we have seen, it is striking in that regard for its lack of moral debate in the shaping of the criminal act. The emotions have been celebrated for their rationality, as in the title of Ronald de Sousa's *The Rationality of Emotion*, as a valued source of perception which is incorporated into our "beliefs, desires, and decisions by breaking the deadlocks of pure reason."[45] But that rationality must include the roots of impulsive violence. The more the philosophical firewall is dismantled that creates the hierarchy between thought and feeling, and this on the basis of genomic imperatives sometimes unleashed, the more tragedy may have to do, ultimately, with the anomalies of evolutionary design.[46] Inversely stated, the inherited aptitudes of an evolutionary brain oriented to the world in its unique and particular ways has everything to do with the stories told about the creatures it guides and determines. Without the intelligence of the emotions as guides, decision-making becomes nearly impossible at the best of times, but when reason becomes disoriented, the logic of the emotions continues, according to its own felt impulses, to fight at levels of raw survival or fatal bluffing. Thus, when tragic actions are based on a reversion to the emotions, our critical attention must turn to the emotions themselves and how their "reasoning" is constituted. Yet what the emotions are and do is a highly systemic and biologically determined matter. Always this crux returns. It may be both self-evident and irrelevant; we have read tragedy for centuries without scientific information about the emotions. Yet we do not cease to try to understand ourselves more accurately both through art and through science in the explanatory terms that trace all human behaviour to its roots in human nature. The argument turns round and round this premise.

Approaching descriptively those decisive emotional states remains inferential work. Seneca argued that anger is a bid to power through the show of the emotions as a marker of the sincerity of intent, an indication through the body of an inward state of cost tolerance to achieve goals, even if the tactic is perceived as temporary madness, precisely because it presents that calculation as a loss of intellectual self-control. Our natures construct this feint on purpose as a strategy of survival. To that we might add Aaron Ben Ze'ev's definition of despair as being "generated when we perceive the situation to be unchangeable, or when we cannot construe our inferiority as unfair."[47] And we might complete the formula by using Steven Pinker's definition of happiness, not only as an ability to alter the environment in ways beneficial to the self but as what can be reasonably attained through a reasonable expenditure of resources in an available environment.[48] The despair that is construed as an irreparable state of damage to self-esteem not deemed one's own fault may initiate a course of action to change the environment that is no longer calculated in terms of cost to the self. Such logic is justified as a future deterrent, or as a matter of pride, or self-justification. It is a gambling strategy, to be sure, because violence also usurps the rights of others and awakens the cry for punishment in ways that the impassioned mind temporarily discounts – all of which may have been more adaptive in ancestral environments. Jonathan Cohen allows for such arrested reactions because the world has changed dramatically – socially and environmentally – over recent evolutionary time, while the limbic system has not. He describes the limbic system as "rapid," "stereotypical," and "inflexible," once it has overwhelmed the cerebral cortex crippled by indecision.[49] Can this be part of the embedded "rationality" of the genomic unconscious?

Given the dramatized evidence in the Yorkshire Tragedy, pure insanity or complete possession appear to be untenable options, taking us back to those intermediary states consisting of radical goals and excited emotions imbued with misinformation which are equally true to our natures, if regrettably so. The Malay gave to the English language the word "amok," which for them applies to persons who become homicidal through a violent or uncontrollable rage or frenzy ensuing from a loss of money, love, or honour. This particular emergent state of the midbrain, in its resemblance to temporary madness, nevertheless conveys the full force of what it *means* for the brain to abandon itself to frenzy. Such a person, in Pinker's words, becomes "an automaton, oblivious to his surroundings and unreachable by appeals or threats. But his rampage is preceded by lengthy brooding over failure, and is carefully planned as a means of deliverance from an unbearable situation. The amok state is chillingly cognitive."[50] Whether this is precisely The Husband's state of mind is open to debate, for it all comes down to an exact hedonic assessment of the organism in its social habitat and the emotional resources for dealing with absent options and perceived hopelessness, but we may be coming close.

Is it then conceivable that such people, striking out through uncontrolled violence, are nevertheless operating according to the logical options allowed by their circumstances? Is it an expression of the abject treatment of cost analysis that comes with despair or devastated self-esteem? Allowing this, amok, as a repetitive cultural practice, may not only be a darker truth of our cerebral architecture but a phylogenetic potential endowed by our genetic design, released through contagious cultural expression – a design, moreover, that would not have been part of our behavioural repertory if it had not proved beneficial throughout ancestral time.[51] Crying signals a state of distress that requests those in the survival group to approach. Frenzy tells friend and foe to read the signs and keep a safe distance. Perhaps we are justified in concluding that, for some, abject misery demands deliverance by any berserker means, and to any delusional extent, from the torment of negative feelings. Such was the apparent economy of this protagonist's brain, less the frenzied exterior. At the same time, such a story is disconcertingly difficult to define in terms of ethics, ethos, or genre: criminal, compassionate, or tragic.[52] The issue might be resolved simply in what it is to run amok, to quell cruel thought through crueler action, as though taken by the surprise of opportunity in a moment of emotional susceptibility which sees its end in the slaughter of all three children and his wife into the bargain.

In the concluding paragraph of his study of "Theories about the Emotion-Cognition Relationship," Tone Roald briefly reiterates the contending views of Stumpf, Sartre, Buytendijk, Hillman, and Funch and concludes that "further phenomenological studies of emotions and their relation to cognition are essential" to resolve the tensions.[53] It is an honest assessment of the crux that inheres in all parts of the preceding argument: reason and emotion are not the natural antagonists of the psychomachia of yore, but they remain, nevertheless, parallel information evaluating systems that function according to contrasting emergent states in the presentation of data and sensation – one reasons through thought and the other through feeling. They are integrated systems, but they are far from homogeneous; if we could quantify their interrelationship, we could finalize the argument. Mention has already been made of the school that presents the relative autonomy of emotions because such emotions predate the operations of the cerebral cortex by millions of years and their opposition to the school that fosters the nearly harmonious integration of the two knowledge sources and the assumed priority of reason.

As an approach to the tragic sense of life implicit in *The Yorkshire Tragedy*, the present study has defined the play as a profile of rising and falling emotions that eventuates in irrational deeds, deeds that, in some manner, have formulated themselves in psychomotor fashion. Our curiosity concerning the foundations of these actions has a double focus pertaining both to the potentialities of human nature and the relationship between crimes of passion and the social order. To recap, feelers have been sent out in all directions regarding the schemata by which we appraise the social world and the recesses of other minds through folk psychology and the further insights that might emerge through more philosophically considered or scientifically tested information. That protagonists may be victims of their self-betraying natures is axiomatic; that is where we began. But insofar as such acts originate in brains, and brains bear their own structural biases, we are entitled to wonder whether the meaning of the present tragedy runs particularly deep because it poses a true (because historical) action that defies self-evident motivational explanation. To find the precise genesis may take us to the fuse of the tragedy, and even to one of the tragic dimensions of our biological constitution. And that in turn takes us straight back to the troublesome interface which Roald designates for more study, because we cannot tie down with scientific precision the degree to which emotions override, ignore, or redefine consciousness and seek to alter the environment in their own image, and we cannot decide the extent to which the primitive hedonic brain must answer to higher ethical criteria. More and more, the approaches to a modular and competitively structured brain weaken the hegemony of reason in all circumstances. From those perspectives our protagonist becomes an exemplar around whom we may build a new analytical vocabulary for the study of tragedy: the thesis yet again.

Turning now to readers and spectators (if the play has ever been produced), in the absence of a persuasive theory of action in relation to a volitional brain and a clear sense of the degree to which the protagonist is deemed morally responsible for his acts, they may well find themselves suspended between a potential for compassion and a potential for condemnation. This too is an established crux, as in all those circumstances in which the misguided who commit crimes may awaken our pity even as we consign them to punishment. Still, we cannot escape the questions pertaining to knowledge, reason, provocation, emotions, the susceptibilities of the mind, and moral responsibility. Martha Nussbaum stated that drama by definition offers a "sympathetic imagining of the possibilities and obstacles that the other person's life contains."[54] Stories catch us out by establishing a point of view, an expository centre, a human portrait, a creature of our own species with beliefs and desires in quest of self-advancement and self-preservation. We are made privileged observers of their lives and of the strategies through which they propose to meet contingency. We acknowledge error, even extensive self-delusion, human frailty, suffering, and tendencies to emotional excess. We imagine the need for closural or finalizing action, for escape, even through acts of atrocity carried out with berserker frenzy, all of which is brought to our understanding through a kind of sympathetic consideration. There is, in effect, no limit to the compassion we might be brought to feel for another, if what is done is in keeping with what it is to be true to the self as a suffering creature.

But sympathetic interest does not mean moral approbation, for humans are gregarious; they live in groups made possible only by cooperation and constraints, reciprocity, fairness, self-control, and some participation in the greater good. The necessity for justice, automatic and uncontested, places a new construction on the play; yet, just as there is no ranting rhetoric over vengeful murder, there is no public outcry against horrendous crimes against family and community. Both are subsumed. Judgment and execution are simply the inevitable aftermath for a man misled by foolish values and ill-formed wishes. The occasional liability of the hedonic intelligence of the species gives rise to a specific theory of the tragic condition, but therein may reside the human substrate of the genre. How, then, do we judge an emotional species? And how do we factor in our own emotional prompts as a part of that species in making the judgment?

Compassion and condemnation are themselves emotional stances, the former attached to the pity that builds communities and the latter attached to the fear among those garrisoned against intruders from without and cheaters from within. We may take the hard Senecan line that harm to others is never grounds for compassion for the perpetrator, and that subcortical instruction through emotions, instincts, and drives will never mitigate the punishment deserved by persons of sound mind whose criminal deeds have been proven against them.[55] Compassion is meaningful only when it is grounded in an ethical vision, only when it has a clear sense of agency and options. If pity is the receiving side of the tragic sense of life, then our protagonist, in his mysterious malignity, falls short of the tragic emotions. What tragedy there is must reside entirely in his repentance and heroic resolve in the face of death. As Martha Nussbaum concludes, "When people commit crimes, and do so with hostile intent, it is condescending not to blame them and hold them fully responsible."[56] To say they could not help themselves is to undermine human dignity and worth. Aristotle did not believe in the tragedy of morally defective characters, even though empathy for the human condition in its entirety blurs that line. Peter Unger is equally decisive. If the hand of an identifiable physical organism in a state of mental alienation commits a crime in an objective and physical world, punishment is nevertheless necessitated in accordance with the demands of society. In this regard, the moral identity does not reside in the brain but in the body. Hence, if the perpetrator, as in this play, goes through a sincere ritual of confession and contrition that alters his identity, nothing is changed. It is the objective body in time and space that must bear the brunt of the deeds it has physically committed.[57] Such an approach to crime and punishment may not only disqualify the dysfunctional brain as grounds for legal extenuation but as grounds for tragic commiseration as well. Errors in judgment by the well intentioned would still qualify if tragedy is anything more than misfortune independent of all but the decisions which place the protagonist in the wrong place at the wrong time. But the tragic sense of life has ever held that the ethical human with a volitional brain must be part of the formula, conditioned by every faculty which pertains to the making of the self and the expression of personhood, including the brain that builds emotions. The sense of the genre slides along this continuum. How should we decide?

Throughout this study, the leading question pertains to the faculties of mind that formulate the volitional energy that issues in a criminal act and spells the play's "tragic" *peripeteia* in a series of murders or near murders. Dramatic narrative provides the analytical evidence, but only after causal and inferential assessments have been made based on what we hold to be true about human motivation and volition, to which the present argument is adding the brain systems that underlie and determine them. There has been a representation in the play of something deemed true, a representation of fatal intent directed at a specific target.[58] In the absence of an expressed rationale, the argument takes us back to the "reasoning" of emotional states embedded in the selective design of the midbrain to explain "a man mad in execution" (vii. 619). The missing link is the level of the protagonist's cognitive participation in this process and thus an engagement with ethical categories. We must fall back on the ethical stances of embodied emotional states in assessing the art forms shaped by the emotional brain.[59] The limbic system, through the "good" and "bad" feelings it generates, is an urgent "ethical" system, and arguably, the model upon which moral binaries are built; it created the oppositional patterns through which we have extrapolated a "higher" ethical system based on reason, empathy, and justice. It is a very old debate, but its origin is in the design of the human brain, where conflictingly emergent evaluations of the social environment struggle for the ascendancy.

Once more, now, around the crux. For many thinkers this will never do. For Alan Palmer, a true action is predicated on seeing the world as it is believed to be and as it is desired to be, with the action serving to narrow the disparity between these cognitions.[60] It is a classic definition of volition and it belongs to the cerebral cortex, to consciousness, provisional drafts of options, with hedonic impetus held to a secondary role. Actions do not proceed without beliefs, goals, and sub-goals, and execution comes about through the will.[61] Cognitivists are in general agreement that under normal circumstances, hedonic information arrives in consciousness; we interpret what we feel and decide computationally (and ethically) what we want to do about it to make ourselves feel better. The present thesis dwells on abnormal circumstances, the potential for emotional tyranny, the prospects of momentary lapses, subconscious urges, genomic dicta and, more challengingly, with subconscious volition. If this is likewise granted to be part of what it is to be all too human, if it is a natural feature of human evolutionary design, then we can reconceive of the species in relation to its stories and, furthermore, identify stories arising from these traits.

The full capturing of hedonic dicta by the conscious brain and its ethical watch hierarchies (a necessary illusion behind the establishment of a moral order) is all very tidy, but if you wish to challenge your mind with the uncertainty of that formula, the study by William G. Lycan entitled "Free Will and the Burden of Proof" will serve as an epicentre.[62] The entire human cerebral engine bootstrapped its way up to its present complexity through the reproductive success of those ancestors capable of reasoning out the courses of action most beneficial to the organism through ever deeper levels of cognitive saturation and computational sophistication. Extreme emotional states, meanwhile, are rarely conducive to considered deliberation. Tragic events may ensue from blinded judgment. Can they ensue when judgment is overwhelmed by emotional imperatives after the magic line of demarcation has faded away? Actions require intentional stances, but the generation of such stances involves the competing judgments of parallel systems, and both systems are genetically composed to achieve the ends appointed to them by the brain's own idiosyncratic architecture. Murder is an achieved fact only with the extinction of life, but it is an act prepared by felt qualities of thought, by a process of choices or eliminations inadequately defined without a theory of the emotions joined with a theory of cognition and action. In the process, there must be beliefs, for without them we cannot explain human behaviour or predict it, although what constitutes belief states is a matter of ongoing enquiry.

With these and related criteria in mind, the reader engages in the evaluation of William Calverley and his fictional double, The Husband, in order to understand or interpret the thoughts that lead to action, in parallel with an effort to determine precisely how the reader's own feelings should confirm through experience the genre of the story. In the middle of the compatibilist arguments for free will and determinism, now as features of brain design, tragedy finds its origins. The question is whether this play about murder is also a play about knowledge without cognition, generating a paradoxically weak intentional stance, formulated by presentiment and executed by subliminal instruction, and thus a tragedy about one of the constituted anomalies of the human brain.

CHAPTER FOUR

On the Systemic Properties of Recollection

Emboxed Narratives and the Limits of Memory in Edmund Spenser's *The Faerie Queene* and Thomas North's *The Moral Philosophy of Doni*

That literature has power to teach we have never doubted. But what literature actually teaches and how that learning takes place is barely understood. It is an amazing anomaly when we think of it. The text would seem to be in control, dictating its lessons by captivating our attention according to its terms. But the brain prioritizes certain kinds of knowledge acquisition according to its own architecture and it designs the stored up memory traces that constitute learning, thereby setting up the curious trajectory that links fictive experience to planning and modified behaviour. Philip Sidney no doubt spoke for many writers of his generation in his claim that among the disciplines, imaginative literature had no peer for its power to give instruction in virtue and the best practices for conduct.

But there is inferential work to be done on someone's part to turn episodic social representations into precepts. Otherwise, memory retains only the action schemata of imaginatively lived experience. The study to follow concentrates on the specific role of memory formation in that learning process and particularly its capacity to transform emplotted narrative into abstract meanings. If readers resist this analysis, it will be less because literature turns objectionably moral before our eyes, and more because the emerging assessment of memory as a semi-autonomous faculty – one which generates more of the meaning of remembered content than should be permitted to a largely subconscious process – will violate our notions of the reading self. Common sense would tell us that *we ourselves* (whoever that is) are the readers and arrive at our meanings cognitively and computationally, and yet some mystery remains concerning the modes and moments at which that understanding takes place, including how and when literature achieves its power to instruct regarding the future promotion and strategizing of the self, attached to a material body.

In his *Apology for Poetry*, Sir Philip Sidney defends the supremacy of poets in Horatian terms; their creations (in poetry or in prose) make learning easy because they mix precept with example, associating the latter with delight and the former with the useful; unwittingly he too is dealing with the story to precept problem. "This purifying of wit, this enriching of memory, enabling of judgment, and enlarging of conceit, which commonly we call learning, under what name soever it come forth, or to what immediate end soever it be directed, the final end is to lead and draw us to as high a perfection as our degenerate souls, made worse by their clayey lodgings, can be capable of."[1] Laying aside his view of our natural depravity, we take note of the faculties engaged in the learning process: wit, memory, judgment, and conceit, allowing that they collaborate in some harmonious fashion to achieve an optimum modification of the self through confrontation with significant and meaning-laden stimuli. But that must be put to the test. Meanwhile, Sidney's emphasis on the training that leads to self-knowledge toward ethical and political ends chimes perfectly with the programs of two of his contemporaries, whose writings will be examined in the second half of this study – writers equally intent on teaching "well-doing and not of well-knowing only."[2] For Sidney, literature must be designed to move and modify memory so that, ipso facto, what activates memory builds the ethically improved subject. That it had the power to impart the felt tenets and templates for the conduct of the political and social life was a Renaissance axiom. The movement from knowing to doing, however, comes down to what they understood of the exemplum.

When Sidney assumes that every story is, in some sense, an example, he recognizes that the nature of the literary creation is, in a sense, allegorical, because examples exist in relation to an informing idea, whether that idea is conceived before the fictional illustration or afterward in the mind of the reader. He shrewdly recognizes the challenge to the mind in making this conversion through his pitting of the philosopher against the historian. His point is to show not only that the former works in propositions and disputations only, while the latter confines himself to dry and factual examples but to set up his celebration of the poet as one who provides "a perfect picture" and thereby "coupleth the general notion with the particular example," finding a new power to "strike" and "pierce" and even to "possess the sight of the soul."[3] This is to add a third dimension to the claim that precept and example form the key parts of experiential memory: the picture has the power to move the emotions which serve, in turn, as felt interpretational prompts and as incitements to memory formation. But the harder question remains: where does the smooth transition from historian to philosopher take place, for by siding with the philosopher, Sidney would have us forget that story and history, initially, are one. How the brain routinely extrapolates the wisdom of exempla is the ultimate point under investigation, together with the limited capacities of memory, which authors must also calibrate in order to coordinate the designs of literature with the competencies and capabilities of the instrument that decodes them.

Here, then, is the thesis: given that memory performs according to its own systemic biases, and that each aspect of the memory-making process contributes to the hermeneutic overtones of the formatted information (now apt for reconstruction at the time of recall), and that these traces form the substance of the literary experience and all that it has taught, there are reasons for anatomizing the operations of memory, because to a potentially alarming extent, what literature means is what memory does. The first half of the following essay seeks to make this clear. Sidney's faith asks for confirmation, and for that we may turn to recent perspectives arising within the cognitive sciences. In the second half, these principles are brought to bear on two literary examples in which the capacities of memory, both exploited and overextended, pertain directly to the efficacy of learning. Such an enquiry must begin with a hasty overview of the idiosyncratic modes of memory in the making of stored experience.

By a simple process of introspection, it is easy to determine that what we remember of moments once lived, conversations once heard, and books once read is far less immediate and comprehensive than the original encounter, even when they are rehearsed or memorized. Hence, one of the largest elephants on the stage of the Cartesian theatre – the illusion that we have limitless powers of recall at our disposal through a mind fully in charge of itself – is the matter of our memory traces, their reliability, "who" edits them down to shadows of former states, and how they are laid down in a recollecting brain and tagged for recall. Such things might hardly matter to us from the perspective of reading stories and reflecting on their meanings were not memory formation so absolute in the shaping of that "core" of narrative events, associations, patterns, and judgments that remains after the reading experience is over. The interpreting of stories hence begins with the design of memory systems; they determine what we learn.

We might have thought that as mental experiences, our memories are merely fading versions of a once vibrant reality (or fictional equivalent to reality), but still in just proportion to everything the original stimuli once were. Then it might be posited that whole-cloth stories remain to be relived in secondary versions, if a little less intensely. But this is a mistake.[4] The products of memory formation are rather more disingenuous, for as an organism ratcheted up to our present level of cognitive acuity by the pressures of challenging environments over ancestral time, our mental faculties and their designs were determined according to their adaptive values. Memory had work to do to the benefit of the organism and it was the pressures of that economy that determined the "meanings" produced by the memory-making processes – meanings which in relation to "truth" were merely good enough. In that regard, memory is not simply fragile but efficiently intentional and proactively editorial. Francis Bacon had an insight into the subjectivity of knowing and learning in *The Great Instauration* when he concluded that "as an uneven mirror distorts the rays of objects according to its own figure and section, so the mind, when it receives impressions of objects through the sense, cannot be trusted to report them truly, but in forming its notions mixes up its own nature with the nature of things."[5] From the general to the specific topic in hand, such constraints upon the memory, both of capacity and reliability, as they pertain to the reading mind, must therefore be pivotal to our successful navigation of plot and the formation of the action sequences, as well as to the values, or abstractions we take away.

If the average information-scanning brain takes in one bit of information per second, then a brain, after a lifetime of gathering (including reading), will have processed ten thousand million bits of information.[6] From that plethora of material, long-term memory bothers to assemble a few thousand recallable traces, and perhaps only a few hundred remnants or schemas of episodic experiences apt for reconstruction, whether lived or absorbed as stories, according to criteria worth pondering. There is, of course, a principle of parsimony at work.[7] Detail is clutter and is hardly worth the effort of processing and cueing; it would serve to no adaptive ends. Too much detail also stands in the way of making the relational aspects of memories clear and meaningful (an explanation will follow). Hence, by what processes are those editorial and schematizing decisions made? We know that memory is a high form of intelligence, brilliant as a mental apparatus, albeit approximate, incomplete, and in some things even treacherous.[8] There is more than the mere shedding of detail. In the building of memory traces, there is also a systemic interpretation of data, a construction of values and patterns, in accordance with procedural design.[9] Memory formation thus becomes a front-line hermeneutist, reading for us and reformatting what we remember according to its own phylogenetic rules.

But when it comes to assessing what memory is and how it operates on data, the challenge is vast and multifaceted. Beginning the story too far back, we would appear to be hopelessly removed from our principal interest, which is the retention of narrative and the stylistic and structural strategizing (and risks) undertaken by authors to enhance the memorial experience of readers so that their stories may have lasting influence. This is particularly apparent among those authors whose designs upon the reader are patently Horatian: to delight in order to instruct by leaving narrative paradigms, templates, précises, exempla – how should we name them? – apt for recall in the realization of the virtuous and urbane life. This was a principal preoccupation for many Renaissance writers, despite a certain reticence to sermonize. But the circle may close sooner than is thought possible since what authors may desire of their stories as "meanings" contributing to the education of a prince or the formation of a complete gentleman must, in some way, work hand-in-hand with the topical tagging of natural memory whereby the subject may have beneficial recourse to stories as they match up with circumstances in present environments (the only survival-oriented value of learning). This will take some explaining, but in brief, it would seem that authors can work to shape memories to utilitarian ends only by putting to use the systemic operations of memory, given that both text, as a form of artificial memory, and the brain are collaborators in the business of creating schematic cores suitable for elaboration at precisely those moments when particular stories, structurally and thematically encoded, become the retained references most approximating an emerging life situation. That is what it means to prepare a creature for a life of contingency by anticipating as many as possible through stories.[10]

Experience then comes without bidding by the brain's own process of analogy association, so that what is taking shape in a present environment is already at least partially or approximately known, together with representative outcomes of similar situations related in meaningful episodic sequences. That, in effect, is what the exempla writers claim to be doing by asking their readers to commit stories to memory. Such experience is literary, but differs in no essential respects from the reader's own experiences, or those of a father or grandmother, passed along as cautionary anecdotes and tales. Such ends may be served, ultimately, by any episodic event to which the equivalent of a "meaning" has been assigned in the form of a cueing abstraction, superimposed value, or action paradigm. For such learning to be applicable to future events, there must be what David Rumelhart calls "content-addressable memory" whereby the brain arrives at its best and most adaptive analogies.[11] This can happen only when the brain manages to locate "the central tendency" of stored patterns by its own systemic inferential powers.

Why this struggle for descriptive wording? In short, the brain must make matches between remembered events and present events for the sake of relevance, and it can do this only by abstracting both in order to discover the common denominators. The question is what that compressing and encoding is like. Memory deals in prototypes or exemplars because only they allow for matching and mapping, once the particularities are shed.[12] Accommodating that mapping procedure is the laying-down of memories in paradigmatically tagged shapes. Or as Chaucer stated the matter at the end of "The Nun's Priest's Tale," "Taketh the fruyt, and lat the chaf be stille," which is to say, take the morality because all is written for our doctrine.[13] That such mental entities require robust networks or connectionist processing to do the generalizing, prototyping, or tagging in the interests of analogy formation (to moral ends) is therefore implicit in the brain's design, making the emergent properties of that design the front-line discriminators of the significant and meaningful patterns in stories.[14] That is an invitation to linger for a few paragraphs over the architecture and properties of retention and recollection, particularly in light of the inclination to call the products of memory "meanings," or at least "meaningful patterns" or relations, which involve inferential evaluations derived from the initial ordering of events. Words are not easily found to express precisely what memory-formation does to the original narrative stimuli to make them apt for recollection, comparison, and application, whether as action templates or schemas, or as ideas or abstractions embedded in patterned exempla, but we can barely perceive what literary experience is or how literature "means" without looking into that process.

Memory making begins with the representation of episodic social sequences singled out by curiosity and attention, elaborated by discriminations and simulation, and coloured by emotional and evaluative responses.[15] The most prominent distinction is the rigour with which we maintain the imaginative provenance of fiction versus the perceptual origin of personal experience.[16] For obvious reasons, our brain insists on maintaining that distinction at nearly all times. Nevertheless, in building our impressions of fictive worlds, we do everything in our power to grant to language-prompted representations all of the required properties of our spatial, temporal, and social conditions, fleshing out settings in the mind's eye, granting personhood or selfhood to characters, and lending emotional involvement and empathy to their plights, in order to gain a felt quality of learning from the outcome of their activities by making their trial worlds as real as possible.[17] We are vitally concerned with who wins and who loses, and with what justification, noting carefully their machinations, rationalizations, incompetency, heroism, endurance, or cunning. This is reiterated here in merely pro forma fashion, for all this we know. It is simply the hermeneutic process in action, wherein we have the illusion of consciously organizing stories in our minds and assigning to them values and patterns. But the crux of this study is that memory either seizes upon our half-formed musings to shape the trace, as we are inclined to believe, or, in the case of the passive reader, labours in the absence of inferential investigation to supply its own principles of patterning and tagging for future recall. Imagining the independence of, or the right proportions of collusion between, these parallel faculties is a mental tease.

A reading experience is, ostensibly, selective. It moves toward understanding in accordance with categories, pursuing a Gestalt formulated in the mind as a completed social episode which is ethically tagged, emotionally charged, visually completed, schematically regularized, and temporally sequenced. In all this, the triage processes of memory formation play a prominent role, thereby challenging memory theorists to calibrate the degree of data framing that takes place before the default data processing of memory takes over. The reduction of information to more general levels and abstract categories – from words to sentences to arguments as minimal psychological or perceptual units – George Miller called "chunking."[18] This is pointedly not a part of conscious critical assessment, but an initial feature of memory composition. Not only does chunking downsize the plenary tale to a schema by shedding detail but it establishes the relational patterning that permits retrieval according to relevant contexts. The conscious reading brain is an allegorizing instrument to the extent we assume that specific meanings are behind the organization of narrative – meanings we are meant to look for, with or without attributing to them authorial intention. The problem is that memory is likewise an allegorizer in patterning the story traces according to their categories of applicability, either as value statements or action templates, insofar as memories without map-worthy meanings are essentially useless to the survival-oriented organism. That bias is a built-in feature of memory systems, which brings us back to the manufactured or extra-textual "meanings" assigned to story Gestalts.

There is some agreement among specialists about how the segments and components of memory should be divided, but less about how conscious recollections are bound together from the component parts distributed throughout the cortex.[19] To account for the operations of memory, there is a tendency to affirm zones, buffers, and centres where they take place – mere metaphors for actions neurologically accomplished. What is certain is that the discriminations we actually perform are neuro-biologically possible, and that hence the parts of memory formation seem to require "places" wherein, and "engines" whereby, the fixing, recursive looping, comparing, and interpreting are done. Teun Van Dijk and Walter Kintsch think of memory as a buffer zone where story propositions are held in suspension, serving as organizational frames for episodic events as they unfold, drawing them into a thematic unity.[20] At the moment of recollection, Forbus, Gentner, and Law introduce a "structure-mapping engine" which permits the brain to superimpose two representations in a "local-to-global alignment process," drawing two action/meaning patterns into a common comparative space.[21] These are but two examples among many.

During the reading process, "medium-term" memory contributes not only to the editing of events destined for long-term memory but also to story continuity even as the events are being read or told. This faculty is responsible for sequencing and contextualizing the reading events in narrative time. Consciousness is a knothole. It has unlimited potential but limited working space. It permits entry to only one formulated thought at a time, if only for a few milliseconds. Typically, while reading, the speed of its operations is programmed by the reading eye leaping forward to add new data.[22] Because consciousness is devoted to immediate considerations in rapid sequence, there must be another "working space" which maintains an accessible background of information about everything that has already transpired in relation to present thoughts. Reading would be impossible without this faculty which, almost imperceptibly, gets on with its interpretive work. This is the so-called working memory in which events and impressions are digested and kept available as intimations of what has gone on before; it is a kind of meta-conscious contextualizing "theatre," residual and latently active, which grounds emerging narratives in the data of their own pasts. Present events experienced in relation to an orienting context is a phenomenon so familiar to us as to hardly bear mention until one tries to construct the cognitive model whereby the brain maintains the sharpness and functional priority of present moments which are, nevertheless, imperatively grounded in an awareness of past events.[23] Very simply, we need to retain an active notion of where characters have been, who they are, what they believe, and how their goals conflict; stories would otherwise lose all coherence – or applicable value.

Paul Churchland speaks of this ability of the mind to make the cognitive past continuously available while processing new incoming data as a product of "recurrent pathways," again in an effort to imagine systems that account for the qualities of the emergent properties we know as past thoughts, impressions, images, and story sequences. In his words, working memory is an adaptively acquired talent for allowing "the creature to represent its current situation in a way that takes into account the situation that immediately preceded it," simultaneously entailing an efflorescent sense of cumulative time. Clearly, all novelistic discourse relies on this modality of cognition, the words themselves not only generating the "now" experiences but prompting the formation of contextual memory.[24] Inversely, if the reader fails to construct an adequate set of working impressions, all subsequent "present" moments collapse into absurdity, as reality does for amnesiacs.[25] This holds true for all stories that fail to generate a sufficient mass of past events and values to keep the present grounded, or that are subdivided into parts too distant to prevent the erasures caused by forgetfulness.[26] All writers are affected insofar as they are strategists who must program the density of their materials in relation to the cognitive capabilities (the working memories) of the generic reader. This is a principal consideration in what is to come, and more particularly in relation to stories-within-stories or inset stories, which may take storytelling beyond the capacities of memory formation.

Returning now to the workspaces of memory, there is not only the contextualizing of successive "nows," but the compounding of episodes into the kinetic groupings that shape long-term narrative memories. Default systems are at work. Attention has done its selecting, the emotions have contributed their colouring, and inference has been at work in assigning value to patterns. But are those "meanings" computational and conscious or passively systemic?[27] Meanwhile, our residual knowledge of the world imposes conformity on new data. Literature may seek novelty, but how we retain it often resembles far more what we already know about the world. (This will be dealt with four paragraphs on.) Schematic reduction for reasons of parsimony is where we began, but paradoxically, the mind will also confabulate information where essential data is scant, filling out settings and temporal sequences to complete the Gestalt in keeping with our minimal criteria for causation and "folk" physics.[28] Memory is a highlights factory that nevertheless posits continuous and unified time, compounding a coherent story, including our "self" story built up from the capricious episodes we assemble into a rational autobiographical continuum.[29] Even concerning our own lives, our minds may be intolerant of inadequate settings surrounding vague and remote events, and thus fill in places in the mind's eye to a necessary level of representational specificity. (Hence the unreliability of some witnesses when details count.) The properties of episodic memory-formation point not only to expedient editing but also to a nearly complete reconstruction of narrative reality, now involving confabulation, chunking, temporal prioritizing, tagging, encoding, meaning infusion, ego massaging, a priori schemas, and sheer forgetfulness.[30] With this quality of product, and for all readers, the critical act proceeds.

Of principal concern are the shortfalls of memory, both working and long term, which challenge authors not only to strategize their narratives in relation to the retention capacities of the "average" or generic reader, but to deal with the "hermeneutic" components of memory: what it allows authors to teach and readers to learn. In this regard, memory has genuinely disconcerting properties. Daniel Dennett's model of the brain as a community of autonomous operations building a diversity of emergent properties which gain a sense of unity only through the convenient illusions of consciousness will meet with resistance. Yet certain studies in the fallibility of memory take us to the brink of Pyrrhonism. In *Consciousness Explained*, Dennett may have taken perverse delight in debunking the infallibility of memory on a variety of scientific grounds. Nevertheless, sober and considered reflections led him to understand that brains work generally on hearsay from their own faculties. Memory is all about access, filters, chemical stability, time-lapse replay, vectors and pathways, and the formation of emergent state cognitions restored by prompting cues, whereby a thing once known can be simulated in consciousness. The phenomenon is so wonderful that we may be astonished that it works at all. Memory cueing flits about by association, implying connectors arranged by the proximities of their respective "contents," so that to think of bullies includes not only semantic categories of conduct but episodic memories of bullying in which the ego may have had a part, perhaps glorified by a sense of strategic intelligence under the circumstances, or flattened by recollections of cowardice and humiliation. These memory cascades are produced by the brain's capacity to make associations of approximate precision on a hierarchical basis. A good memory will protect the fragile "I" from unflattering memories as an essential operation of well-being by finessing and massaging both facts and impressions.[31] There are many causes of error worth showcasing from simple data reduction, often to make events conform to expectations of the world, to misremembering, to ideological filters, and sheer inattention. These factors take place both at the time memories are formed and at the time they are reconstructed from rudimentary traces.

Meanwhile, our sense of certainty about our own experiences and beliefs Dennett refers to as subjective "incorrigibility," including our tendency to overestimate the reliability of our memories.[32] He reflects on the chunking and editing as a dismantling and reconstruction of the original stimuli, corrupted even by what we know and impose on the world to make sense of it. Thus, there is no final draft, no canonical view of reality past or present, and no definitive experience to base memories on. Mere probability is constantly involved in interpreting the world, and particularly so at the time of recollection.[33] Meanwhile, narrative itself has disappeared or else "has been digested or 'rationally reconstructed' until it has no integrity."[34] Such memories consist of abstractions and episodic impressions far removed from the events that led to their formation, and in time, even they go wonky because memory is spread throughout the workspace of the brain and makes use of a quilted work of neurons across which multiple memories are imprinted, using different configurations among the same neurons. Hence, there is a constant threat of new memories contaminating and corrupting those that were laid down earlier and are less often activated.[35]

So much for the retention of stories, the scattering of their parts, and the fidelity of their reconstitutions. So much, too, for the social and political agendas perceived in stories as unassailable truths. It is difficult to imagine that all we sense, know, experience, enjoy, and remember is derived from mind states systemically generated to produce qualia existing nowhere else in the world except in our minds. That is as bad as it gets. As Michael Gazzaniga summarizes it: "the human brain is built in a way that ensures our past memories are faulty."[36] Where can we go with a faculty that files memories according to self-generated meanings, binds them to old feelings, enhances them by way of accommodating the ego, while filling in the missing contexts, and drifting toward cautionary resumés?[37] The fact is, quite a distance, for we are a surviving species, so far, and our memories have been crucial to that success; we are planning animals making up for our many other deficiencies by basing decisions on the experiences we can recollect. Such emergent properties of mind have proven useful in explaining dilemmas, correcting errors, and clarifying situations; we profit by the manufactured cues, images, and action sequences generated by long-term memory-formation, both despite and because of the editing and redesigning.

Stated otherwise, if our orientation to the world of stimuli and changing environments really is, in terms of our perceptions and instincts, that of a survival machine (our own survival or at least that of our offspring, according to the theory of the selfish gene), then memory making must also be a part of that machine. Victor S. Johnston asseverates, further, that "only biologically relevant events, defined by our feelings, initiate the formation of memories."[38] That is radically concise and waltzes over those more recreational or aesthetic interests that also focus attention and incite memory. As Brian Boyd points out, planning the future through provisionally drafted option scenarios from which a best choice will be made employs the same areas of the brain as memory recall, enlisting "the hippocampus and prefrontal medial temporal and parietal regions to promote a form of 'life simulator' that allows us to test options."[39] In sum, as a faculty reacting throughout prehistory to ancestral environments, memory achieved its design in relation to its capacity to store past information in the interests of solving future problems. We have information that is no longer available to us in the immediate environment. This axiom puts us in mind once again of the authors who go so far as to append epimythia to their stories, stating forthrightly the intended morals whereby tales counsel conduct through abstracted precepts, or schemas of significant and meaningful actions.[40] But we are now challenged to know whether we can speak of the parts of an illustrated conduct book and the parts of episodic memory in synonymous terms. If the moralist teaches by exemplum and the biological value of recollection is the ability to plan against a background of experience, there is an apparent coalescence. But the differences in their respective "ethical" biases as "conduct systems" are difficult to quantify, even though they must share common operations for the laying down of memory traces. Arguably, the moralist reader becomes a walking repertory of exempla designed to prepare him or her for all eventualities in the public and private spheres through the values illustrated and learned. Yet memory completes this work in its own hermeneutic fashion.

Regarding the regularization of actions according to pre-established schemas, the point has been proven many times over. Sir Frederic Bartlett's experiments in the early 1930s involving reading groups presented with strange mythological stories revealed how their reordering, through recollection, favoured established expectations and explanations.[41] There are many scholars to consult on the matter of schemas. Michael Tye endorses them as essential to the consolidation of data and the efficiency of learning. "Were all the things we can discriminate retained in memory, we would quickly suffer unmanageable information overload. Limitations on memory are necessary for information reduction," in the direction of familiar patterns.[42] Ornstein and Erlich confirm this in *New World, New Mind*.[43] Narrative events that do not match conventional expectations tend to be normalized during the process of recollection. The brain's architectural biases include exaggerated discriminations and binaries such as the category of good and evil because they give classificatory sharpness to our understanding of social actions. These are the schemas at work and they are of self-evident interest to any student of memory and recollection in relation to stories as experience, for stories are formed in the memory and later recollected according to schemas, which by definition carry their own organizational saliences and editorial powers.[44]

Students of literature might well have an interest in what prompts recollection through tagging and relational mapping. This is particularly so if the moralist intends her stories to foster informed ethical judgments by example and circumstantial recall. Daniel Dennett spoke about an "if-then" prompt system that constantly invigilates new stimuli. By some such system, the brain knows when it wants to remember something and how to take a little piece of the whole to incite its reconstruction – a phrase or notion, a fragment of plot, or a conceptually modelled character. Bernard Baars speaks of a vast library of archived memories "triggered" when the "calling conditions" appear in working memory, aware, a few pages later, that the story files may be vast but the access lost, as in the tip of the tongue syndrome.[45] When the brain recognizes things known which match present forms, it reacts with a "that reminds me of" connection – or so we must express the processes of a complex neurological triage operation.[46] This matching and mapping is essential to the writer who would have her stories used on cue in the conduct of life. Clearly, analogy-making is key to the entire procedure, and Douglas Hofstadter among others has led the way in studying the epistemology of comparison and how the brain might perform the discriminations pertaining to a propinquity of forms.[47]

With this we come to the heart of the matter pertaining to stories as operative forms of wisdom made useful upon recall. Finding the vocabulary to describe the phenomenon in all its pliancy and plasticity makes for good work, for it is another cognitive crux pertaining to memory. Stories may be lodged in the dispersed reticulations of neurons as distinct potential states apt for restoration, but the brain must also recognize them as something typical that can in turn be identified with similar forms and events. How does the brain decide how things compare and how closely, and is it carried out as subconsciously as it would appear? Consciousness can hardly afford the luxury of performing such associations as these, any more than matching auditory stimuli or printed signs to words. Such things must happen by processes of their own in order to keep consciousness fully invested in current events. Stories stored must likewise be evoked spontaneously to feed into conscious concerns, prompted by patterned circumstances, an image, a precept, an action sequence, a moral value, or all of these, in which case, "who," in the memory-making process, assigns those perceptual or moral values that become the call-up cues? Knowing that we must avoid homunculi in the brain as intelligent agents making choices, that is to say, creatures who must contain more homunculi to make choices for them, we must fall back on arguments from design, eliminative systems that work by category and value assignments tantamount to ideas or abstractions acting as titles, catchwords, or prompts.

Because new stimuli are investigated in the prefrontal cortex, this area of the brain is involved in memory retrieval by analogical association. Connections move outward through layers of closer and further associations which also, by efficient degrees, come to a point of saturation in order to prevent the brain from losing itself in dreamy distractions. (Our brain has perhaps reached its maximum allowable evolutionary capacity in terms of association, metaphor, and analogy because preoccupation with more remote matches would be less adaptive.)[48] Returning to basic principles, in the words of Dedre Gentner and Arthur B. Markman, "we store experiences in categories largely on the basis of their similarity to a category representation or to stored exemplars."[49] This "practice" takes us directly to the operations of case-based reasoning, similarity processing, and analogical mapping, all of which are critical to human intelligence. The challenge is not one-to-one correspondences that show likeness, but analogies that permit comparison in the absence of surface similarities. What remains are the relational properties and the ability "to match connected systems of relations." Gentner and Markman call that ability "systematicity."[50] They can only conclude that analogies are not about coincidences but perceptions of likeness that are systemic and coherent only at relational levels.

For those familiar with the events pertaining to Kosovo, the idea of ethnic cleansing will often come into mind, even if the concept is never mentioned, and those reading an account will later remember the use of the term even if it never appeared in the text. This must occur because the Kosovo story has been "chunked" down by memory-formation processes to become an illustration of that concept, so that, inversely, mention of the concept will bring Kosovo to mind as a prime example. Only then can the Kosovo story be mapped onto other cases of genocide as practised in different historical contexts.

Stories likewise gain their predictive value through the achieving of relational meanings connected to summary abstractions acting as cues. Thus they become base domains prepared for alignment with target domains by analogy, taking us back to memory, ideas, and action templates, leading to "spontaneous analogical inference."[51] These are the "meanings" made in memory whereby this faculty supplies restored data by contextual prompting, and that takes us back to the "structure matching engine" alluded to above. We can see the system at work when current planning draws on recollected literary exampla. It is another matter altogether just how often we actually do this in moments of social decision-making.

The creations of authors must in some sense be held hermeneutically hostage to the memory- and analogy-making processes whereby story digests become templates for planning. Memory thereby becomes a builder of allegory in reverse by adding the precepts implicit in the unfolding of episodes in the form of memory prompts, a process authors can frustrate only by imposing qualifications, dilemmas, reserved judgment, even indecision with regard to the narrative, or by appealing to aesthetic appreciation beyond precept. The moralist, by contrast, may rely on memory tagging to turn simple forms, such as proverbs, beast fables, related apologues, and exempla into wisdom literature apt for the direct instruction of life. Memory is at the procedural crossroads between how we learn and how stories teach, and how episodic information is patterned and converted to the "symbols" of our understanding for storage, association, and comparison. It is by no means the only reader in the machine. We can computationally infer meanings from memories, such as they are, through conscious reflection of many kinds, constituting the whole of the critical enterprise. But memory remains a default server, a medium that constructs a disconcertingly large part of the message.

And now for the challenging part: how we deal in literary terms with the memory that assigns meanings. If we acknowledge the role of schemas in furnishing expectation templates, and the role of encoding whereby memory chooses an "essence" by which a Gestalt is restored, and the mind's ability to make analogies by distilling common properties from two or more action structures, we must seek to imagine the laws of distillation whereby forms are perceived for their "moral" utility, and then profile the entire operation as an idiomatic feature of human nature on a largely phylogenetic basis. The crux points us not only in the direction of the "artful" transformation of chronological events into narrative, and the representation of fictive worlds according to our categories of knowledge of the world,[52] but in the direction of the transfer from narrative complexity to inferential understanding. The more categories we have for stories in anticipation of hearing them, the more firmly and extensively we remember, because categories help to particularize memory traces.[53] (Serious students of literature should read lots of stories!) Inversely, the longer the perceptual tagging of a story is delayed by ambiguity, undisclosed matter, or alternating points of view, the longer the memory trace is in the making. Fiction can be drawn out to those ends, or collapse through the mind's incapacity to complete the exercise.

Other constraints have to do with similarity, for the more things are alike, the more they interfere with one another and perplex the making of precise memories.[54] Some stories form quickly, and particularly those contributing to an allegorized agenda of expressed values in the form of exempla. Concomitantly, authors must design their narratives unwittingly, intuitively, or by introspective insight, in relation to the capacities and biases of the human brain, for texts are nothing before they pass through brains, and they are nothing if not fitted to what brains do best. Memory does what it does and narratives take their chances both in sustaining context and in teaching precepts. Arguably, this is a vital perspective for assessing the power of stories, and those authors who perhaps overreach for having failed to intuit the limitations of human faculties may provide the clearest examples.

Memory, under the most positive of constructions, abets in the metamorphosing of narratives into pedagogical entities whereby stories may be said to teach virtues by illustrating them in action (or in decline). The literary work, through its strategic designs, authorial commentary, illustrative redundancy, and power to arouse emotions may seek not only to instill exempla quickened for recall but even expose the fallacies of the human faculties in order to sharpen the critical acuity of the self-examined life. The exemplum recalled is a form of reasoning, or logic, because as a narrative it is experienced by the ethical self and forms the basis for social understanding and planning. More fundamentally, much as we are inclined to think of the brain as a rational computational instrument that must work in formulae such as syllogisms to make up its mind, it is the pitching and comparison of micro narratives reduced to perceptual essences that are far more frequently employed.[55] They are the life-orienting traces that remain of lived experiences, including all those that have been absorbed through reading or hearing stories. Two of the Renaissance writers whose literary programs come most readily to mind as overt expressions of the relationship between story, memory, and virtue are Sir Thomas North and Edmund Spenser in their very distinct capacities as translator and author respectively of putatively comprehensive manuals of ethical instruction taught through stories – stories knowingly and strategically submitted to the operations of memory.

One of Tudor England's more surprising translations – a conduct book of sorts which has achieved far less notice than the translations of

Baldasare Castiglione or Stefano Guazzo – is Sir Thomas North's English rendition of *La filosofia morale* (Venice, 1552) of Anton Francesco Doni entitled *The Moral Philosophy of Doni* (London, 1570). What North had in fact landed upon was a book of fables that constituted the *Kitab Kalila wa Dimna*, a work still well known in the Arabic-speaking world, itself the eighth-century translation by Abdullah ibn al-Moquffa of Burzoe's sixth-century Persian translation of the first and longest section of the Sanskrit *Panchatantra*. In that original work Vishnu Sarma promised to deal with the ignorant princes by distracting them over a period of time with a carefully compiled complement of engaging tales, recognizing in the medium itself the full potential of the messages, through the process of memory formation, to secure the overall political fitness of the royal youths. In a general sense, it is a variant on the "dial of princes" or *Fürstenspiegel* genre with its declared intent to serve the reader as "a looking-glass ... wherein thou shalt most lively behold the daily and present dangers and deceits of man's most miserable life ... [wherein] the eyes of thy understanding shall be made open to discern the flatteries of deceitful men and the wisdom of the most guileful world."[56] The difference is that the boys are so dull that default memory and default association will have to accomplish all the benefits on their own; this is the first trial run of pedagogy made easy, or "statecraft for dummies." Default memory would do it all, including discerning all the subtleties of *Realpolitik* by forcing the memory to cross-pollinate and cross-tag interwoven stories.

North's "Prologue" proceeds with one of the most succinct statements imaginable concerning wisdom, the vital role of an active memory, and the concentration of the reader in building long-term mental defences against the miseries and deceits of the age. The entire rationale of the work is that it serve the reader as a complete survival guide in a treacherous world by telling enough stories to cover all contingencies, each serving as a template for reasoning in a changing environment, largely by alerting the understanding to the hidden motives and desires of rulers, courtiers, and sycophants: the ambitious, the insecure, the fickle, and the opportunist. North assures his reader that the medium is far from tedious, but rather delightful, and that the wisdom it dispenses is ancient and proven, as are the literary methods for its dissemination. Everything has been designed so that such doctrines should be "imprinted in the Reader's mind" for his or her profit. This must be the work of natural memory, concerning which North can make few observations apart from its offices in committing transient narrative to permanent doctrines. He is fully aware that his book has become a mnemonic instrument, "for it may, in a manner, be called an artificial memory to benefit themselves at all times and seasons and in all arguments, with every particular thing that these wise and grave men have invented, shadowed with tales and parables."[57]

The reader is invited to align this statement of design and purpose with all that has been set out above concerning the phases of working and long-term memory. To be sure, these are merely subsumed in North's description of the work's value. But at the risk of redundancy, it is worth pointing out that the design of the Doni collection of interleaved or story-within-story or nested narratives – some forty of them incorporated into an extended beast fable concerning the treacheries of court life – is intended to educate three dim-witted and recalcitrant young princes by depositing into their memory banks the stories required for their future reasoning concerning the multiple dimensions of human nature as they are manifested in the public sphere. The wise sage who has undertaken their education treats the stories as philosophical touchstones; they have proven longevity and a demonstrable power to align themselves automatically with current and future political circumstances. The entire point of the collection is that the princes do not have to do this for themselves by conscious interpretation; it is done for them by the inherent qualities of the stories as meaningful tales, or tales made meaningful in the process of being remembered. Again and again, the reader is assured that the book is a jewel, the most valuable of scriptures, the essence of the self-examined life, for "this precious gem of knowledge whoso shall lodge it in the secrecy of his memory shall never lose it, but shall rather augment and increase it with age in such sort that he shall win a marvelous commodity to him."[58] If the princes are under-motivated, the stories must, in essence, be that much more able to make default impressions that will resurface according to the systemic operations of recollection at those times in which they are the most suited guides.

Nevertheless, North evinces a sense of uncertainty concerning this automaticity in his caveat that full memory formation requires concentration and careful discriminations, for he admonishes his readers to call their wits together to understand how the work was "framed" lest they read as the blind. Moreover, readers must not be in haste, but must take the time necessary to ponder, reflect, relate, and reinforce all the parts from beginning to end, undercutting Vishnu Sarma's easy confidence in the meanings made by memory. In this he hints at the "making difficult" of the work's basic design that serves to retard the forward movement of the narrative in the interests of building a complex web of associations leading to hard-earned wisdom. In the language familiar to his day, he cautions the reader to read synchronically, as it were, in associating beginnings with endings and in holding the middle parts constantly in mind. In this, North may even hint at doubts concerning the offices of working memory, of the making of chronological sequences, of causal sequences, and the critical association of parts. He recognizes that in the telling of many exempla, each containing a kernel of generic understanding in relation to a representative event, there is work to be done, not only in fixing each as a precept in the memory but in understanding their intertextual resonances. He was a great believer in the formula and sunny in his promise of delectation, but skeptical of the facility by which the full profit might be gained. North, of course, had no inkling of the debate he has been drawn into here, but his dual approach to memory is a reminder that the brain is constituted of parallel systems, and that while we persist in employing mechanisms shared with our mammalian ancestors even to the establishing of the simplest of Pavlovian responses, we nevertheless have a cortex which can be brought to the enterprise with its capacities for reflection, inference, and rehearsal. Memory making, for all its systemic autonomy, to some still-to-be quantified extent performs in relation to a larger cerebral economy.

As stated above, in discussing memory, we are, in effect, discussing learning, and in discussing narrative techniques, we are discussing the disposition of narrative materials toward the achievement of an author's declared pedagogical ends. The simple crux of the matter with regard to individual fables is how the brain turns events into examples and examples into cued precepts readymade for environment-driven instant recall to self-promotional, collective, or moral ends. There has been speculation on how stories retain their particularity yet, by dint of their plasticity, achieve a *passe partout* profitability through mapping or comparison at relational levels. We are back to the mind that functions by analogy, progressing "outwards" from closely proximate to less complete fits according to brain design.[59] The architecture of the memory induces prototypes or exemplars to interact "with any other data in the system with which it is capable of interacting."[60] This inventory of data on comparative grounds may constitute the most fundamental "wiring" of the cerebral cortex: the brain that knows in accordance with what it already knows. We can even be a menace to ourselves when the enthusiasm for analogous connections attains conspiratorial levels, or inferential links are allowed to drift into irrelevant or ideologically driven associations. That too is the work of a brain obsessed with establishing causal interconnections now based on overheated probability schemas of the world, but a topic for another time.

Concomitantly, a kind of allegory is born of the working codes which memory assigns to stories. Consider the twelfth tale in North's collection, that of "An Ape Meddling in That He Had no Skill." The title is already pitching the story in moral terms, but the brain is pretty good on its own in making morals out of examples, suggesting a simple experiment: ask readers unprompted by a title to say what the story "means" as they remember it. We are hoping for a high incidence of correlation. The tale is told by an ass, the aunt of the mule she is admonishing for interfering in affairs that do not pertain to him. She has chosen the story specifically because it illustrates unambiguously the benefits of minding your own business. This becomes one of the high order caveats to those who would lead successful political lives, especially where the affairs of the prince are concerned; the point is made repeatedly. The codicil that "whoso is given to be a searcher-out of other men's doing, he can never be reckoned good nor honest" confirms that purpose. Her tale is of the ape that observes a woodsman splitting logs with an axe or beetle and wedges. The beast is keen to try his hand and waits until the labourer takes a nap. But in removing the axe from the cleft, he succeeds only in having the log snap closed on his foot. The consequences are dire, for the woodsman, now styled a "churlish clown," wakes up and clubs the animal to death, presumably for his folly in meddling in things that did not pertain to him, although the magnitude of the punishment seems in excess of what has already been a self-punishing error in judgment.

The action is sequentially matter-of-fact, yet it is spontaneously recognized as an emblematic schema that converts readily to a cautionary notion. This the brain does as readily as it decodes metaphors or detects symbols. We are barely able to imagine it as devoid of such intentionality. Yet the question remains where this transformation is performed, whether by a conscious inferential process, or by the memory processes which package such traces for cue-activated recall. Remarkable about the aftermath of this story is that the stubborn mule avoids any application to his own situation by denying the analogical associations between judgment errors in the manual crafts and ostensibly similar errors in the arts of the courtier. Self-serving denial defeats the analogy-making of memory prompts and association networks.[61] By refusing to transfer the story template and its meaning to a relatable sphere, he, of course, beggars Vishnu Sarma's entire pedagogical project, reminding us, at the same time, of a critical underlying principle: that learning from narrative entails a capacity to reduce data to essences simply because lived episodes are never identical.[62] What the reader has gained, however, is a pictorial emblem, a micro-tale, and a moral, the universal truth of which is proven by the mule's cruel demise. He who maps not, perishes. But for others, the story, once it is known, abides as a warning schema generically labelled by memory and ready for an "if-then" recall. Such utilitarian recall is frustrated only by ideological interference and denial or by obtuse exempla imperfectly formulated or eroded by prolonged delays in their completion. This is particularly the case when stories are interrupted by the telling of one or many interim stories. Strategic narrative organization calibrated to the designed capacities of memory systems is our principal preoccupation over the next few paragraphs. Authors dispose in relation to what brains can do, and when they push the margins to cognitive and pedagogical ends, they do so at the risk of obscurity, confusion, or overload.

The techniques which excited the makers of the *Panchatantra*, but which gave pause to North, were the *entrelacement* or interweaving of stories, by which is meant either nested or inset tales interrupting one another, or the interleaving of tales, one following the other in close succession to illustrate a single point. There is nothing mysterious in these practices. The stories-told-within-stories technique was used by the ancient Greek novelists as well as by medieval romance writers, and it was passed along to Ariosto who made it a fundamental procedure in his *Orlando furioso*, as did Edmund Spenser in his designing of *The Faerie Queene*. But whether the practice constitutes a taxation of attention and memory in a way that enhances learning or that impedes it depends on a sliding scale of factors, the foremost pertaining to the distinctiveness and urgency of the overlapping narratives and the ability of attention and memory to keep track of them. Readers have no choice, in such cases, but to enter into the spirit of compound narration for its playfulness, complexity, interrelated meanings, and novelty. As Lisa Zunshine points out, writers "have always experimented with the palates of their readers," and the leading example she cites is Heliodorus's *An Ethiopian Romance*, a work dating to the third or fourth century AD, which she rates as "profoundly experimental in its handling of causal sequences and stories embedded within other stories." Instructively, she concedes, on the basis of her own teaching, that modern readers find the work baffling, yet confirms that the Heliodorian prototype fostered the conventions of romance that prevailed throughout centuries.[63] Interlaced stories call for sustained curiosity and suspense. They promote rehearsal to fortify the working memory with contextualizing facts and values. And they induce a search for comparative relations even among the incomplete parts. The case against them, however, pertains to the brain's incapacity to retain the suspended parts. Quite simply, experiments involving retention and interference demonstrate that "children who had heard only one story were much better at answering questions about it than were the children who had heard two stories."[64] No surprise there. Similarly, interrupted stories may actually quell active suspense, suspend emotions, and tempt the overcrowded working memory to begin the systemic discarding and erasure of its deactivated files. Moreover, the more the stories resemble each other in shape or precept, the more the second will interfere with the first in deleterious ways.[65] Lapses in time are yet another factor. North recognizes that dilemma, even in placing the onus on the reader to ramp up levels of concentration. The paradox merely confirms that the technique was a calculated risk, a challenge to orientation and memory, yet a stimulus to the "cognitive fluidity" required to disentangle complexity, generate compound analogies, and infer precepts in order to make a cognitive investment in the procurement of wisdom. Curiosity about narrative endings may heighten concentration, compartmentalization, and comparison. Literary works may be characterized by the trials they impose, and memory may be investigated in relation to its levels of performance.[66]

If the emboxed story is intended to incorporate propositions within larger and larger propositions, to nest information within other forms of information, then its processing bears some resemblance to the operations involved in recursion, which is the mind's capacity to compound points of view in causal and chronological sequences. This takes us rather close to certain limits of human intelligence. We are good at recursive challenges only to about five layers, just as we are good at holding seven digits or bits of information in our short-term memories, plus or minus a couple of digits, but no more. "Fred told Mary about Sarah's encounter with the zebra after it had leapt over the zoo's fencing. He was particularly struck by the story because it reminded him of an experience with a black bear in a national park, and it was Mary who told me all about what I'm telling you now." (Who encountered the zebra after it leapt over the moat? No cheating.) Add another narrator to this and we begin to lose track. Just how the brain manages to remain oriented in recursive structures has exercised cognitivists, leading some to develop models of neural fixing and feedback. Steven Pinker opines that we do not create discrete networks for each component of a recursive structure, whether about propositions within propositions or points of view in sequence, but store all these elements *once*, while "a processor shuttles its attention from one structure to another, storing the itinerary of visits in short-term memory to thread the propositions together. This dynamic processor [is] called 'a recursive transition network,' one which in turn enables us to coordinate thinking when problems split themselves into smaller and smaller parts."[67] To be sure, some "connectivist" system does the work in accordance with selective design, however we choose to name it, like a machine in its own little room. Marvin Minsky deals with recursion in his chapter on memory entitled "Interpretation and Recovery" in which he likens the capacity to someone packing a suitcase in sporadic fashion who stops to pack smaller items before returning to the larger, smoothly recovering the first operation exactly where it was left off through a memory "control system" that prevents us from starting all over again from the beginning.[68]

Recovering partially completed stories, arguably, works along similar lines, particularly when those story parts are assigned to different narrators within sub- and sub-sub-structures – following each of which, we are expected to recover previous stories right where we left off, context and all. Not surprisingly, writers have experimented with our capacities for "multitask" reading, both in the emboxing of narrative voices (recursion) and in the emboxing of story units (*entrelacement*). The technique of episodic *entrelacement* may have originated as a suspense-building delay tactic, and in the paralleling of materials it can also serve to reify pedagogical ideas through multiple narrative elaborations. Or it may have been offered as a source of mental recreation, a little cognitive festival or mannerist extravaganza. But success, at the same time, is predicated on intuiting limits and in delivering precepts against the default working capacities of the mind.

In *The Moral Philosophy*, when the lion at last slays the wise and good bull, traduced as he was by the mule, the ass begins to berate his brother (the mule) for his treacherous and hypocritical role by telling stories, the third of which is "The Ungracious Traveller." All of these tales, in a sense, place the political framing tale of the lion and the bull on hold, which is itself enclosed within the relations and points of view of two of the founding translators, Burzoë and al-Moqaffa. Two travellers come upon a great treasure, and the more naive of the two is prevailed upon to store his new-found wealth in a common place where each might go to furnish himself according to his needs. The first, however, steals everything so that when the second goes to retrieve a small portion of his money he finds nothing. Accusations then fly in both directions, the thief playing his mendacious part with verve, so that the magistrate can make nothing of the matter. At last the magistrate decides, in Solomonic fashion, to consult the tree under which the pelf was once hidden, and to the amazement of all, the tree actually begins to speak, accusing the innocent man. But the wise magistrate smells a rat and builds a fire at the base of the tree, whereupon the thief's father, hidden within the tree to play a villain's part, is smoked out, and the entire amount is transferred to the falsely accused. Yet the story does not go straight to its conclusion. The father, before accepting his part in the trick, launches into a story of his own, that of "The Bird and the Snake," in essence usurping the ass's place as narrator in order to dissuade his son from attempting so hazardous a ploy. (We might debate whether this inset is a break away from, or an extension of, the tale of the travellers.) He tells of the bird, beset by the egg-snatching serpent, which goes to the crab for advice, leading to the offices of the mongoose, which in turn eats both snake and bird. The moral of this inset piece is difficult to nail down, and one may well ask how memory will tag it for recall. "A wise man will look ere he leap," concludes the father. But "so it happen not to me as it did to the Bird that would kill the Snake, I am contented" is an equally prominent pronouncement by the son who, thereby, in essence, is rejecting the moral of the tale. All this the father pronounces to his son before assuming the role of the witness in the tree – as though he had learned nothing at all from his own cautionary tale. The lesson was about the son's risks, forgetting his own. The defeat which the father fears is perhaps to be generalized to the level of the bird's demise, which alone would permit some form of complementary mapping in the mule's mind. If both stories are about calculating risk, as well as not meddling in the affairs of others, little can be made of the irony, in the case of the beleaguered and bedraggled bird, that she is consumed by the instrument she had employed to protect her from a ravenous enemy against whom she had no other recourse.

Matching that with the story of a man who would cheat another of his rightful wealth is the kind of work that worried North. Reading too closely may only make it worse. His concern was establishing the ideal reader in the context of what minds do with analogous materials, tagging, coding, and mapping; sometimes it is not enough to seize the worth of the exemplum, and sometimes responsible discernment overshoots the mark. Hence, if we remember this tale-embedded-within-a-tale at all, we may do so without generating a moral orientation or an ethical tag. What moral does the death of the bird convey? The crab was an untrustworthy counsellor for recommending the mongoose, and the mongoose was a treacherous friend in light of its feeding instincts. Thus in the admonition to "look before you leap," the fault lies with the bird whose knowledge of the nature of things fails her, though the blame seems harsh because her struggle for survival, caught between snake and mongoose, was foredoomed; how do we moralize that? Memory may not fail the story as episode, but may fail to supply it with the coding that makes for a more closely moralized mapping and application.

Or perhaps we should not give up so easily. It should be possible, by introspection, to determine at least the circumstances under which this story might be recalled as a prompt for future planning, in keeping with the father in the story who carried it in *his* memory and produced it as the foremost exemplum for enlightening his son with regard to his present dilemma and choice. That in turn was the work of an author who assigned it to him on the understanding that the memory might format just such a tale, to be recited under just such circumstances, in accordance with what a "natural" memory had made of it as an ever potential object for recollection. The question is whether the author has made of it what the systemic operations of memory would devise during the process of chunking and dispersal for storage and eventual reassembly. This is the problem in hand, namely, how author's meanings through stories recollected align themselves with what memory and conscious inference together make stories mean. The problem is both cognitive and hermeneutic, shifting like a Necker Cube in the mind's philosophical eye, now seeming to face one way, and now another in accordance with how the brain sees it. This is analytically challenging, and yet Vishnu Sarma took it as child's play. The way memory works should take care of the problem.

Returning to the matter of interlacing and the recursive handling of voices in *The Moral Philosophy*, there are several other sequences of stories with their changing narrators likely to defeat the attention of the casual reader, among which, this further example. Even to construct the following resume, I had to reread the entire section quite carefully, doubting the terms of the opening by the time I reached the end. The ass tells to his brother the mule the story of the holy man who lost his treasure to a notorious robber who posed as a sincere disciple – one who, for many days, fasted and prayed with the master in order to make away with his treasure. This holy man, ostensibly concerned over the soul of his erstwhile student, sets out to find him. To this matter we know we must return, but it is suspended by an interim quest constituted of several discrete tales. Along the way the holy man meets the fighting goats who inadvertently kill the not-so-wily fox who sought to lick their blood by stepping between their clashing horns. Hold that moral. Then he spends the night in a brothel where the old bawd seeks to slay the lover of one of her "girls" because he is interfering with business, but is killed instead by the belching (eructation) of the lover as she seeks to blow poison into his mouth through a tube. Seize the sense. Thereafter he makes his way to the house of a jealous husband who thinks he has cut off the nose of his unfaithful wife, but has only disfigured the "bawd" who stood in her place. The bawd or maquerelle, in turn, falsely blames her accident on her barber husband's mishandling of a razor. Thus the cuckolded husband is bamboozled by both wife and bawd. Turn this to wisdom. The quest now comes full circle as the last tale morphs into the first. The holy man, a privileged observer, is outraged by the accusation of the bawd against her husband and follows them to court to witness on behalf of the husband but there finds his disciple, the thief of all his goods, "newly punished for an old offence." Instantly he forgets the barber, wife, and bawd and cries out for justice against the double thief whose dear but wayward soul had been his initial concern.

In this sequence, it is not the nesting of stories that confuses so much as what each story means to each of the parties involved: the listening mule concerning his crimes at court, the holy man in relation to the quest for his disciple, and the characters within each of the respective stories. This is not recursion, but multiple audiences and points of view which contextualize each story differently and point to complementary or contrasting readings, yet it resembles recursion in demanding of the reader a capacity to arrange and retain each of the contributing parts before attempting to blend them into a working set of values. If you have come away confused from this paraphrase of events and sequence of stories, the point is the better made. Together, they entail somewhat more than even several readings can provide, despite our faith that they were perfectly complementary to the mind of the collector who perceived in them the common denominators that brought them together.

We are reminded, meanwhile, of the intent of *The Moral Philosophy* to relate the critical number of stories needed to complete the political and public educations of the two young princes – to produce, in essence, a full set of applicable memory traces. Would it be fair, then, to hold the collection to its promise of educating the complete prince through its forty-one stories, including the framing tale of the lion, the bull, and mule? Probably not. Yet it would be an exercise of a certain magnitude, nevertheless, to attempt to quantify precisely the kinds and categories of wisdom explicitly stated or implicitly delivered through inference by these stories, now conceived as a plenary theatre of behavioural intuitions and prompts both consciously and systemically supplied.

We must now formulate a moral philosophy in its totality through some version of all the operations of memory and cognition outlined in the first half of this essay. We might reflect on the nature of story precepts, how clearly they were perceived, how concretely remembered, and even the likelihood that they would ever be recalled and applied as practical guides to future decision-making. Or to put this into the language of Herbert Simon's model for learning algebra, to what degree have these stories complied with the expectation of an "adaptive production system" by finding their stored equivalents according to the symbolic language the brain happens to be using in order to create new symbol structures and thus make them part of the operative information that is constantly running within the system?[69] Such stories achieve their effects only if they succeed in running the long-term memory formation gauntlet and are recollected as applicable patterns of action in problem solving through comparative association. North was clear about this in his instructions to the reader and, as a member of the Leicester Circle, he may have been more sincere than we might imagine in offering a guide to political conduct to one of the forces of the realm. But we still seek to know in hard experiential terms or through rigorous analysis whether the collection can or ever has come close to its promises, and that can only mean determining what the book has done to the memories of its most attentive readers. After all, by the Victorian period this collection, best known as *The Fables of Bidpai*, had been translated into thirty-eight languages in some 112 versions and 180 printed editions, making it "one of the most enduring works of imaginative literature of all time."[70] What are those recollected traces like? What are the allegorical equivalents assigned by the memory formation processes? How available are they to the subject on an "if-then" basis? Perhaps appraisals of this nature can be made anecdotally without consideration of the operant cognitive processors, but willy-nilly, though reason is the father of wisdom, memory is the mother of meaning and will have played her decisive and systemic part – and arguably of the two, memory is by far the greater factor.

Spenser's grand design in *The Faerie Queene* shares with North's *Moral Philosophy* the cultivation of memory in the learning of virtue through the technique of interlaced or nested storytelling, but Spenser ramps up the level of hermeneutic expectations to even more challenging levels. For him, the interweaving and juxtaposing of stories becomes thematic. Ostensibly, Spenser, in his grand program of the virtues illustrated, builds toward the completion of a sprawling episodic yet thematically unified essay in which each sub-narrative performs and illustrates a sub-motif. His overarching theme in Bk. IV is friendship, which, in itself, is a companionate virtue, one that necessitates sharing and trust, most readily expressed through an exchange of privileged information and confidences. Those who are the enemies of this virtue rarely have stories of their own to share and are described in their deeds by others. Those who strive for the paired or communal good, meanwhile, endure crosses and thus their stories grow by episodes and are cut off by intrusions. Because the courses of love never run smoothly (and such courses are pursued by many within the "community" implied and represented in this complex book), stories open and close at a rapid rate, and yet few find true closure. The communities of our own lives are like this, creating an imitative form for Spenser to fall back on, if you will, insofar as our social existences are made up of many half-completed actions or conversations, of plans in progress, and hopes unrealized – stories opening and closing with the ring of the phone, the rounding of a corner, or a meeting over coffee. The book self-consciously lingers over unfinished, long, and indeterminate episodes, some stories being partially repeated when they are resumed, or told to other characters, all of which aids the information-challenged reader, but rarely to full satisfaction (if I dare leap ahead). Indicative of the many cases of inset materials and displaced points of view that occur throughout this book are the six cantos which come to an end with suspended actions, each time interrupting the working memory, placing on hold its temporal sequencing and its fund of transpired facts, as well as feelings about the characters and their moral statuses.

In the nineteenth stanza of canto eight, Spenser employs the little transitional formula, "till on a day." It marks the moment when Arthur discovers Aemylia and Amoret, the one half-starved and the other near death from her wounds. Their plight is not only an occasion to reintroduce this emblematic king into the action, if only for a brief time, but to place preceding episodes on hold, resolved or unresolved, and call back to mind the expansive group of interconnected tales that leads up to the present moment for these two characters, backtracking to a subset of narratives originating in Satyrane's jousting for the girdle of Florimell, following which Amoret and Britomart set off together on their respective quests for Artegall and Scudamour (v.29), and to the passages in Book III in which readers are introduced to Timias's ill-fated love for Belphoebe. Memory, at this point, must also supply a clear and instantaneous impression of Amoret's meeting with Aemelia and their close encounter with intended rape and cannibalistic murder – sufficient, at least, to make sense of Arthur's administration of a restorative liquor and the service he seeks to render them by placing them on his horse to lead them to safety, albeit ironically toward the cottage of Sclaunder (viii.34). It is as good a place as any in this compound narrative to take stock of what has been remembered of previous episodes and what has been forgotten, even catastrophically forgotten, given that any meaningful forward movement depends entirely on a sufficient recall of critical past events. Amoret and Aemylia's sojourn in the Ogre's cave was itself an inset to explain Amoret's disappearance during the encounter between Britomart and Scudamour. There was also the betrayal of Timias, followed by his self-exile, and the rediscovery of Belphoebe's proximity through the story of the messenger bird. Only a lengthy summary of the many plots could show the degree to which delay and suspension mark this book, coupled with the implicit or ostensible grid of memories such events were to have created in the minds of the most alert and receptive readers. Such a narrative technique brings to awareness more critically than ever the matter of reading and memory, of the continual presence of the past in its ongoing cumulative state as an essential condition for the comprehension of present events, and of the ways in which narrative strategies can place memory under stress, strategically or inadvertently, through an excessive employment of recursive strategies and emboxed stories.

The challenge of making meaning from these episodic designs becomes particularly perplexing with those stories that refuse to close because the characters are themselves so psychologically incapacitated that they cannot achieve their happiness, but find themselves in prolonged states of trauma, thereby making nightmare out of story and obstructing natural desire for completeness both of the narrative and of the bonding that leads to companionate marriage. In this way, Scudamour's courtship of Amoret becomes a narrative that "harder may be ended, then begonne" (x.3), while, in a similar fashion, the union between Britomart and Artegall is repeatedly forestalled by fate, error, and intervening adventures, even after their meeting in Canto vi in which they share a nearly complete moment of wooing, vowing, and consenting until "mariage meet might finish that accord" (41). Even then, pressing affairs separate them once again and the marriage is indefinitely delayed. To be sure, this is the way of romance – the episodic trials that prolong the desiring emotions, strengthen loyalty, and put pre-matrimonial mettle to the test. All along, memory makes virtue of experience by storing up the trajectories of persons seeking mutuality, influence, adulation, to love and be loved, hence loyalty, and a common cause in the world. But the fragmentation of desire through the conventions of emboxed narratives frustrates the coherent retention of story and contributes to a growing sense of alienation, solipsism, or malign fortune. Given the fragility of thought, the impositions of fate, and the misreading of other minds, the rivalry over rare resources, the resilience of disliking or disdain, the mishaps of friendship and the tribulations of love, we see forces abounding which are able to delay closure sometimes indefinitely, thereby tripping up the Gestalt formation that constitutes closure in a stasis of failure or success. This is business as usual in the creation of the social binaries that animate these human "scriptures." But social process replaces the tagged learning of finalized events. We are back to the two kinds of memories outlined at the outset.

Perhaps Spenser's fragmented design functions thematically in inverted or negative ways by exposing hope to contingency and incompletion. Nevertheless, hope is essential because friendship at work in an extended community is a matter of ongoing and interrelated stories as guided by principles of goodwill and cooperation among associates. Multiple and interwoven stories are merely an artistic manifestation of the social kaleidoscope that we know in everyday life in gregarious situations – those in which persons, each with a story, enter and perform within the contexts of their respective identities: past histories, present desires, and symbolic messaging. The interlacing of stories expresses our cognitive plasticity in shifting from centre to centre in life situations, just as we must do when characters enact their quests and endeavours in a social economy in which barbarism can be avoided only through acts of self-regulation, loyalty, and solicitude for others. In both circumstances, we are challenged to remember identities, incomplete actions, and proclivities in order to add meaningfully to emerging events.

Reading Spenser is good cognitive practice. The interlacing of events is part of the illusion of that greater fictional community in pursuit of the happiness that comes through friendship in its many categories and traits. Interlacing is merely a correlative to the social life. Meanwhile, from a perceptual point of view, the many sub-features of friendship between men, between women, between blood kin, and between men and women, including lovers, may, in part, have led Spenser to his seemingly endless juxtaposing of uncompleted stories. Friendship has its many contributing parts and negative counterparts, from adoration to jealousy, or from trust to suspicion, or from transparency to hypocrisy; to name them is to return, arguably, to the text itself to read each inset tale, each micro-romance, each relationship, whether in binary, quaternary, or other groupings of persons, for its unique informing idea. We are back to the domain and offices of memory to track, contextualize, chunk, tag, and fix the recall traces – all as ineluctable operations in achieving the Spenserian mimetic, experiential, and pedagogical program.

But to say as much is to resurrect the problem of allegory in its more and less naive manifestations, for we are altogether less certain with Spenser than we are with North that story is tantamount to exemplum, that plot equals symbolic episodic action or a template of understanding *tout court*. So we must be content with our thesis for as far as it can take us and no further. Spenser was vitally interested in memory, with the ways, according to Renaissance faculty psychology, that information is fetched in the library of the brain by a clever young runner named Anamnestes (one who reminds) and presented to the aged Eumnestes (one who remembers well) (II.ix.45–60). This has a lot to do with the entire work as a vehicle of communal memory concerning the ancient heroes and founding events of the English nation and the Tudor dynasty, but the mechanisms of recollection, drawn principally from Aristotle's *On Memory and Recollection*, have overtones for the thesis in hand. Spenser mentions those things which are mislaid or lost and the effort required to retrieve them, which brings us back not only to the work as a vast manual on the actions of learning in the making of a complete gentleman or courtier with timely recall and contextual recollection as his supreme objective but to the work as a product of stylistic choices. The two function hand in hand, and to make an evaluation of the efficacy of the Spenserian design in relation to the thresholds of memory formation is by no means straightforward.

The problem relates to points made by Aristotle in the *De anima* (3.3) concerning the offices of the imagination and the conversion of direct sensation into mental images which the memory can in turn process. The secret to memory was the striking power of those images and the logical, emotional, and verbal prompts which could bring them to recollection. Maurice Evans makes the connection between narrative memories and artificial memory and "the ways in which writers organised the knowledge which they wished to imprint firmly in the memories of their audience." For Spenser, the art of memory and the narrative techniques conducive to memory came into close association.[71] This leads to the orator's employment of memory places or "loci" and the association of ideas with images, turning *The Faerie Queene* into a vast mnemonic theatre of story pictures, so that in travelling through those pictures in ordered fashion, the attached precepts can be recovered one after the other. The ultimate product is prudence, or wisdom – a comprehensive plan of behavioural options. Spenser's work is, of course, more than a picture gallery of mnemonic devices, but in the language of cognition, including the formation of memory traces, there is only a short distance to go to draw Renaissance (Aristotelian) vocabulary into the most recent analysis of working and long-term memory. Then the patterns, pictures, symbols – whatever they are to be called – must be built up through stylistic choices to a necessary critical mass, achieving an intensity that withstands delay and the proactive work of oblivion in clearing out the working memory. All these factors impinge on authorial choices and on reader success.

I do not wish to make more of the role of memory in forming the reading experience than the evidence bears out, or to reduce the reading experience to the accidents of a neurobiological system, being only too ready to plug for the rich layers of cultural understanding, the computational checks and balances, and the critical reflections of the conscious mind. But Spenser himself attested to the central role of memory in achieving his ends as a writer, and to the fact that *The Faerie Queene* was a giant national and ethical mnemonic, a memory function that could be taken right back to the faculties of the brain and their roles in retaining and retrieving whatever they could of the data and design of his great work. That is to say, he too was mindful of the organization of his work in relation to the systemic capacities of memory formation and recollection, thereby allowing for the two questions raised in the present study: whether the reading experience has done its work in relation to authorial aspirations in building a network of potential recollections through the encoding of episodes and precepts; and whether Spenser's interlaced and thematically interconnected storyettes have served him well in constructing the value prompts pertaining to an ethical schema in the interests of performing what he deemed critical to the education of his ideal as well as his "common" reader. Such a project can be realized only through the offices of memory formation whereby stories are simultaneously turned into "places" of meaning and their interconnectedness into a community of interrelated and self-cueing parts. Where such automaticity stands in relation to reasoned inference will never be settled with mathematical precision, for minds work in parallel, systemic, and adaptive compartments, and in proximities based on schemas which define reality, out of which we build the "storied" precepts which invigilate our social world on a partially subconscious, habituated, and competitive basis. That is the kind of animal our brain systems have made us. At this point we move beyond the fallacies and biases of a fickle, interested, schematic, inferential, and confabulatory brain at work in preparing memory cores for future recall, to ask whether, in any age, Spenser's readers have routinely returned to his stories as they are remembered by way of informing the planning that meets the vicissitudes of the active life. That is a difficult question, answered, nevertheless, by the faith both Spenser and North held in literature's power to extend experience within an ethos of ethical consciousness toward improved performance in the active life. Memory by design is utilitarian, or at least its faculties were selected out exclusively in terms of the benefits they provided to sustain the organism long enough to permit the genetic replication of its traits. That is the brain we inherited, and storytelling of diverse kinds will continue to serve those ends. Surely with Spenser, as with North, pragmatic learning for mastering the contingencies of the urbane life was part of the formula, in the pursuit of which they made or transmitted the narrative choices calibrated to those ends. These are matters for considered critical evaluation.

Further to that critical work, the case has been made for a memory that reduces matter to paradigms suitable for comparison as a prerequisite to mapping base patterns onto target patterns. Only then can subjects learn from the stories they know. Likewise it has been argued that memory meets the data of stories to blend them with what it already knows of the world, and deposits those memory configurations in relation to inferentially devised abstractions assigned to them. The critical but elusive question has been posed concerning the compatibility between the conceptual tagging of allegorists and the recall coding of default memory and how the hermeneutic brain deals with the dissimilarities. Spenser's "essay" on friendship calls for a profiling of the sub-motifs of the central virtue through a proliferation of stories illustrating cautionary actions or inferentially drawn moral precepts, not necessarily teaching the same thing. In light of evolutionary design, memory-making faculties might be altogether more inclined to fix pragmatic lessons for a survival-oriented creature than to install the abstract theories of a moral conduct system, but such discriminations are difficult to make insofar as the Spenserian virtues are likewise adaptive in those social millieux in which altruism, cooperation, mutuality, and bonding define group cohesion, family formation, and communal survival. Allegorists seek to control connectivity according to precept – it is the definition of their craft. But the narrative episode is caught in the middle in attempting to serve both its author and the default habits of reading brains. But by the time Spenser reaches chastity and friendship, he may have found himself telling stories best left to the devices of memory on its own, rather than to over-discipline them as part of a philosophical program driven from the top down. That is open to ongoing discussion.

The technique of *entrelacement* or the emboxing of stories is a touchstone for bringing memory studies to bear on literary practice. The full capacity of working memory for the retention of interlaced data is a moving target, dependent on latent talent, attention, rereading, rehearsal, and literary experience. Nevertheless, its limits have been demonstrated by experiments employing lists of words, a diversity of objects, or unfamiliar stories in relation to distractions, delays, or interference. Insert stories likewise have the effect of distracting, delaying, or interfering. Those constraints and limitations must similarly circumscribe the number of discrete social contexts we can manage, their ongoing density and sharpness proportional to the quality, length, differentiation, emotional content, and personal stakes associated with each emergent situation. In literary instances, such capacities and limitations may be tested on an anecdotal basis by simply asking first-time readers of Spenser how often they had to backtrack to restore critical data in order to continue with a meaningful pursuit of the story, or how often they plunged on in unfamiliar circumstances hoping for hints that would restore the past. Inset tales would appear to be of utmost importance to Spenser's literary and didactic endeavours, but we can never know his mind in the matter with certainty. Convention, humanist imitation, and sheer novelty may have been paramount, or more strategically, he grasped the cognitive challenges and literary play of juxtaposed episodes and comparative impulses, or the value of making the reading experience difficult in order to whet attention and exercise our computational habits. A defence of Spenser is not difficult to build. But *The Fairie Queene*, however we wish to profile it, is not only a memory-making quilt of stories, but a memory-testing quilt of stories.

In Bks. III and IV, Spenser brings us into the realm of romance and away from what has been called naive allegory, introducing a quality of storytelling that encourages, arguably, a greater investment of empathic concern for heroes and heroines in their struggles to find union and stasis, and thus a more emotional involvement generally which should, by rights, contribute to a fortification of the memory traces these stories generate. Does it therefore become a meaningful question to ask whether the stories in Bk. IV enjoy a greater emotional weighting in our memories because of the kinds of stories they are, and whether that effect has been dissipated by the fragmentation of their presentation? Memory research has demonstrated that emotionalized memories come up first, bearing the more urgent record of themselves as felt experiences, reflecting in their effects all that the limbic system has done for mammals and primates in saturating their involvement with environments through feelings.[72] Because feelings are polar and graduated between the hedonics of pleasure (well-being) and displeasure (adversity), notions of the desirable and the undesirable are not far behind, connected in turn to what is deemed good and evil, both for individuals and for the "survival" groups with which they identify (socially, religiously, or nationally). The cognitive and cultural implications of a brain that imposes binary values on the natural and social worlds, beginning with first myths, such as the tree in the Garden of Eden, holds enormous significance for the nature of the human and the ordering of their collective experience. Emotions arise in relation to emergent actions arousing desires and apprehensions plotted out experientially against the passage of narrative time and coherence. The memory of emotions can be sustained, but the experiencing of emotions cannot. New data decontextualizes them and reorients them, and the interlacing of stories, with their conflicting emotional registers, cannot but have an effect on "felt" qualities of the reading experience.

North agonizes over a perceived obligation to intertextualize and computationally investigate the potential thematic relations among the interleaved or nested stories that constitute the Bidpai tradition. In the interest of extracting all possible hidden meanings, North allowed for the mental surcharge ensuing from the technique of multiple or overlapping stories as a by-product of extracting regulatory adages, maxims, and mottos from often trivially moralizing fables – all in the interest of maximizing the pedagogical efficiency of the collection. But the default moralizing assumed by Vishnu Sarma fell short of North's humanist and political vision. Such cross-readings represent a rare if impossible achievement simply because of the inability to consciously reduce discrete stories to relational levels within other stories. But the most troubling implication is the departure from conscious reasoning or forms of cognitive and intellectual play. Arguments based on the evolutionary development of our data processing faculties always point us in the direction of ancestral values and lessons because the conditions that promoted the adaptive design of our brains find their rationales in the environmental challenges that affected survival. Memory studies emphasizing the roles of "conceptual" tagging, binaries, and action templates thus have an awkward way of imposing reductive and deterministic elements to parts of the mental life we prefer to think of as elective and intentional. Yet the implications of the systemic data processing of our inherited faculties cannot be wished away. Their roles are pervasive and the value patterning is evident in the mechanics of trace formation and recollection. Here is a final witness to that effect: "Learning … is not a general-purpose mechanism that allows all environmental relationships to be acquired with equal proficiency. Instead, it is a constrained mechanism that depends upon an affective value system that provides an immediate appraisal of only those events that have important reproductive consequences ... It is the 'omens' of life and death, positive or negative feelings, that direct the learning process (the inner genetic algorithm)."[73] This "value system" of designed priorities and the concomitant biases of learning, chunking, cueing, schemas, confabulation, and probabilistic recall is hardly "red in tooth and claw" in plotting for number one, but reveals its proclivities in the priorities of retention. In some fashion, the effective moralist must work in philosophical alignment with what memory faculties do.

Play is always serious, and particularly so when its mimetic or representational substance is social, volitional, and consequential. We may have the impression that we can choose to remember any whimsical thing we might wish, and with effort we can, but under other conditions the editorial memory is still at work by its own criteria, organizing our past as new moments preoccupy our conscious minds. Stories mean as the memory reads and formats them. It is the first and most critical dimension of all hermeneutic endeavours. The counterpart to memory is, of course, forgetting, and that too must enter into all aesthetic, structural, and moral calibrations pertaining to story.[74] Fabulists with their glosses and allegorists with their guiding precepts are, among writers, the most committed to fashioning the memory to premeditated pedagogical ends. Hence their stories are typically geared to moralized matchmaking on a base-to-target application through the "systematicity" of analogical thinking.

Paradoxically, the moralists examined here are among the writers most disposed to test the limits of memory either through comparative complexity or through interlaced and recursive episodes. North and Spenser were cognizant of their roles in establishing stories to live by and hence for stylistically maximizing their efficient and utilitarian recall. At the same time, there was enormous ground to cover to create the complete conduct book, given the complexity of the social life with all its exigencies and quandaries, tempting Spenser to become a narrative over-reacher and North to debunk the simplicity of apologues by embedding them in an enriched ethical program through hyper-reading.

At the same time, North and Spenser illustrate the involvement of all literature with memory. Idiosyncratic, yet quasi-describable and quasi-quantifiable, memory becomes a working measure of all literary purposes and possibilities because its operations are universal to the species. All hermeneutic impressions are based, not on works, but on what is remembered of them. Memory is learning's matrix and maker. How it works is the result of adaptive design over eons of evolutionary time; it is not absolute, but purpose-oriented, and approximatively efficient. Realizing its nature and limits may, thus, entail a major overhaul of the critical enterprise, and before those insights can be very good, the science must be very good. There is work here for the future.

CHAPTER FIVE

Crying and the Ambiguity of Shakespeare's *All's Well That Ends Well*

Shakespeare is one of the masters at styling the "infinite variety" of the human – the impressive array of the ontogenetic traits of which our species is capable. But to be a species, we must also be defined by the phylogenetic bedrock of our natures whereby, through a consistency of design, we presume to know a great deal about others because they are so much like us. Literary characters, in their fundamental behaviours, must operate cognitively and emotionally in chorus with our entry-level understanding of what it is to be human, or we could make very little of them. That, in turn, suggests that both authors and their readers know a great deal more about human nature than what they have learned through experience; a common genetic inheritance has wired all our minds alike when it comes to such matters as emotional production, because the ancient design templates for the limbic system have been uniformly disseminated through the genome. Without this species-wide repertory of traits, we could barely imagine the development of a theory of mind whereby we devise reliable insights into the intentionality of others. Each person would be a species unto herself. In particular, for the present study, we would have no idea why others frown, laugh, shout, pout, or cry if we did not share a midbrain design that produces similar emergent hedonic states in response to similar generic stimuli, and we would have no grounds for judging others for laughing or crying out of place.

Arguments in support of this simple truth may luxuriate in many directions, but they always return to the human capacity for understanding other minds, a topic dealt with in the MIT *Encyclopedia of the Cognitive Sciences* under the heading "Intersubjectivity."[1] Implicit in this research is that Shakespeare, by dint of his membership in the species, possessed at least a competent, but arguably impressive, ability to deal with interpersonal interactions through the same principles of "folk psychology" that unite us as a species of mind readers. In the words of Daniel Dennett, we "go about populating each other's minds with beliefs" in order to explain and make rational their conduct. José Luis Bermúdez in "The Domain of Folk Psychology" profiles the skills involved in predicting and interpreting the psychological dispositions of others based on reasoning about their mental states "both occurrent and dispositional."[2]

Reading and creating fiction depends on the ability to ascribe mental states to others, including the elements of personhood that make such ascriptions possible. We believe in the rationality of others in seeking their goals and presume, within limits, to intuit their causes and motives. This practice is tied up, in turn, with the phenomenology of personhood and the full volitional and intentional distinctiveness of persons as "rational beings striving to satisfy their desires and aspirations in light of the information they possess about the world."[3] Even of fictional characters we must believe they have minds and that they are very much like our own, insofar as their makers project upon them the same notions of time, causation, feelings, basic desires, and instincts they themselves share with others of their species. Thus, cognitive investigations can be extended to include both fictive minds and reading minds, not to mention the minds of authors (all of whom were alive and thinking as they were writing – not one of them dead), and the ways in which humans characteristically deal with theories of knowledge and the states of other minds.

Moreover, Martha Nussbaum states that "neuroscience, when not wedded to a reductionist program, can make richly illuminating contributions to the understanding of emotions, their intentionality, and their role in the economy of animal life." [4] That is reassuring, because the specific human trait under investigation in the present study is crying – a trait which is generic to the species, deeply embedded as a prompt-response mechanism in the hedonic midbrain, and constantly available as a contextually equivocal signal of distress, sorrow, frustration, or sudden joy. Crying is therefore part of the repertory of human emotional expression, with Shakespeare's mimetic intuition on the one side and neurobiological investigations into the meaning of crying on the other. The present task is to bring them together. Our reaction to emotions, both our own and those instilled in fictive characters, is a matter of vital social concern; we are obsessed by what others are feeling, especially about us, and we scan the world for signs, rather confident of our abilities at psychological mind reading – Shakespeare included.

Paradoxically, however, what crying means and how it interprets our world according to the ancestral values that made the cry display an adaptive feature of the human limbic repertory, and the degree to which it remains an ambiguous and contextually specific signifier, are matters still open to basic investigation. That makes the term "reductionist program" the key concept in Nussbaum's observation; it is the bugbear to be faced en route by all observers of culture simultaneously concerned with the implications of brain science and evolutionary psychology. No serious scholar can buy into radical reductionism whereby Pleistocene circumstances constitute a linear interpretation of modern cultural diversity, any more than any serious scholar can deny the tilting of the brain's predilections through adaptive design. The work in hand is in making careful discriminations concerning the melding of ontogenetic and phylogenetic properties in human behaviour whereby, in the present case, crying is limbic in origin, yet subject to the misconceptions of those who cry and the misconceptions of those who interpret. Crying nevertheless remains semiotic bedrock for the species, despite our hermeneutic challenges in reading its meaning in all cases and under all circumstances in literature as in life. After all, humans do cry, and they do so first and foremost because the response is an autonomous limbic reading of crises in the environment controlling the life of the subject, and secondarily because it conveys a sense of that emotional state to those in proximity. Crying demands attention and response because of the adaptations to ancestral environments that designed the emotional brain. It makes for caring communities by inducing us to acquiesce to its involuntary sincerity.

Through simple introspection, we might be brought to understand important new first principles, such as that while humans are experientially and constitutionally diverse, yet in matters of our emotions, we share a common neurological architecture and thus a phylogenetic as well as an ontogenetic way of seeing the world.[5] We do not learn to cry, but do so according to the constitutions of our midbrains, as our first and most heart-stirring rhetorical gesture in life so amply demonstrates. Post-natal crying is an interpretation of our condition in the world long before we are equipped to reason about it. Upon that single axiom, there is a thesis to be elaborated about the nature and meaning of this behaviour. It is a reading of the world and a message to the world, and it wells up through mental machinery far below the level of volition and reasoned reflection. We do not make crying, but are made by it when it sweeps over us; we become whatever it is that is exposing us to the world involuntarily, and we try to reason through the causes in order to quell our own alarm.[6] This pertains not only to the way in which we appear to and are interpreted by others in real life but also to one of the ways in which fictional characters reveal their internal states of alarm or excitement to onlookers. Moreover, to cry is to have the biological capacity to cry, and that capacity has an originating and potentially explanatory history. What does it mean for a species to share a common capacity, and will that understanding depend, in turn, on knowing what crying was for in our ancestral environments? Traits belonging to an entire species entail a genetic coordination made possible only through adaptation; hence, crying must have value in relation to common environmental hazards.[7]

The examples chosen to illustrate these problems are two moments of crying in Shakespeare's "problem" comedy *All's Well That Ends Well*. I foreground them as cogent representations of the social and aesthetic emotions and as loci for literary interpretation. But given the complex explanatory connectivity between neurobiology and textual hermeneutics, those two literary moments will at times slip temporarily into the background. Very near the play's opening (I.i.35–96), and again a scant few lines before its closing (V.iii.314–16), two characters are brought to tears in circumstances of stress, loss, or rapture sufficient to incite this very particular and expressive psycho-physiological reaction. Moreover, those stimuli are sufficient to engender public emotions displayed through the body as distinct from the feelings perceived only by the individual who experiences them. That threshold is important, for it distinguishes between the private hedonic instruction which the midbrain constantly supplies as feelings concerning low-grade physical and social events and the physically expressed emotions which follow from more radical changes in the environment, as when sorrow or overwhelming joy brings constrictions to the throat, tears, and even uncontrollable sobbing. It is telling that in both dramatic moments, Shakespeare's characters do not wish to make public spectacles of themselves, but they are betrayed by their reactions – betrayed, in a sense, by the midbrain processes that have deemed such displays to be adaptive conduct and advantageous to those who display them to others.[8] Crying in company makes public that which begins in the deepest recesses of the mind, despite the reticence and embarrassment of those so overcome, and presumably to adaptive ends.

In the first act, the Countess of Rousillon cautions Helena not to make such a show of her tears, whether for overindulging her grief over her father's death, or shedding manipulative and essentially forged tears in order to solicit more sympathy than is her due from the social entourage.[9] At the play's end, Lafew, by contrast, in witnessing the promise on the part of the recalcitrant Bertram to truly love Helena after her seemingly miraculous return to life, complains with a touch of embarrassment that his eyes smell onions in an attempt to dissimulate his joy over the reunion. He too would avoid a show of sentiment that makes more emotional claims on those around him than he intends. He would prefer not to be taken for a sentimental old fool, after all, who cries at weddings and happy endings. Yet both characters cry, willy-nilly, and in doing so, may be providing us with two of the most nuanced signifiers of the play, precisely because they are subcortical evaluations of the social environment beyond the control of self-imposed facades, game plans, and constructed selves. Emotions thus become displays that invite interpretation, both by those who are embarrassed to show them and by those who register their high truth value in reading other minds. Emotions are semiotic and no doubt have been so from ancestral times as instruments of group cohesion through displays of fear, joy, compassion, and grief. Tears, by definition, though a sign of vulnerability, also make claims on others within support groups. Simultaneously, emotions read our world through an ancient brain system and impose their judgments on consciousness in the form of pleasant and unpleasant sensations. The evolutionary backstory whereby this trait became part of our genetic design is implicit in every act of crying.

The Countess tells us through her candid character profile of Helena to Lafew that there are bad minds in which virtues are pitiful because abused and minds such as Helena's which have no such duplicities, for she is merely honest, simple, and good; this is a revealing introduction to a character who is taken by many to be a devious little schemer out to get her man by any means, fair or foul. This compliment is paid in Helena's presence, however, and when Lafew attributes her tears to her vanity, the Countess is quick to reply that tears are rather a gesture of humility, and that the memory of her father's death, in any case, continues to incite her feelings of grief. Both take themselves for competent readers of other minds, but in this case, both are in error.[10] What is clear is that the manifestation of her condition has taken priority, for tears impose their demands within societies where members are expected to care for those in distress. The Countess and Lafew had been speaking of her father's memory, perhaps tactlessly so under the circumstances, and now Helena is mildly rebuked for a show of emotional excess over which she is expected to maintain rational control. Society also has its interests in monitoring indulgent feelings by which they feel manipulated. Yet the empathy of the Countess and the suspicion of Lafew (products of imperfect folk psychology) are wide of the mark, as Helena tells us in the confidence of a short soliloquy. Bertram, indifferent to her suffering, is preparing to leave. Once she informs us that love is the occasion of her grief and that Bertram is the object of her desire, we realize not only that the crisis of his departure is the cause of her tears but that her corporeal signing of that grief has been destined for him alone. Her tears are seals upon the sincerity of her affections, or at least their affective potency. Shakespeare knows that. We must assume, as well, that in making her cry, he allows her to signal to us something of her true interiority. Girls weeping over hidden causes are not crocodiles.

The language of the emotions, attached now to amorous longing, registers the grief of separation. It pronounces, in its way, not only the degree to which Helena's prospects for enjoying the mate of her choice have been diminished by his imminent departure but, by the logic of biogenetic design, all that such a departure signals by way of her reproductive fitness as a mother to future generations. The question is whether there is such a plot built into the impedances and finishing lines of the human midbrain, the neural "cyclotrons" or "amplifiers" that form and dissolve these limbic sensations. How, in effect, does the limbic system formulate its judgments on the emerging contingencies of social environments? How do those involuntary reactions contribute to life goals, such as they are interpreted by these "amplifiers to specific motivational systems?"[11] To be sure, Helena may have preferred to hide her tears, but a primal assessment of her falling fortunes welling up through emotions has made her distress public. Moreover, the backstory reflected in the present time of Helena's world must pertain to the hedonic instructions felt by our ancestors in matters of mating and parenting whereby such values fostered the future of their genes. Helena's eventual offspring are thus crying through her eyes. Her ambition to love and be loved by the object of her choice has itself become an environment, a realm of strategic mishap and disaster under the full invigilation of an emotional brain. For onlookers in the know, tears become the physical seal of her commitment, the confirmation of values and desires almost beyond words, and the markers of a life imperative, a goal endemic to the species. Misappraisal may confuse others, but Helena knows why she is crying. The argument now embraces not only crying over an immediate stressor, but the emotion itself as the fuse of romance, confirming her place in the genre. We may dislike her means, but her motivations are deeply engrained and tears confirm a destiny in her own eyes.

When Parolles enters, weeping is replaced by banter, but their debate is precisely about those things that are expressed through the genes. Most pertinently, they discuss Helena's clarion desire to dispose of her virginity to the partner of her choice without squandering it on sexual opportunists. She is, in this, equally true to her liminal instincts. Her emotional life, according to a genuinely Darwinian script, has already been transferred from father to prospective lover, from the old generation to the new, as she prepares her best options for pursuing her destiny as a wife and mother. In Helena we see not only the hedonics of a secret moment but Shakespeare's emotional orchestration of this character's psyche feverishly at work in the pursuit of its biological future. Brain designs carry their own plots and, through the emotions, make our life-stories conform. Meanwhile, Parolles brings out a reasoned confirmation of her archetypal priorities: to manage her chastity toward long-term interests, computationally strategizing her powers to attract a steady and protective mate. Reading Helena comprehensively through the priorities of the genome that designed the psychological tilt of her mind may be as archetypal as we can get; there are the telltale signs, allowing for any and all deviations she brings to us. Determinist it may appear, but no more so than in offering some predictability because she is a member of the human species. What can we do? She makes her choices, but with a phylogenetically designed instrument, and so far she seems right on track.

The conclusion to the play is rather drawn out as Diana plays at riddles with the king, equivocating over how she came by the ring. But with Helena's final entry, following a lengthy plot constituted of the tortured tribulations of predestined lovers, the play proposes one of the hastiest of denouements on record. That she is alive appears miraculous to those present, but despite this *coup de théâtre*, Helena does not waste a minute before getting down to her entitlements, whereby, through a minimum of talk, Bertram is brought not only to contractual acquiescence but to a declaration of eternal love, provided her claims are verified and she can prove that she is more than a wraith. Hardly has there been time to form a rational assessment of so much new social innuendo, but there has been sufficient time for Lafew's emotional sensibilities to form a decisive and demonstrable evaluation in the form of irrepressible tears. Midbrains can be fast and efficient. Making light of the matter, he turns to Parolles and asks for a handkerchief and prepares to go home, no more no less, and thus the play comes to its end after a few matters of business mentioned by the king and his allusion to the new "flow" of pleasure. These are the few inflections we possess concerning this micro-world's return to wellness and homeostasis, a world barely clear of the troubled thoughts we have been encouraged to hold with regard to both protagonists.

We might have made even less of the matter had Lafew indulged in the laughter of the incongruous, or exclaimed over the unexpected, or wondered aloud over the enabling circumstances yet to be divulged. But now, even at the folksiest of psychological levels, the play asks for an alignment of events with limbic reactions, to make an assessment of what we know about adolescents suddenly united by an unaccountable pledge of mutual devotion and a well-wisher's empathy. Shakespeare made his choice of character and reaction for this specific and critical occasion by making the tears of an old man the sole signifiers of closure. The question is where to set or limit the terms of an interpretation, whether social, affective, or aesthetic.

Moreover, we apportion our emotional resources according to "in" groups and "out" groups, the former as survival companions or extended families or tribes and the latter as alien tribes or competitors for rare resources.[12] Familiarity, negotiation, or intermarriage may adjust those lines, but the emotions make quantitative distinctions concerning the numbers among whom they can invest their most supportive estimates. Ideologies, too, may compensate (such as the brotherhood of man or the global village), but paleo-cerebral conditioning determines levels of compassion and indifference by an intuitive cost analysis of its own. These factors, for better or for worse, define communities, which would seem to spell death to the global village. As Steven Pinker points out, realistically, "group selection ... does not deserve its feel-good reputation. Whether or not it endowed us with generosity toward the members of our group, it would certainly have endowed us with a hatred of the members of other groups, because it favors whatever traits lead one group to prevail over its rivals."[13] Empathetic emotions are reserved for and shared among members of self-interest groups in relation to dependency upon that group. Tears of joy felt on behalf of others within the extended family as signs of a discrete emotion may be more ambiguous signifiers than the fear felt by every member of a platoon about to rush out of the trenches, but such signs remain cogent by dint of the very constitution of the limbic system as a reliable witness to urgency or radical change in world circumstances. A part of that reaction is the constitution of the in-group interests that have occasioned them. Lafew's tears are about "something" important not only to him but to the group with which he identifies; its generic story is also read through the design of his brain. The truth perceived by his emotions is always open to interpretation; he may be misinformed. But the question remains concerning what there is before his eyes that has turned his emotions into a public instrument, and what it is that pertains to the survival of men and women in groups through the empathy he has felt for them.[14] He identifies with Helena's interests so that what happens to her is not only about her but about the interests of the survival group which, as its single nubile female, she represents.

The story of Rousillon, the dynasty of the Countess, and the interests of the older generation in its reproductive potential may also have an ancestral lien on limbic design; Lafew's tears may tell an archetypal tale. This is a level of the play which Shakespeare authorizes through the gnomic running commentary of the household clown Lavatch, who chatters on about the "getting of children" and the "loss of men," and the whole undercurrent of death and rebirth, and the inversion of "dying unto life," and of a Helena who, "dead though she be she feels her young one kick." Bertram and Helena are wilful players in a larger drama, concerning which the oldsters can merely wait for the time and good fortune which will promote their union; dynastic emotions must be placed on hold until "unlicked" youth has time to mature. The problem is that many readers think Lafew's limbic system has fired prematurely on wishful thinking and incomplete evidence.

By dint of his constitution, his investment in dynastic concerns, and human make-up, Lafew may be allowed to weep for joy, that weeping arising from the architecture of his brain, as all spontaneous and irrepressible weeping must. As an emotional response, we are to conclude that it has been incited by an adequate stimulus in his social environment, touching a production mechanism adaptively designed to make such a display. If there is little rationale for such a response in private, there are grounds to conclude that its underlying purpose is to communicate to his community a sudden awareness of an exceptional group benefit following suffering and anxiety. This is to make explicit the argument implicit in the design of the cry mechanism. The body, through the endowment of a feeling brain, calls for communal rejoicing because all, according to Lafew's limbic evaluation of an unexpected change, is suddenly well. This is the point at which readers may object, and it is unfair to reason them into corners if they do not feel the wellness. Lafew is an audience of one, yet for him the sequence from design to display at this precise moment seems binding. Harking back to Nussbaum's caution against radical reductionism, it would appear, nevertheless, that Lafew's emotional response, predicated on his kinship with the household, including its prospects of renewal and continuity, is signalling one of the salient goals of the race. In the words of Pinker, "some parts of the mind register the attainment of increments of fitness by giving us a sensation of pleasure."[15] That may come close to the definition of comedy. Lafew is signalling in his systemically human way a sudden opportunity for reproductive prosperity, thereby supplying the (Aristotelian) emotional quotient of romance comedy – still allowing that he may be deceived by the illusions of hope in what he witnesses.

Still, from his perspective as a well-wishing insider we can elaborate. His allegiance to the clan has been established from the outset when he enters wearing the black assumed by a family in mourning over the death of its patriarch; he is one of them. Throughout the play, their collective destiny has been concentrated in the options of its single heir, Bertram: to accept the most logical candidate for an endogenous bride; to go in search of the more risky solution of a foreign bride; to confirm his bachelorhood; or to throw himself away on the wars. For those left at home, the ideal scenario is his capitulation to the wishes of the king and the acquiescence to an enforced marriage. That is rechannelled into their grief over the ostensible death of the unique eligible female within their community. These assumptions fall into place as prerequisites to the formation of goals in relation to an emotional finale. As spectators, we are granted privileged information, that Helena is not dead, that Diana is playing her enigmatic part, and that counter forces are in motion whereby patience, planning, and tasks fulfilled might eventuate in a form of mutuality, if Helena does not overplay her hand. Affective misalignments between the principals have constituted the tribulations of romance, while in the final scene, the only missing component is a word of acceptance. When Bertram pronounces those words, wherever his heart may have been, Lafew's sensibilities are overcome by the witnessing of an irrevocable contract, a reversal of fortunes (the tell-tale *peripeteia*), a putative resurrection, homeostasis, and, dare we say, wellness. He interprets the play through what it is to cry, insofar as the feeling brain, by its inherent design, is particular in its interpretation of the world – ancient, archetypal, and fixed on the interests of survival groups with which the feeling individual aligns his own more generalized reproductive continuity. Just as compassion requires an understanding of the suffering of others as a significant part of one's own scheme of goals and ends, so joy includes a sense of personal benefit in the advancement of the fortunes of those upon whom the continuity of the social group depends.[16]

What we, as spectator guests within that community, are intended to feel in relation to this emotional prompt becomes the next challenging question, for tears have now orchestrated the romance factor of existence, which is clearly one of the most hedonically engaging aspects of human social life. Characters within the play have been taken in tears, thereby justifying an argument on phylogenetic grounds of what it means to cry in relation to environmental origins. The challenge is whether we too are to be drawn into the ancestral values of courtship, death, competition, deliverance, and reproductive opportunity which are so deeply engrained in our own emotional design, and in the genre of romance. Typically, this is the case in less troublesome circumstances. In such instances, we are entirely willing to embrace the semiotics of crying, almost contagiously, whereby our sub-volitional sensations respond to shallow joys as well as to deeper values of the genome. Such hedonic codes as these, linked to the themes implicit in the mechanisms that produce them, correlate closely with the genres of literature. This is what it means to experience the worlds of others through empathy and fellow-feeling elicited by the design platforms conducive to caring, cooperation, and the projected feelings which alone make identity, suspense, and cathartic pity possible – design platforms pertaining to the collective success of the species.[17]

Nevertheless, we may be close to one of the presiding cruxes of the middle comedies. Shakespeare may also be inviting observers to object that Bertram and Helena have not performed a ritual of redemption that satisfies the criteria of their own respective limbic thresholds; they too may resist membership in Lafew's hedonic community, as many readers have done. After all, we have our own feelings about what we have seen, and our own eyes may be dry as cork. We can only discuss and share in the dilemma that may constitute the quintessential problem of this "problem" comedy. Our own experiences and reservations can never erase what it meant for Shakespeare to have Lafew confirm his experience in tears, or to set him up as a putative choral prompt, despite his inability to generate a universal emotional contagion.[18] He patently performs for us the emotional catharsis of romance comedy. Quite pointedly, however, Shakespeare does not tell us whether others in the entourage joined in. Lafew is an isolated witness, and in that Shakespeare does little to stanch the flow of doubts over Helena and Bertram coming to a meeting of minds, much less going forth to be happy ever after. So what of the final "cathartic" reading of this complex comedy, which, for many, frustrates both laughter and weeping, and barely authorizes pleasure? Our dilemma is that in dismissing those feelings, must we also opt out of the Darwinian plot attached to the rising and falling emotions of survival or defeat in the near background of every romance? Without Lafew, the *All's Well* plot, in falling below the level of limbic registration, has no unifying emotion, and thus, in Aristotelian terms, has neither grounds for a generic classification nor closural meaning, both of which are derived through spontaneous emotional confirmation and truth: laughter, pity, and fear, tears of joy, or at least deep contentment and pleasure. Emotion alone creates membership in communities and rewards the empathy that accompanies romance.

Wrapped around this hedonic midbrain, meanwhile, is another, more recent and altogether different kind of reader of environments; I'll call it the "cortextual" brain because it looks like a good pun. The hermeneutic self is a compound machine, not only saturated by hedonic background humming or urgent intrusions but computationally preoccupied with the logistics of persons in groups with regard to merit, entitlement, resources, cheating, and the sinned against and the sinning, down to minute discernments – a parallel Darwinian aptitude. Problem comedies may earn their epithet by pitting our diverse hermeneutic readers of social representations against each other. Pinker makes much of our capacities to keep track of those who give more to the group than they need to (altruists), and those who take more than they give (cheaters), yet who often contrive to give just enough to make it worthwhile to the group to tolerate them. It all has to do with our antennae regarding reciprocity, favours, debts, obligations, and justice. Pinker illustrates our obsession with social computation through a series of game scenarios, for example, the one in which a person is given a sum of money to share, all of which she would like to keep, yet who is destined to lose everything if the sum granted to the other through sharing is not accepted. Each computes his own benefits, the one calculating the least that will be acceptable and the other deciding upon the sum he is prepared to reject out of spite.[19] Arguably, reading other minds in these computationally negotiated terms may have been the single most influential pressure in ramping up the reasoning acumen of the brain, capable as we are in discerning cheaters and cooperators down to the finest levels of memory. That *All's Well That Ends Well* engages our attention in this way surely competes with our emotional celebration of fortuitous reversals. We need not rehearse here the many entitlements the principals assume for themselves and our respective weightings of their antagonistic claims. But we do keep score and become computationally absorbed in a manner that leads to gossip within groups (and seminars) in terms entirely oblivious to the teleology of romance. We carry out such "social work" on a presumption of expertise in matters pertaining to worth, honesty, honour, and morality, which are themselves the products of emergent mental properties prioritized by design. The emotions may dictate archetypal wishes, but intellects will simultaneously posit their assessments of probability, fairness, and deserving.

Many a discussion of this play has followed this line of analysis, unable to go beyond the respective entitlements of the two genders and the moral weighting of their mating stratagems or evasions, leading to polarized readings of the protagonists in moral terms. This play is not about good feeling, but about gossip. Brain design, in this, has an equally weighty part to play of an explanatory kind. It would seem that readers must settle matters between the two principals, often by selecting evidence and confabulating flaws – by making Helena a trickster before her time, or Bertram more than the unseasoned adolescent he was. The play is true to our species in that regard as well.[20] Someone must earn our approbation in the "he said/she said" battle of interests that constitute this play, and seminar participants are far more interested in scoring their respective claims than in reading closely or reserving judgment. We are subtle and committed adjudicators and rarely confess to our incorrigible biases.

Our dilemma may be restated as follows: in the head-over-heart debate, the head loses out when it comes to defining genres, for it is in the communal homogeneity of emotions that the defining properties of literary kinds are most easily found. Minds may read environments in many idiosyncratic, rationalized, or ideological ways, but feelings are systemic and group oriented; they tell ancestral truths as they are designed to see them, and group values, in all societies, depend on them for cohesion. Aristotle certainly thought so in his *Poetics*. Yet even as conscripted participants in the emotionality of art, readers and spectators may remain sufficiently detached emotionally to give priority to cerebral calculations. The question pertains to aesthetic distance. We may be drawn into the logistics of plot and character without taking up honorary membership in the survival community of the characters. Lafew's contagion of tears can succeed only if what moves him enlists spectators with an equivalent compulsion. Tears are not to be constrained, for that is precisely the basis of their semiotic validity. In that regard, we are left not only with rational reservations about characters, motives, and generic emotional thresholds but with dramatic structures. Why should Shakespeare's programmed reversal – with its celerity and archetypal rectitude attached to the design of romance in which two attractive if difficult individuals are surprised by love – not grab us as it did Lafew? Failing this, we are left in the murky world of Shakespeare's intentions and whether they even matter, the title included.

Tellingly, Aristotle made his mark in the world of literary criticism by linking tragedy to the successful creation of a homogeneous and collective emotional state among the spectators, which in turn defines the genre. His precise description of this state will always remain a topic of investigation, and particularly the so-called "cathartic" effect, for emotions are readers of the status of things within represented environments and not the pathological elements within spectators which art is designed to expel. But in global terms, Aristotle understood clearly the correlation between types of stories as stimuli to types of emotions, and that the "release" of those emotions in spectators is a kind of ultimate verification that the story has achieved its ends and may so be classified. These are, as Aristotle prescribes, protagonists of merit defeated by adversity, stories of contingencies unmet, which not only send heroes to their graves but their communities into lamentation. Such stories, properly told (according to the terms and conditions outlined in the *Poetics*), will play to the limbic brain, to the end of eliciting the approved hedonic responses. A work of celebrated proportions thereby judges the emotional competence of spectators and not the inverse, because the design of the emotional brain is common to the species, and thus the instrument of communal response. Stories thereby become builders of experiential communities and the vehicles of exile for those unable to sense the contagion of collective joys and sorrows.[21] Dry eyes may demonstrate a fatal detachment that makes outsiders of those unfeeling spectators. Arguably, Shakespeare, in his announcement of wellness through the semiotics of tears, at least pays teasing lip service to the joy associated with the realized imperative of romance, but always amid lingering cerebral doubts that what Lafew sees is what he will actually get.[22] Such are the hermeneutics of indecision pertaining to problem comedies, pitting the computational primate against the feeling mammal. Yet other matters pertain.

The emotions in readers are routinely exposed to orders of narrative that deal in net gains and net losses along the life trajectories of imaginary persons.[23] Just how readers are brought to invest real emotions in the airy insubstantiality of literary characters is a question always good for a philosophical debate, but the experience of readers will never allow for categorical denial that our minds are so engaged. It is an old problem, which the current enquiry into the tripartite structure of the brain may help to resolve. That problem can be posed in many ways. We might ask what constitutes the reality of fictive characters that we should care about their destinies. Or we might ask how many of the conditions of reality can rightfully be imputed to characters who remain mere constructs. The first pertains to the psychological depth and complexity of the characters themselves. The second touches on the difficult matter of "belief," or as Coleridge would have it, "the suspension of disbelief" through a mental investment in fictive worlds.

Fortunately, in dividing the brain against itself, critics may have it all ways. It would appear to us one thing to say that Helena had feelings about Bertram, which she expressed in tears. Shakespeare shows and states as much. Yet it seems quite another to say that Helena must enjoy all the rights and appurtenances of having a material brain apt for producing such emotions. We are anxious about failing the hallucination test by taking Helena as the girl next door, because it would appear absurd to say that by dint of her tears, Helena has an amygdala which has engaged with her hypothalamus, that her midbrain has experienced an efficient environmental cause in relation to primal goals, and that her weeping is an expression of anxiety over the reproductive criteria inscribed in her brain as a guardian of her "selfish genes." By other criteria, however, we lose entirely the substance of her functional personhood. We tell ourselves that Helena is nothing but words, a name, and a series of literary functions and effects, an agent in her story having no phenomenological substance. So how do we decide? After all, our minds stand on guard over the radical and essential difference between reality and fiction. To lose that distinction is to lose everything.

But now for the paradox concerning the reality discriminations of the human brain. The peculiar value of the limbic brain is that it makes no distinctions between real and fictive persons. It invests itself with equal intensity in both modes of the representation of personhood, expressing its hedonic evaluations indiscriminately. The emotionality of the imaginary supplies the "gut feelings" around projected scenarios; it enables us to plan according to our hedonic intelligence in completely imagined and provisional ways. If this were not so, any emotional evaluation of the counterfactual, as in comic or tragic catharsis, would be unthinkable. Fictive characters and their worlds are played to our feelings on an equivalent-to-reality-basis. This may be the byproduct of an adaptive arrangement between the feeling and thinking brain, or merely a residual fact of the early design of the limbic system, long before the cortex learned through a conscious self to discriminate between that which is perceived and that which is merely imagined.

More controversially, our confabulatory skills in creating representational Gestalts that meet with our schemata of reality must also include the representation of characters as persons. The feelings of empathy, hope, or grief they arouse can only be predicated upon entertaining them as full-fledged members of our own species. If Shakespeare makes Helena capable of tears, by implication she possesses the internal workings necessary to their generation. Just as much is implied in her capacity to see or hear as a parallel form of perception. In short, she is either a person with all the attendant properties of personhood, or a mere zombie who occasionally evinces certain hominoid forms of behaviour. Arguably, without our belief in Helena's selfhood, including a capacity to dream, desire, make vows, hold convictions, form speech according to the principles of syntax, and make love, we have no felt interest in her, indeed practically no interest in her at all. We are back to the opening argument. By faith and without prior evidence, we grant an interior life to all persons as part of their categorical entitlement until otherwise disabused. Helena is to our imaginations a very person of very persons, as emotionally and as cognitively competent as she must be for us to care about her and feel for her aspirations and failures. That characters can be made to sweat, feel adrenalin flow, blush, have muscle spasms, sob, and feel their heart rate increase is merely further inducement to accept the package of personhood in its entirety. It is not hallucination, but profound play. Even the jagged, promiscuous, and diversified contents of the "stream" of consciousness are seamed together in an illusion of smooth, continuous, deliberative, and harmonious selfhood. It is all an adaptive illusion which Coleridge could only call "belief," but which has its full evolutionary backstory. It is part of the way our brains do business with the world through the attribution of properties in relation to kind, crossing the reality/fiction divide in operative terms. But that is another matter.

Such elaborated representations of persons are preconditions to exercising our folk psychological analysis on a fully functional basis. Thus it is that hearts believe and respond even while heads may register reservations. But the problem with Helena and Bertram is not that they are mere characters, but that as representations of persons, we do not believe in the emotional substance of their enforced declarations. The argument, at this juncture, moves into options: that we disengage with the protagonists because we do not believe in them fully as persons (which would limit empathy with any literary character whatsoever), or because we simply do not believe on social and psychological grounds that they have earned our emotional confirmation. The first option would seem worthy of dismissal on purely experiential grounds, but it will stage a strong comeback in succeeding paragraphs.

The discussion thus far has dealt with our hedonic emanations as signals pertaining to events and stressors in social environments through what Colin McGinn has called "fictive immersion," that such emotions are real, social, empathetic, archetypal, and communitarian. But there is a last condition: what if those emotions are different in kind because of a lingering computational awareness of the fictional, no matter how invested our emotions are in the play? What can our emotions be attached to if not to the social and environmental values that selected them and confirmed them in the genome to the benefit of succeeding generations? On the premise that we have but one single emotions-generating system, one that makes no qualitative distinction between percept-generated representations and imaginatively generated representations, then the emotions can be attached only to values known to generate limbic reactions in ancestral environments. But can the more recent developmental awareness of imaginary worlds as phenomenologically distinct from the perceptual world condition and dilute those emotions? Can it be that in that knowledge alone we fall short of the emotional challenge of the play? It has been suggested, but I am skeptical. If the emotions are not invested in characters as equivalent to the real, then they are invested in the play according to other criteria. Is it that by changing the spotlight of our attention, we can bring our midbrain to feel less immediately about social values, and more intensely about artistic values and accomplishments – the making special of the aesthetic gaze? This we would hardly wish to deny. But how then can we separate the aesthetic thrill of consummate artistry from the so-called cathartic emotions arising from social outcomes for the characters? Must we not have a double set of emotional criteria evaluating the artistic medium and its messages in radically distinct ways, even if we can discern only a single state of emotion at the play's end? But then which criteria should prevail in the interpretation of that single feeling? Is what we feel at the end of *All's Well* anything more than a response to a story well told, independent of the destinies of the characters? Cannot "wellness" pertain to the supremely conceived artifact?

These matters have been touched on in many discussions of aesthetics and the feelings evoked by art, including a challenge posed by McGinn that brings the question to a very particular impasse. He returns to the "as if" basis of fiction and the matter of belief as a prerequisite to the engagement of emotions. But doubt remains for him over whether the emotionality aroused by fiction is the real thing.[24] First there is the artistic artifact, the print, the words, the book, all of which anchor the reader's senses in the process of creation and transmission, so that the illusion is never complete and the intentionality of converting words into their alternate-world Gestalt is never absolute. This remains a distinct prophylactic against the contagion of choral emotions on the part of characters within the fiction; that it is just fiction always remains an active part of our experiential reception of the social world. Only in dreams does the perceptual dimension of reading disappear, allowing the images a complete freedom. Yet even in dreams a knowing mind interferes with belief, for even the terror of a dream tiger is insufficient to make the dreamer leap from the bed and run. Thus, if the illusion is never complete – a Gestalt our conscious analytical minds will always frustrate – then the emotions it arouses must likewise be provisional and intermediary. McGinn will go only as far as to allow that the writer is like the hypnotist in placing the brain in a state receptive to illusions-as-real by occluding the checks of perception.[25] Only because our logical minds are diminished are the emotions consulted, which points us in the direction of Plato's objections to art in general. Thus for McGinn, "dream fear is not *quite* the same as real fear; it doesn't have quite the same *clout* as real fear."[26] It is not as committed, not as intense. There must be intermediary or artistic-grade emotions. This follows from McGinn's central axiom about the brain: that the distinction between percepts and images, between perception and the imagined, is absolute.

But are these discriminations within the capacity of the "primitive" midbrain, which appears to deal with provisional drafts on the same basis as real drafts of events? It is an intriguing question. The axiom that I continue to hold is that the emotional brain, originating among our mammalian ancestors, has encountered no data necessitating the development of uniquely aesthetic emotions. What possible Paleolithic conditions could have existed to give constant reproductive advantage to those whose feelings were attached to purely fanciful worlds? Such an evolutionary backstory can hardly be imagined. McGinn, nevertheless, challenges us to accept that the fictive representations submitted to the emotional brain for evaluation openly conveyed that fictional condition. Thus, in reading fiction, if the emotions are aesthetic and removed, they are diminished in relation to stressors in the environment and as signals of group identity and empathy. Can that be the explanation for our indifference to Lafew's tears? Probably not, and yet the question in hand should leave no options unexplored.

With that, we return to our engagement with characters as persons and the genesis of the social emotions. The problem of this "problem" comedy is that Shakespeare proposes a romance sealed by the feelings of a choral character through tears of joy which are, nevertheless, rarely confirmed through audience empathy. The question is not whether Shakespeare missed the mark in designing the play, but whether the audience dissonance arises within the depths of human nature whereby we make contending judgments about social data. If the play is problematic, it is so because of the emergent properties of *our* minds, which have imposed their idiosyncratic evaluations on the presented data. Lafew cries as a sentient being within the action, and by all the working indications we can imagine, we seek to read the significance of his emotions in social terms. If we shed a tear with him over the ostensibly happy lovers, we do so either because the environmental stimuli he sees we also see because we have adopted his communal perspectives as our own, or because we apply the emotions "felt in ordinary fictional immersion" tantamount to "*quasi*-belief and *quasi*-emotion," but of the same social order as those felt by Lafew.[27] Alternatively, the emotions of spectators are entirely different in kind from those represented in the play, being essentially aesthetic in nature, and we are left to ponder just how our ancestral environments could have given us reproductive advantages by developing such parallel feelings pertaining strictly to the bright and beautiful. McGinn has not gone so far, however. By the diminished hedonic qualities associated with fiction, he is not positing the felt qualities of the aesthetic in their place. In effect, we are looking at three questions as one: whether we fail the Lafew test because we cannot computationally imagine the bond between Bertram and Helena that Lafew thinks he sees; whether our emotional attachment to the lovers is not as vital as Lafew's because his world is, for us, ontologically fictional; or whether we have divided our felt responses between the cathartic potential of social representations and the felt potential in pleasingly realized artistic designs.[28] Ultimately, McGinn agrees that the emotions of fictional immersion are "real enough, but not with quite the sting of real sadness" with regard to his specific example.[29] It is a potential but as yet uncertain distinction, which yet does not gainsay the fundamental involvement of the emotions in the evaluation of characters and their worlds on an equivalent-to-reality basis.

Bookending the play is the weeping of Helena and Lafew. With Helena, crying is a deeply felt emotional display with adequate cause that is undisclosed to those in her circle, but revealed to the audience. With Lafew, weeping follows logically as a fact arising from his immediate observation of events. Shakespeare made his choice for the character's brain and for the play; can we trust him to accurately present the human potential called for by that moment? As previously allowed, Lafew's choral prompt may not move us: because we do not see events as he does; because he is deemed an unreliable or overly sentimental witness; because we are not involved in his community with sufficient immediacy; because our fiction-generated emotions are different from his in kind; or because the products of the imagination are tagged as falling short of reality. There are several ways out. By the logic of empathic communities, the power of emotional contagion, the *peripeteia* factor of romance by which those predestined to love have been delivered from adversity (even the adversity caused by ignorance of themselves), and by the realization of social goals deeply engrained in our limbic fostering of reproductive success, our emotions are certain to be aroused in some capacity by the social content of the play. Shakespeare may befuddle us with a character who shies away from his emotions as a public transgression, yet he knew as a close observer of human nature that tears belong to contexts and arise according to apt stimuli, making tears the biogenetic proof of a completed tragicomic ethos. That conclusion takes us back around the circle to accusations of reductionism and deterministic bullying. Is this not another elaborate rationale that denies the precious incorrigibility of our native psychological intelligence that tells us in no uncertain terms that between these two, nothing can yet be well? The problem of "problem comedy" may well inhere in this very impasse, arising as it does in the conversation between the hedonic and the analytical brain.

So once more back to the basics. Tears require interpretation by those who experience them, those who witness them in the play, and those who witness them as readers or spectators. Tears make their demands on each in accordance with the urgency and the strategic meaning of the situation. They are of such a nature in themselves that little can be done to hide their appearance and little more to mitigate or fashion their effect. That has been good for the overall fitness of the species, despite the embarrassment often felt by the signalling agent. Because humans live with knowledge and contingency, they have become emotional beings; emotions are part of our strategy for dealing with random causes in fortifying attention and physical readiness. Feelings are open to assessment, yet they are judgments and evaluations of the world in their own right. They may meet with rationalization or denial, but they remain operative and instructive.[30] As signs, tears call for public grief or public rejoicing. They may be discounted, as by the Countess, or by Lafew himself, who jokingly attributes them to smelling onions. But the playwright, all the same, is speaking through the language of the emotions and the neural systems which generate them. To ask what that language means in the context of the play is to ask what it means to the species bearing the trait. And that takes us back through the long, long trail of the present argument. Lafew cries as a member of a cry-programmed race; if we cry with him we confirm that fellowship in relation to a mutually perceived social order; if we do not, there are reasons enough, but an opportunity lost. The cortex speaks and Plato's horses are reigned in. The play will fail in its bid for emotional confirmation and the experiential reification of its genre.

There is no intention here to force a reading that goes against deeply engrained social wisdom pertaining to the dysfunctional psyches of otherwise designated partners. W.W. Lawrence sought to resolve the matter in favour of the literal promise of the title by asserting the priority of folklore and the allusions to such devices as the bed trick and the fulfillment of the "impossible" task as sure signs of a live-happily-ever-after conclusion.[31] The play is about the transitional and transformational experiences over time whereby the camp follower Bertram might be modified to the paternal Bertram anticipating his imminent fatherhood. G.K. Hunter, in his introduction to the Arden edition of the play, keeps his distance from ultimate interpretations, yet allows that Bertram was young, a mere "boy," and that things happened to these adolescents beyond their control, that time, destiny, and justice followed schedules of their own, and that faulty as they were as humans, we should not deprive these eleventh-hour lovers of their good fortune.[32]

The archetypes of romance continue to make their claims, and those archetypes can now find a new source of confirmation in the survival biases of our emergent properties of mind arising from the tilted design of the brain. All that was selectively designed was designed to a purpose, and that purpose had to make its contribution to sealing that trait in the genome responsible for replicating the design. Reproductive success is the fuse to it all. That kind of scientific substrate will no doubt remain unwelcome as a means for fixing the truths implicit in story structure to the truths implicit in cerebral design. But the more we know about the origin of the brain, the more we are going to be stuck with the social and hedonic implications of that design and its role in the genesis and shaping of culture; to stare that fact down is mere intellectual bravura, or whistling in the dark. This is not a campaign in reductionism, but if we are to have any interest in these two characters, their behaviours must conform to our expectations concerning the themes of human survival. Emotions are calibrated to the rising and falling fortunes of those who produce them strictly in relation to reproductive fitness and all its attendant complexities. That master narrative is now more than archetypal; it is the only backstory we have for gaining insights into the origins of human nature.

We can nevertheless imagine players shadowing the structures of that great story without possessing the qualities and emotional commitments that will make it work. *All's Well That Ends Well* is a telling case study in that regard and ultimately one of the messiest in setting out the imperatives of romance, a story that, through trial and torment, arrives at a stasis signalling the happiest prospects of mutual love. Meanwhile, what it means to cry at the level of the species brings the nature of the emotions into the reading of social environments and thereby into the interpretation of imaginative worlds, providing much of their ontological bedrock. Nevertheless, there are interferences that are also to be traced to the composition of the brain and its information-processing nuances, interferences that may detach spectator emotions from the immediate experiences of the characters, thereby diluting the power of the emotions to define genres on hedonic grounds. This principle pertains to all narrative expressions. And yet, equally pertinent is the argument from design, that emotions are calibrated to the crises of the species: they classify the nature of experience; communicate themselves to groups; solidify communities, including theatrical communities; and place their imprimatur upon all manner of human experience, acting as prompts at the most phylogenetic levels of being.

If Shakespeare forestalls those felt associations in this play, their prospects in other contexts are hardly diminished: *All's Well* is merely negative proof. Emotions and the ontology of the imaginative is a philosophical and hermeneutic crux. It brings to the foreground the constitution of the limbic brain as a front-line reader of environmental conditions through the empathy and alarm these conditions bring to expression toward all agents to whom we attribute personhood. There the discussion must be brought to a halt.

CHAPTER SIX

Toward a Cognitive Theory of Proverbs

The Dialogue of Solomon and Marcolphus

The now little known medieval *Dialogue* between the ancient king of Israel and a traditional German peasant is a work worth reading in its own right as an early, anonymous sampler of the proverb culture of the Middle Ages. As with any such collection, it challenges its readers to decode its sometimes cryptic or gnomic sayings, thereby setting up the larger question pursued in this study, namely, just how we deal, cognitively, with the mental operations by which the proverbial is decoded and rendered meaningful and applicable. As it turns out, even such "simple" forms can lead to complex explanatory systems, centred in the brain's capacity to perform analogy and inference and, inversely, in the kinds of cultural artifacts we generate because brains do business in these systemic ways. (That hermeneutic loop will turn up several times along the way.) Always there are the challenges of texts and how our brains are equipped to deal with verbal play and the equivocal, now in the dual registers of the high and low cultures alternating throughout this collection.

The problem of brain-generated meaning is hardly new: philosophers in the eighteenth century were as cognizant as we are today that the world is full of stimuli that are something in themselves, but "something" that we can experience only as the properties of the cognitions that those stimuli produce. Our brains, according to their design, generate such meanings largely in their own image. Even Francis Bacon, in *The Great Instauration*, intuited that the brain "in forming its notions mixes up its own nature with the nature of things."[1] Stated more polemically, as a survival machine, the brain has its biases: it furnishes functional representations of the environment that are adaptively sufficient. Such biases pertain to the qualities of systemic knowledge endemic to the neural design of the brain that enables and partially predetermines the meanings of which we are capable. One interactive complex of those design potentials gives us access to proverb structures as encapsulated bits of wisdom that can be applied in appropriate contexts. Hence, there is a knowledge component to proverbs, again based on how the brain interprets word usage according to the emergent experiences that usage is capable of producing. All cultural creations "play" to the capacities of the brain requisite for decoding them. That factor thereby defines the meanings of proverbs as being "out there" and "in here" – as things made to mean, and as things reduced to meaning. In sum, proverbs submit their characteristic semantic, formal, and propositional demands in search of the neural arrangements apt for their decoding. That is the quality of cognition under investigation, and clearly it corresponds to a universal capacity, given that proverbial structures are culturally ubiquitous. These are the bookends: sentences to be read and their achieved mental Gestalts. That is as much as I seek to assess here, aware that such a discussion will intrude upon the extensive work of the paremiologists who have scrutinized the proverb in and out of cultural contexts in terms of rhetoric, semiotics, style, world view, folklore, and wisdom literature in ways too complex to deal with here in comprehensive fashion.

The proverb pairings anonymously assembled in the names of these two combatants, King Solomon and the peasant Marcolphus, represent the Old Testament (Hebrew Scriptures) and the early medieval proverb tradition respectively. They promote a simple literary form in contrasting registers, relying not only on those mental operations requisite for the interpretation of metaphorical commonplaces but on those enabling comparative assessments of statements deemed to share in some often nearly occult common property – the presence of which we can intuit more readily than name.[2] That is to say, as transparent as they look, the disparate features aligned to form the metaphor, the sense of which is seized through an operation passed over simply as wit, can easily slip into the obscure – a failed inference regarding juxtaposed semantic fields. Proverbs at their best make very real demands on the neural networks that produce semantic meaning; in pairs they sometimes jolt us into metaconscious consideration of what the proverb is as an ambiguation and release format.

As with any such inquiry into the ways brains interpret data, we are entitled to ask just how much we really want to know. That we *can* arrive at adequate interpretations of proverbs, for some readers, may be entirely sufficient. Verbal meaning as an emergent property of mind is like the picture on a television screen; neuron circuitry or electrical circuitry are entirely optional matters. Indeed. But for the curious, further introspection can shed some light on metaphor recognition and the modes we employ in turning these specific verbal configurations into wisdom coloured and perceived as proverbial. Such an inquiry deals, self-evidently, with cognition simply because metaphors do not exist in nature; they are mental properties which arise because the size and production modes of the cortex induce comparative thinking at many levels by combining and associating convergent elements – despite Paul Thagard's reminder that "there appears to be no evidence to date concerning the neurological basis of analogy making."[3] That is a serious reservation regarding an explanatory approach to a computational feat we nevertheless perform and experience. Yet, Daniel Dennett concludes at the end of *Consciousness Explained* that we can never escape the production of the metaphorical. Metaphors remain fundamental tools of thought: "no one can think about consciousness without them," he states, in defence of their presence even in his own scientific discipline.[4] Always we come back to the constraints placed on thinking by design. We may turn to syntactical, cultural, propositional, and formal conditions, but all of these are likewise properties of brains. The elephant in the room is the systemic operation of the cerebral cortex in discriminating meaning: how we "do" analogy and how these patterns of meaning relate to the categories of knowledge that filter and frame our understanding of environments.

The proverb is a particular case in point, for as a *forme simple* it appears among the verbal creations of all races and cultures. In his list of human universals, Steven Pinker includes "proverbs, sayings – in mutually contradictory forms."[5] Of course, such a phenomenon may be downgraded to a mere ethnographic truism by dint of cultural contagion. But universals have something tellingly particular about them if all cultural groups in all places practise them, because universals must flow from some principle of nature which creates categories of universal sameness: in the world of extension from the laws of physics and chemistry; in the world of the intellect from the common properties of brains inclined to create or achieve meaning according to their phylogenetic categories. The proverb, I argue, conveys information in ways curiously resembling some of the foundational schemata of our defining cognitive processes, even while reflecting a huge range of cultural values and referential diversity. There is something to suggest that as a species, our overall adaptive potential has been enhanced by what proverbs do as mental experiences in parallel with what knowledge schemas do in the manufacturing of cognitive meaning. Each proverb represents the economy of a single generic statement about the world that is widely and efficiently applicable, serving as a point of interpretation for many potential contexts in rapidly changing environments. It is not hard to see their usefulness as a cultural extension of the ways we interpret data in order to improve our reproductive fitness. The cerebral cortex gained its girth by supplying quality mental experiences that gave our species an edge, and "proverbs-think" is undoubtedly a contributor. Back to Solomon and Marcolphus.

In 1492, Gerard Leeu of Antwerp published an English translation of a work formerly produced by his shop under the title *Collationes quas dicuntur fecisse mutuo rex Solomon ... et Marcolphus*. Of this single English edition only one copy survives, found today in the Bodleian Library in Oxford. Leeu was following in the path established by Caxton of arranging for the translation of continental favourites in order to supply a growing market for English books. These were commercial enterprises, the successes of which were determined by the approbation they received from purchasers on a title-by-title basis. Concerning the *Dyalogus or communing betw[i]xt the wyse King Solomon and Marcolphus*, however, there is little indication that it met with wide approbation among English readers, perhaps causing Leeu to regret his efforts. But the success of the work on the continent is another matter altogether, for the Latin and German editions of this centuries-old work number in the dozens, some of them elegantly illustrated.[6] And before those printed editions, there was a manuscript tradition that filled out a stemma with variant presentations dating as far back as the first years of the second millennium.[7] Just how this collection of diverse materials initially came together can only be imagined in accordance with a careful consideration of its component parts, for it begins with a mock proverbs contest between the great king and a Germanic low lifer that comes to an end only when Solomon grows weary of the "strife" and Marcolphus turns instead to the playing of practical jokes. This latter section forms a miniature trickster cycle, which ends, as they typically do, in the death or exile of the protagonist. Clearly these two parts were independently devised and developed before they were sewn together.

Our preoccupation here is with the ninety-one pairs of proverbs (in the English edition) that constitute the opening section.[8] This recreational enterprise, no doubt the work of clerics, is based on the humorous juxtaposing of sayings and maxims from official and folk sources. The early dating is corroborated by the subsequent inclusion of approximately one-seventh of Marcolphus's proverbs or dicta in the *Fecunda ratis* of Egbert of Liège, a work composed sometime before 1024. The direction of the borrowing might have remained uncertain were it not for the fact that Egbert's teacher, Notker, had been complaining about paired proverbs in the name of Solomon and Marcolphus at an even earlier date (he died in 1022). Egbert remains, nevertheless, a most important witness to the cultural enterprise represented by both collections, for he states in his introduction that his intent was to create an "opusculum rustici sermonis," a collection of rustic wisdom which he claims to have gathered directly from the people for translation into Latin.[9] As a teacher at the cathedral school in Liège, he was convinced that such proverbs would facilitate Latin instruction for students drawn from the lower social echelons.

Arguably, by the year 1000 Marcolphus was already performing in his comic peasant role as the baiter of the world's wisest man – in itself one of the simplest forms of cultural play based on the contrasting echelons of cultural production: biblical wisdom literature associated with the world's wisest man and the traditional sayings of the local German peasantry. The nature of the exercise was to show that for every wise saw of ancient Hebrew culture there was an equal or even wiser saying held in common among members of the local population – or so it was made to appear. Notker, the celebrated monk of St Gall, wrote a commentary on Psalm 118 in which he made a detour into the matter of secular stories and their capacity to obstruct spiritual teaching. He called them "fabulationes" told by loquacious heathens and worldly raconteurs, and in passing he states: "was ist es anderes als dass man sagt dass Marcolf gegen die 'proverbia Salomonis' streite? An alledem sind nur schöne Worte ohne Wahrheit."[10] (What else do we have [except lies] when Marcolf [Notker's spelling] strives against the proverbs of

Solomon? These are but attractive words, totally devoid of truth.) That Marcolphus was striving against the proverbs of Solomon can signify no other work than the present *Dialogue*, which Notker loosely associates with the "lies" of popular folk tales. In saying as much, Notker may also be signalling his concern over pedagogical devices that diluted the sobriety of Church teaching, polemically asking whether the present collection was a profanation of official learning in mixing the high with the low. We can easily appreciate his clerical position and the desire to inculcate the "truths" of biblical and Latin culture, but the fictionality or mendacity of rustic proverbs creates an odd position for attack. Either cats are or are not instinctively attracted to milk, but that is beside the point for now. Cutting Solomon down by matching him with the wits of a yokel and giving him the worst of it is *le monde à l'envers*, a carnivalesque inversion, a potential pot shot at those who excluded laughter from the world of religion, and a touch of rustic chauvinism. It was both a pedagogical recreation and a social leveler. But whatever the subversive overtones, a literary formula had been found that was destined to find favour in the European imagination for half a millennium: wisdom literature delivered through a ludic frame.

At the risk of opening a lengthy detour, it may be said in passing that contests in defiance of Solomon's wisdom were by no means new to the year 1000, although that would appear to be the moment in cultural history when the wisdom contest between Solomon and assorted pagans was reshaped into the proverbs match as we know it in Leeu's edition and its German, French, and Latin antecedents throughout the late Middle Ages. Polemical debates between Solomon as the defender of Christian values against the assaults of alien philosophical positions were of far greater antiquity. Pope Gelasius, pontiff from 492 to 496, in expressing his concern over the intrusions of heresy and the vestiges of pagan religions, mentions a *Contradictio Salomonis*, which, among several options, Erika Schönbrunn-Kölb believes to have been a wisdom contest involving a parodic handling of ancient Hebrew proverbs, although the substance may as easily have been theological or arcane.[11] Markolfus is first associated with these debates in two short Solomonic dialogues of 164 and 331 lines respectively, written in a West Saxon dialect by unknown clerics around the year 900.[12] At line 180 of the second poem, Saturn, the pagan opponent to Solomon, is described as the Prince of Chaldea, a man who understood the wisdom of the East, had disputed with the sages of the Philistines, and knew the history of India and the sciences of the Greeks and Libyans. This Saturn had likewise travelled through "marculfes eard," the Land of Marculf, thereby linking the name to the East and to pagan lore, potentially fastening the entire tradition to Byzantine sources as well as Talmudic materials. By all indications, this group of sayings and protagonists represents the oldest surviving collection of paired proverbs in the Western world, the origins of its materials buried in an archetypal past – materials which were collectively shaped by many minds choosing and rejecting the component features. Accordingly, these sayings survived through generations of oral transmission before finding a place in the written records, thereby representing a principle of "natural selection" by which the "fittest" sayings were retained on the basis of their relevance and "truth" to the conditions of the material and social worlds that fostered them. It is a principle that pertains as well to the stories in the oral cultures practised in all parts of the world.

A comprehensive analysis of so simple a formula as oppositional proverbs nevertheless remains elusive insofar as the relationship between these sayings is constantly shifting. In many cases, Marcolphus is not in disagreement with Solomon, but provides, in his own register, specific examples of the principles set out in the initial proverbs (without exception offered by Solomon). Nevertheless, both participants are intent on creating communities of the wise around their contrasting mentalities and views of the world. Every retort by Marcolphus, no matter how apt, oblique, or irrelevant (and some are), is a deliberate estrangement in tone if not in content, and a confirmation of his own community of wisdom or counter-wisdom.[13] Such maxims are provided with the syntactical designs of their authors, each stylistic manner, in the words of Donald Wesling, giving rise, nevertheless, to "the rhetoric of the small deep wound."[14] Such proverbs not only contain precepts and counsel but little emotional insults, tactics which create "in groups" and "out groups," those who know and espouse these things and those who think like fools or aliens. Proverbs thus contain sundry forms of collective wisdom but also prods to conformity. The binary design creates antiphonal communities and their concomitant mentalities: mealy-minded platitudes and urban commonplaces versus hard-won rustic insights.

Marcolphus, in this regard, can be conscripted as another Germanic rebel resisting official culture in the name of the folk. Conrad Celtis, in the humanist educational program described in his *Oratio in Gymnasio Ingelstadio publice recitata* of 1492, expressed his desire to eliminate Germanic barbarism yet maintain the discipline, manly virtues, and essential wisdom of the German race.[15] Solomon's proverbs are thereby made to appear equivocal, perhaps banal, or perhaps sanctimonious.

There are certain broader perspectives that characterize the two contestants. As James G. Williams points out, Solomonic proverbs are based on a notion of retributive justice: as one gives to the world so one receives.[16] Actions have consequences, as they do for Marcolphus, but for Solomon they are adjudicated by divine principles operating in the world. His views are often stark and absolute – "collect worldly goods and get nothing; live in righteousness and escape death" (Proverbs 10:2) – as determined by the omnipresence of a God who watches and guides, whereas retribution for Marcolphus follows from the act of stupidity itself. Solomonic man is often blind so that his quest for autonomy leads to disaster unless he is guided by a greater force, whereas Marcolphus is a philosopher of self-reliance, who is, at the same time, far less confident of the world's bounty and thus more fatalistic in his outlook.

Yet both frame their worlds in terms of causes and consequences constituted of mental categories promoted by a designed brain. Both the Solomonic and the Marcolphian world pictures create little social scenarios involving specific attitudes and acts followed by stark social consequences. Such proverbs are calibrated to the mental processing that constitutes learning by proposition and by example – such learning in itself a matter of complex computational operations including chunking, abstraction, integration, and recollection. This is the essence of the form that, on the basis of traditional wisdom, sets up, in minimalist terms, its prescriptive insights into the nature of the social and natural orders. Putatively, the one who knows all such precepts and takes them to heart has an adaptive edge in the struggle for self-advancement in challenging environments. In this, I think it is safe to opine, the platforms came before the proverbs, enabling them, even though it would appear that proverbs themselves are adaptive in extending the cognitive prospects of the human enterprise. Dubious assertions along these lines have gotten evolutionary psychologists into trouble.

In order to be of value, the proverb, as a simple form, must first of all release its data for lexical disambiguation followed by conscious reflection and intimations of significance. Ultimately its applicability must be openly inferential and appropriately assigned to immediate social or volitional conditions. Proverbs exist within a specific frame of propositional formulations whereby knowledge of the world is reduced to its most elementary and absolute terms, employing implicit logical schemas of cause and effect in chosen registers of activity which are sometimes specific and concrete, but are more often reduced to an optimum degree of generality, from which the specific must be inferred according to the situations on which they are to be mapped.

This is merely an attempt at precision in saying the obvious: that proverbs are sometimes literal, as when Solomon says, "With bawling people hold not company," and sometimes tending toward metaphor, as when Marcolophus says, "Who shall find a cat true in keeping milk?" The former is literal only because the target of this cautionary observation is quite precisely contentious persons, whereas the target application of the generic cat drawn irresistibly to milk pertains to more specific forms of human behaviour, to be supplied by the reader, guided by his or her systemic computational aptitudes. Somehow, the human brain is good at mapping generic orders onto specific situations through reductive analogy, and your story is as good as mine in accounting for the environmental pressures which put this cortical wiring in place. How we may presume upon these capacities is made clear in the proverb pair, for it is Marcolphus's reply to Solomon's recitation of Proverbs 31:10: "A woman strong in doing good, who shall find?" that illustrates our skills. Marcolphus makes a metaphorical association between the instinct of women to do ill and a cat's instinct to lap milk: it is in their natures. These may at times be combined, as when Solomon says, "Against a strong and mighty man thou shalt not fight, nor strive against the stream": stream, current, futility; strong man, fight, futility. Marcolphus seems to concur in terms of vultures that strip the feathers of strong birds, a curious corruption of the popular saying that whoever skins vultures, skins a tough bird, as in "Dur oisel peile qui escorce votur" which also implies, perplexingly, that vultures prevail where others fail.[17] But this need not detain us here. The point is that despite their observational truth at one level, proverbs imply other targets, and those targets entail a form of cognitive mapping that is so familiar to us as to pass unobserved, but that requires a remarkable degree of computational plasticity in reducing two explicit social or propositional frames to the exact level at which one is seen to pertain to the other. How we do that neurologically is moot, but that we do it is bedrock to human experience and our ways of seeing the world. Proverbs play to that capacity and the characteristic levels to which we can map concrete structures onto provisional structures.

When Marcolphus observes that "The shepherd that waketh well, there shall the wolf no wool shit," he poses a statement of logical consequentiality in the form of an integral bit of folk wit and wisdom. It is perfectly clear that vigilant shepherds increase the survival chances of their flocks, but is it useful only if Marcolphus is giving advice to shepherds? In fact, we automatically escape the literal by assuming a conceptual metaphor which translates the signifier into multiple signifieds according to matching orders of causation and result. Mark Turner describes this cognitive transfer according to a "generic is specific" rule.[18] (I might have said a "specific is generic" rule in that cats and milk is the specific which stands for any general expression of the powers of instinct, or shepherds and wolves stand for any relationship of vigilance in relation to harmful effects, but Turner saw it the other way and thus it must be kept: from the all-purpose proverb to the specific application.) Such equivalence schemas in the human reasoning processes are, for Turner, by-products of "right-left" body symmetry, from which the embodied brain takes its instruction for viewing the world. This doesn't seem quite right as a complete explanation for the default categories of thought requisite for knowing our physical and social worlds, but it is a point well taken, for the body is a part of our most immediate materiality and requires categories of intuition for appreciating its design, by analogy with which we may extend certain categories of analysis to the greater world (the following note is recommended reading!).[19]

But symmetry seems qualitatively different from general to specific, and an understanding of the nature of the material world requires many other categories than X is a mirror image of Y. Analogy-making of many kinds, in any case, seems absolutely basic and essential to our computational processing, even if the precise neural connections remain obscure whereby this deeply programmed pursuit of meaning moves outward by association to levels of saturation. The vigilance of shepherds regarding wolves is a behavioural relationship, the principle of which can be extended to all situations defined by that same generic relationship. The difficult part to imagine is how our cortex is organized to pursue only those relevant relationships after assessing in each potential match its substructure of comparable meaning. The proverb, in this way, becomes a schematic action-plus-value-template submitted to a massive repertory of adequately similar schemata in search of those matches according to the "laws" of analogy, in a sequence of diminishing comparability to a point of saturation and irrelevance which is systemically determined. The process must work in ways similar to the investigation of the human lexicon whereby discrete verbal stimuli are granted meanings of greater and lesser pertinence – the implicit operation of hearing or reading words. Stated simply, if generic is specific in the case of the proverb concerning the shepherd and the wolf, then shepherds, flocks, and wolves form a little beast fable apt for superimposition on like domains. It may apply to parents protecting their children from pedophiles or kings defending their realms against invaders, but far less to children learning to swim where a "try, try again" metaphor would be more apt. Of this mapping transfer, we are as certain as can be, and although scholars of the proverb (and of metaphor in general of which this ability is a subset) have described it in many different ways, all their analyses point to a fitness-designed neurological pattern drill that makes the operation possible. Our cognitive apparatuses, quite unsurprisingly, encourage us to see metaphorical associations and to draw their precepts into the circle of personal benefit. That this mode of pattern transfer according to abstracted levels of analogy is a prime bit of cognition and thus a way of seeing the world – and clearly the prompt for creating pieces of our cultural world in its image – need not detain us here, but gives pause for further reflection on the universals of the species through ancient design. Making proverbs is an extension of our inherited capacity to decode them.

Proverbs featuring conceptual metaphors appear first as language symbols recognized and interpreted as images in a propositional bit of syntax. They resonate at nearly the same instant as little judicial scenes based on causes and anticipated effects. They often create miniature scenarios in a material world such as hillsides, flocks, and wolves (Neil

R. Norrick's "species-genus synecdochic proverb").[20] But the formula is foremost a set of propositional values, for at the same time that we envision a sleeping shepherd and the aftermath in the form of wool in wolf excrement, by all that is implied in between we generate an action schema involving the conditions of a guardian and the actions of a predator. There is the archetypal story. We may pause to reflect on *topoi* such as human fallibility, the instincts governing initiatives along the food chain, or generic forms of domesticity and survival, but the reduction to an action schema alone makes the mapping from the generic to the specific possible. The leading question may not be how we carry out this computational performance, but how compulsively it figures in our mental activities. The nearly automatic or default mode by which we see a specificity of application in generic schemata may, in itself, reflect a brain systemically designed to draw all relevant patterns of the world into the survival perspective of the self. Analogy may be one of the nearly run-away compunctions of our cognitive operations – a leading adaptive trait genetically confirmed.

This principle seems so important to me, and has taken such pains to work out that, with your indulgence, I want to run through it one more time (you can always skip this paragraph). Douglas Hofstadter, in a most engaging essay, put forward analogy as the "core" operation of cognition, insofar as every concept entertained by the mind releases a flood of associations beginning with semantic contextualizing by comparison and finishing with the loop through long-term memory that brings to consciousness in the short-term memory, on a hierarchical basis, those elements deemed connected to the topic, precept, or idea then held in the foreground, on a prolix analogical basis. Analogy is the selective and opportunist coupling of everything perceived with everything known – perhaps tantamount to a definition of wit. Concepts trigger interrelated thoughts, and thus, on a pop-up associational basis, the mind builds, fills out, juxtaposes, illustrates, and thus saturates itself with meanings. In this way, concepts are explored, expanded, classified, personalized, applied, and troped according to all the relevant modes of knowing which the brain systemically applies. Metaphor for Hofstadter, as for Lakoff and Johnson, is the default operation of all thought and interpretation. He speaks of concepts hitting the neurons to form association "clouds," characterized by fuzzy limits where associations drift into irrelevancy.

Add to this the meeting of incoming concepts with the definable cores which are culturally and semantically determined, largely by the default operations of cognition coupled with personal experience, in turn modified by logical and poetic reflection. In this habituated sequence, embedded in a big brain with its abundance of in-depth connectivity potential, we must assume all the analogy making necessary to decode and apply proverbs. They are merely provocative concepts, which get the fullest of such treatments on a dissonance-resolution basis.[21]

That is, of course, to dodge the question of our neural competency for doing so, except to presume upon a scientific just so story that by selective design, we have ended up hardwired as an "analogizing" species. This is perhaps one of the ultimate emergent properties of human consciousness. By one set of analytical terms, such properties are the products of a massively modular or connectivist organ of assimilation that searches for levels of meaning through simultaneous parallel systems, associating like-with-like according to the knowledge schemas that organize our understanding of the environment.[22] "Generic is specific" (species-genus synecdoche) defines one of these operations, the word "mapping" pertaining to an ability to "trope" or to pursue communities of connotations according to the data of the verbal stimulus and the constraints of cognitive processing. For us, generic *is* specific because the principles of interchange among mental domains is an integral part of brain architecture. Such mapping is compound, often imprecise, and sometimes wishful or opportunist, but as a feature of mind, it has clearly withstood the test of evolutionary stress as a systemic source of new information. Proverbs are precisely conceived to set up the search involved in drawing inferential meanings from pictorial axioms. The level of the generic is calculated specifically to that end as a form of mental play. By mental mapping, we draw paradigms into the circle of personalized precepts at the systemic insistence of the cerebral cortex.

Proverbs in pairs extend the ludic profile because just as we are invited to detect double meanings within single proverbs, we are invited to evaluate the common values linking paired or contrasted proverbs. Dialogic response, however, admits of many nuances, often playing at the margins of our comparative capacities. There are several potential meaning schemas that can be drawn upon, one of which perceives A as the opposite of B, or as countermanding or denying B. Pairs may indeed be in opposition; Marcolphus would appear to be in that mood most of the time. But when Solomon pronounces that "All right paths go towards one way," and Marcolphus concurs in observing, "So do all the veins run towards the arse," the antithetical meaning is abandoned, even while the stylistic registers are contrasted. The oppositional becomes, in a sense, aesthetic; Marcolphus agrees with the precept but drastically lowers the register, poking fun at the hoity manner of declaiming truths. The first is a moralized commonplace – the convergence of the paths of righteousness. The second is an anatomical observation of no immediately apparent "generic is specific" application except, perhaps, for the medieval rustic mind. But in response to the wise king, Marcolphus is being sarcastically deflationary: righteous paths and rectal veins are one. This is a challenge in ideal readership. How do we know which schematic restraints to place on our reading and still satisfy ourselves concerning a purposeful statement, for in our cognitive processing we are seeking to replicate the processing attributed to the mind of Marcolphus by the cleric who first compiled the pairing. What did he presume to exist as a quality of relation between the two? Proverbs in pairs are also little dissonances, little cognitive jags, esoterically conceived and posited, even while we understand that they are never random, or private, or riddles without solutions. The goal of the maker is to seize attention by posing information that temporarily threatens for as long as it refuses to reduce itself to the stasis of familiarity.[23] This prospect looms often in the collection. Marcolphus's adage about the shepherd and the wolf is in direct response to Solomon's asseveration, "Though it be so that thy wife be sour, fear her not." (When shepherds are not vigilant, the wolves shit wool.) Readers may spend a long moment seeking to align these two through a common schema.

But why all this talk of schemas and biogenetic neural systems designed to see data according to predetermined categories of cognition? The answer is that proverbs seem to perform around a select subset of such mental categories, taking cues from them for their own characteristic forms and practices. "Kant was the first theorist to think of the mind as a system of functions, conceptual functions transforming ('taking') precepts into representations."[24] The brain works by making discriminations concerning both externally and internally generated information, and those powers of discrimination must be in place before the information arrives, some of it through a lifetime of learning and experience, but much of it by dint of having a brain designed by the same genetic coding that builds the limbs and organs for specific problem-solving reasons. We know the brain by the functions it performs, enabling us to retro-speculate that what we experience as unified sensations and perceptions in consciousness are reassembled discriminations and evaluations of their constituent parts by a brain that binds them all together, including word configurations that follow a trajectory toward simple, literal, and compound, associative, personal, historical, and idiosyncratic meanings, only when each element conforms to a currently held sense of the world.[25]

Connectionism is one neural scientific system for modelling the implicit operations at a neural level. By a system of "channels" and "impedances," blind neural networks contribute to the only final product that can matter to our brains: an emergent property that constitutes a meaningful representation to consciousness. But the tricky part is that meaningfulness is determined largely by the schemas of knowledge about the world systemically embedded in the discriminating operations of these neural networks. Because they have been designed by natural selection to correspond to the constants that pertained to ancestral environments – gravity, an overhead source of light, oppositional forces, the need to nurture offspring, and much more – they are designed to understand and predict the world in many generic ways. Violations to these expectations produce immediate anxiety and alarming cognitive dissonance.

Moreover, concerning these schemas of knowledge, Jeremy Campbell points out, in *The Improbable Machine: What the Upheavals in Artificial Intelligence Research Reveal about how the Mind Really Works*, that survival reasoning trumped logical reasoning in the selective process. Failures to replicate through computer programming the many puzzling nuances of human intelligence and meaning production have brought a clearer understanding of just how messy and biased the brain really is, yet how efficient it can still be in its averaging and approximations, its shabby logic and imperfect reasoning, because the brain has been programmed to interpret not in the interests of logical, mathematical, or scientific truth with computerized thoroughness and accuracy, but to make inferences often from slight data, which (Sherlock Holmes style), through rapid multiple and concurrent multi-system processing, lead to amazingly useful information within seconds. Ironically, computers can do the kind of thinking that humans find difficult, such as mathematics and complex problems in logic, but they are poor in making the simple observations of our everyday lives about who is doing what, how to read looks and glances, or how to eliminate masses of irrelevant data in a single sweep of attention.[26]

The study is full of examples of the brain's idiosyncratic forms of "pre-knowledge" and the stories of how its aggregate of aptitudes came together. Schemas are the result of what Campbell refers to as "the worldly machine," by which he means the real world environmental pressures that placed their demands on a design that was adapted to self-replication. The resulting brain is characterized by a remarkable degree of "economy" in its efforts and "an easygoing tolerance of inaccuracy" by making use of opinions, stereotypes, theories, and other kinds of schemas to cut corners. It makes sense of the world by building models out of fragmentary evidence, applying probability and confabulation as necessary, "fitting morsels of new information into rich structures of preexisting knowledge," and making strange things familiar by reducing them to known categories on the assumption of regularities and expectations concerning the world. In this way we make usefully approximate predictions of the future. This is how schemas work. Moreover, we bypass logic, which, in its formal terms, is fairly useless in everyday life and not the way the brain works anyway, usually because it doesn't have enough information to set up the requisite conditions; to create a syllogism, all the parts must be known. Rather, the brain thrives on poor-quality information and does not hesitate to offer interpretations by "mobilizing enormous amounts of relevant world knowledge, all at once, to provide a spacious context to the words we hear or read." That is as relevant to the multi-layered meanings of proverbs as it is for any form of verbal communication.[27] Proverbs may be reduced to syllogisms, but that is to artificially replace the missing parts to the formula which the brain does not otherwise require to come to its "reading." In a word, "*schemata* are the psychological constructs that are postulated to account for the molar forms of human generic knowledge."[28]

This kind of messy, associative thinking is precisely the kind of connectivism that works through the data of proverbs and sets up the meanings we ultimately entertain. Stated a little more carefully, these schemas are latent systemic memories, invigilating hierarchies, or emergent properties in their own rights that function in a highly distributed fashion to settle on the correspondences pertaining to stimuli in order to characterize them according to their own properties on a match/mismatch basis with information retained in our long-term memories. Hence, much of the intentionality of mind is generated by the biases and weights of these modulators and activity patterns in making their contributions to the formation of conscious experiences, including our beliefs and desires with all they entail regarding a sense of deliberative self-identity.

One of the basic categories of thought production involves oppositional binaries, the kind of patterning that occurs both within and between paired proverbs on a regular basis. This is a production of the human brain, because such binaries do not occur in nature. Up is the opposite of down because the contrasts in nature have induced us to see these values in more absolute terms. The remainder of this paragraph illustrates the point in fascinating ways, but it may be leaped over by those keen to get on with proverbs. Victor Johnston provides a telling example of the way in which natural selection has not only ignored those sources of information in the environment that are not pertinent to our biological well-being but has enhanced and amplified minute distinctions into major categories of experience. Colours are emergent properties arising through the brain's capacity to discriminate the sensational equivalent of light rays. Much has been written about the way in which we experience colour, and about how we could ever know from person to person whether red is a precise and universal cognitive sensation.[29] Presumably it is, given that this feature of our perceptual powers is very old and given that all members of the species have, with notable exceptions, inherited the same genetic design regarding colour. But that is pure speculation. Johnston's point is that where discriminations are of prime value, natural selection has ratcheted up and amplified them, as in the case of red and green, which are, from a human perspective, very nearly opposites, but which, in terms of wave lengths, are "essentially identical," differing by only "150 billionths of a meter." He explains that "for most animals, visual perception is limited to detecting and discriminating within a small range of electromagnetic frequencies centered in those reflected by the leaves of plants. This is not surprising since the survival of most animals depends on a food chain based on the ability of plants to convert solar energy into sugars using photosynthesis."[30] The human visual spectrum likewise places green at the centre and arranges its forays into the blue and red sectors around it. The survival benefits of wider discriminations were largely irrelevant and so the "gods of evolution" simply left off where they did (such as ultra-reds and ultra-violets). Revealing in all this is the degree to which nearly identical properties are perceived as widely divergent based on adaptive functionality. The brain is, in fact, defined by schemata which favour the conscious subjective experiences of greatest benefit to the organism, often by amplifying their experiential quotients such as sweet and sour or bad and good. In that way, oppositional perspectives arise from the schemata of our "worldly" oriented perceptions, but we can take them for granted only among members of a species with the same genetic coding.

This is a rather long detour to account for opposition as a function of mental categories and of knowledge schemata that present themselves as absolutes in an eternally relative world. The oppositional binary has passed into consciousness as a "natural" way of viewing the world, leading to values that are themselves schemas revisited upon the world as good and evil, safe and deadly, loveable and vicious, always to be done and never to be done, wise and foolish, obedient and disobedient, heaven and hell, righteous and iniquitous, beautiful and ugly. It is a prime feature in the world of the makers of proverbs, for axioms are often intransigent and decisive, categorical and sure, and thus precepts of unqualified value in the making of safe and ready decisions. Proverbs may help in the face of ambiguity, while at the same time making us wary of overly refined descriminations. Binary logic impinges upon consciousness as a schematic way of knowing the world in the interests of conformity, obedience, and proven wisdom. This is so only because natural selection has improved survival rates by scaling critical kinds of knowledge down to stereotypes; proverbs perpetuate the effect culturally by reducing the world to binary propositions that have proven their efficacy over time as functional truisms.[31] This evaluative and constructionist property is a streamlining schema that allows us to organize space and time into pairs of opposites, enabling us to see what things are in relation to what they are not.

Proverbs rely on this neural procedural classification to create their habitual dual perspective on social and natural phenomena, whether within the proverb itself or in antithetical pairs of proverbs which pronounce upon behavioural options from positive and negative perspectives.[32] Opposition gravitates easily in social settings to moralized observations: "A wise woman buildeth a house, and she that unwise and a fool is destroyeth with her hands that she findeth made." Thus sayeth Solomon (139).[33] This represents the conventional positive-negative formulation of ancient Hebrew wisdom literature, which in turn relies on binary conceptualizations of the world; the good life is all about the collective wisdom concerning what thou shalt and shalt not do.

Schemas, in terms of their applications, are therefore paradigmatic in their cores, but they have soft edges, pertaining to as many specifics as conform to the template. That principle alone accounts for the possibility of pairs, or of lists bound by a single value and plan. Proverbs are open to textual, contextual, and material reconstruction, so long as they say the same thing, seeking through the hazards, adjustments, and collective experimentation of transmission, whether oral or written, their optimum level of novelty around the axiomatic precept. Among the commonplace schemas is the observation that volitional creatures are limited in their activities by mental and mechanical constraints, that in many categories of activity they cannot do two things at once. They may sing and tap their feet, or drive and reflect, but they cannot sleep and eat at the same time, and they should not drive and text message or use other hand-held devices simultaneously. But why would the makers of ancient proverbs think to include among them that "As a man playeth upon a harp he cannot well indite"? And does Marcolphus match tit for tat, according to the precise value of the schema, in observing that "When the hound shitteth he barketh not"? Both "general" formulations provide a common schema, that two all-absorbing activities must be practised sequentially. But the parallel constraint pertaining to defecating dogs remains ambiguous. (Can dogs bark and shit concurrently?) Marcolphus pretends that the schema is the same – that creatures cannot do two things at once – but the question remains open whether, as a stand-alone general precept, the dog's limitations can be remapped into cautionary human specifics. The answer is that it can, but only if constraints upon the diaphragm are viewed as equivalent to the constraints upon the mind. Such pairings are altogether more challenging than those encountered in the book of Proverbs, which is the working prototype for the format. "He that covereth a transgression seeketh love; but he that repeateth a matter separateth very friends" (17:9). The formula is widely employed and sets up expectations of a single precept modified by the double perspective. Marcolphus plays with great subtlety on this formal convention by undermining the symmetry in many subtle ways. Has his shitting hound confirmed a precept or connotatively destroyed it? And does his perfectly just observation of this fact of the animal kingdom lend itself to metaphor? What is the specific target implicit in this general set of terms and to what form of potential behaviour does this schema serve as an admonition? We prefer to assume a positive answer than to admit defeat because we are wired to believe in the limitless possibilities of analogy if we can force our powers to find them.[34] (A diatribe here against the free-for-all analogizing and inferential "worm pulling" of some modern criticism and most conspiracy theory is, of course, out of place!)[35]

As stated earlier, metaphor does not exist in the world. Yet the brain is equipped to catch out analogies by leaping registers and associating similarities, or discerning common properties among unlike things, or detecting common action schemas governing otherwise divergent objects. In light of the nature of the world we are justified in wondering why the selective processes that built the human brain installed this literal to figurative mode of discernment.[36] Once in place, however, the species learned to exploit it by constructing metaphors as a means for inducing auditors or readers to test their cognitive acumen. These verbal artisans became specialists in making strange, in posing riddles, for as properties not in the world, metaphors are novel, and as such impose themselves upon our computational curiosity.[37] Mark Turner, among many others, has observed of consciousness that it is not only a limited one-thought-at-a-time processor but that it habituates quickly with the effect of speeding up and diversifying the content that passes through.[38] Promiscuous attention is part of its adaptive design, for even ten seconds is a very long time to hold the attention of consciousness with a single unchanging precept or picture. New material invariably rushes in, working its kaleidoscopic effects, linking up associations, pursuing related connotations. Much material is simply too banal or uncaptivating to achieve the mark, the weighting, the critical mass to win a moment in consciousness. For nearly self-evident reasons of survival advantage, this "theatre" of competitive awarenesses has been designed to give tonal salience to the present, thus distinguishing it from all past and future thoughts.[39] Moreover, consciousness is readily bored and gives priority attention to things innovative, novel, threatening, counterintuitive, absurd, or profoundly relevant, for it is in our interests as a species to pay closer heed to radical changes and the unexpected in the environment than to things habitual and unchanging. Our unconscious mind, with its trained habits and instincts, can take care of the rest.

Proverbs are wonderfully minimalist on this score, yet effective none the less, for they play closely on the Necker Cube effect of opening simultaneously on the literal and the metaphorical, depending on the level of saturated inspection. "Eat that ye have and see what shall remain," says Marcolphus (147). The intimations of ambiguity in the most subtly prepared of literal statements are charged with cognitive suspense. Consumption and surplus, survival and desire, act and consequence; why should a man eat that he may take cognizance of what is left over? The leanness of the saying is part of its arousal effect. There is an action schema, cause and effect, the eaten and uneaten binary, and an option to elaborate food into a metaphor for all things consumed and remaining. But the sum effect is perhaps to take no stock of possessions, of portion and residue, but to live by the simplest of urges and economies – to live like the lilies of the field. How are we to know? No computer can help us – only our sense of the world and its realities and our sense of the Marcolphus world vision and the double meanings of language. To achieve this depth of reflection over such incomplete data is not only a feature of dispersed modular thinking but of an intentional strategy for pressing the brain to the edges of its operational capacities to generate a final, unified meaning. Failure to achieve a complete cognition results in a fractional hedonic demotion, and when we achieve cogent cognitive appraisal, there is a sense of satisfaction or reward. The brain has a sense of its possible Gestalts. We might even wonder whether "getting" proverbs finds a little dopamine reward in the nucleus accumbens, a micro-orgasm of momentary satisfaction.[40]

Marcolphus states in one such pair: "What the stone heareth, that shall the oak answer" (147). This is Marcolphus making strange in perceiving in the dumb conversation between a stone and an oak tree the low levels of meaning emanating from a low level of mentation. The wisdom of a fool is to that fool what the oak's answer is to a stone – or so we would assume in failing to make much higher sense of what these two might discuss. The metaphor makes absolute the stupidity of the fool by equating his levels of understanding with that of an inanimate object. Yet even without Solomon's prompt concerning foolishness, our minds assume that specific is general and transfer the schema of dumb dialogue to higher forms of life. Stones and oaks must be about something else more relevant. We will enhance our fitness by making every nuance of the material world that sustains a trope an aspect of our own well-being because we are cravers of success and conspirators in verbal tweaking. Analogies are our unique several. Not all the world is metaphoric, but all has metaphoric potential once that "level" of discrimination is alerted as a meta-category of "perception," just as nothing is beautiful in the world until it is made so by "making it special" in our minds.[41]

We have come now to the difficult crux of the argument, whether the schemata of the neural networks, which are reputed to guide the formulation of conscious perception and thought according to their salient designs, provide the explanatory model for the mental operations that include metaphor transfer and binary opposition as well as the many other more particular schemata described by paremiologists in the formation and interpretation of proverbs. That probability comes close to necessity, yet we find it difficult to contemplate because we are dealing with a theory of mind and how its Gestalts are formed. Proverbs present themselves as ideal laboratories for such an investigation because of their minimalist nature and limited set of component parts. Normally, in dealing with literature, our concern is with social knowledge through the representation of characters as simulations of minds in action. Wisdom literature works differently in its assessments of the social order through pithy, sometimes witty, slightly troped propositions, offered as experiential truths in moderately hyperbolical terms. They appear as forms of popular wisdom often typical to a social group or nation. That is already a lot to account for in so minimalist or "simple" a form. Because proverbs thereby entail a particular and limited set of mental operations, they provide a repertory of elements pertaining to a theory of mind. As Lakoff and Johnson have shown us through the schemata associated with the body, prior information both inheres in metaphors and serves to interpret them. There in a word is the challenge; both notions of schemata, those in brains and those in things, are or are not of a common order of cognition. But if they are – and so it would seem – then what the brain pre-knows and how it is disposed to know is what analogies and hence proverbs are, both in what they are made of and how they perform.

More difficult to determine is the degree to which the proverb as a precept for self-advancement is a symmetrical cultural continuation of the systemic information we possess through our emotional prompts and preliminary categories of understanding. The challenging part of this argument, however, is that proverbs *are* cultural extensions of the "wisdom" imparted through the genome because culture takes over when conditions in the environment change too rapidly (or recently) for natural selection to install genetically transmitted solutions. Hormones instruct boys' bodies how to become men's bodies, but rites of passage devised by the elders define how they become integrated into men's culture. Proverbs operate on both sides of the equation. This principle is central to the argument by Paul Hernandi and Francis Steen concerning the adaptive value of proverbs. They contain a "multiplicity of behavioral options faced by human beings in our relatively recent evolutionary past." Proverbs take over to guide populations concerning the most volatile aspects of the environment, thereby supplementing our instincts.[42] If you wish to know in brief what the chances are for an individual to break with her usual habits, remember the leopard and its spots. Whether this is subtle or blunt instruction is beside the point; it represents the experience of generations and errs on the side of safety. It streamlines the decision-making process in the interests of efficiency, much as the schemata of cognition do in polarizing and categorizing many of our discriminations in conjunction with our hunches and feelings.

Moralizing from exemplum to precept appears to be one of the foundational orders of cognition, a fundamental way of making sense of the world. This is in perfect keeping, generally, with the pragmatics of fictional representation described by Raymond A. Mar and Keith Oatley in their thesis title, "The Function of Fiction Is the Abstraction and Simulation of Social Experience."[43] They evaluate stories as useful social guides. Of course, one must know how to fit the moral to the context, for it is sometimes right to "look before you leap," while at others "he who hesitates is lost." Proverbs, nevertheless, offer to extend the wisdom of the genome by adding sure and proven culturally defined precepts to the repertory of survival information. They seek to stabilize the categories of experience, reducing the amodal flux of events into immutable precepts, and they seek to make themselves memorable by achieving optimum verbal formulae with properties sized and fitted to what the memory most easily retains. Their defects are in tending toward sharper distinctions than social reality admits. Their strengths are in identifying the generic causes and generic effects "situated in the enduring present."[44] For Roger Schank, they become "situation labels" or ready-made configurations of language apt for mapping by analogy onto epitomized social circumstances. When authority is absent, "the mice play." When the person under discussion suddenly appears, "speak of the devil!" With astonishing facility, our brains perform the match.[45]

Arguably, proverbs are time-honoured formulations for packaging fixed categories of experience. They are verbal consolidations of perceptual circumstance, which facilitate easy reconstruction in the minds of auditors. As insights into categories of experience, they communicate common understanding and thus express cohesive thinking in groups. As independent spokesmen for their wisdom circles, Solomon observes that "Need maketh a right wise man to do evil" (we can imagine the extremity that might corrupt the highest principles), to which Marcolphus observes, "The wolf that is taken and set fast either he biteth or shitteth" (wild animals captured and tied, and hence made desperate [and humans in similar straits] turn savage or craven). Mapping proverb with proverb is the harder part. Mapping to the world is quick and easy.

At a different level of computation, that of encrypting and decoding metaphor, the neural circuitry of the brain has achieved a remarkable if inexplicable mastery. Some proverbs are built through the inventive reversal of this decoding capacity. Lakoff and Johnson, in keeping with their study of metaphor, concluded that "our ordinary conceptual system, in terms of which we both think and act, is fundamentally metaphorical in nature."[46] This too implies a theory of mind to the degree our linguistic environment is composed of buried and overt metaphor, although to assume as they do that "the categories of our everyday thought are largely metaphorical and our everyday reasoning involves metaphorical entailments and inferences" may be slightly overstated.[47] The term becomes less useful if all domains of perception are treated as tropes, for clearly the brain does not convert all of its presentiments into the something else with which they may be associated.[48] When the brain is doing metaphor, it is conducting a "supra" or "meta" operation, simply because big brains can do that, whether for fun or profit. In this spirit, Samuel Levin recognized that "metaphor comprises meaning in excess of, or differing from what can be deduced from its sentence vehicle," an excess which must be reduced to a common equivalent through inference.[49]

Nevertheless, the brain is an inveterate and compulsive pattern association machine, the degrees, kinds, and proximities emerging from the multiple neural dimensions available for the simultaneous processing of stimuli. Just how the brain deals with incongruity, similarity, and metaphorical expression, and with their respective powers to captivate attention as a problem-solving domain, has been little examined empirically.[50] Paul Thagard set the challenge: "both the generation of a metaphor by a speaker and its comprehension by the hearer require the perception of an underlying analogy."[51] That much we know, and that much we can retrofit upon neural capacities. There must be relational similarities well beyond associations of language and syntax, but the mechanisms are the domain of neuroscientists. This is where the entire argument began and there we find ourselves again. Proverbs not only judge and recommend based on collective cultural experience. They also rely for their fundamental operations on subsets of an "architectural knowledge" universally experienced throughout the world, accounting for the stability of the generic form, the decoding of metaphorical language, the antinomies of folk physics and of the emotions, all the universal categories of human experience, and heuristic operations of mind – all of them variously pertaining to the cognitive processing of proverbs. Also as stated, proverbs are designed to halt the normal rhythm of cognitive processing by throwing up little computational obstacles in the interest of making mental effort a part of the learning and retaining process. In all such reasoning about the "literary" mind, the goal is not only to understand how literature works but to formulate the templates of semantic production for the brain attempting to think about itself. As Mark Turner stated the matter so succinctly, consciousness is wonderful in many of its discriminations, "but it is a liar about mind."[52] Analysis of literary forms may impose a bit of discipline. There is progress. But Paul Thagard's view still pertains: how brains actually perform "analogy" remains occult.

This brings us full circle to the simple forms and universal practices recognized by André Jolles. He was decidedly on to something in anatomizing the similarities of the formal practices defining proverbs, jests, riddles, and parables over wide geographical and cultural areas by seeing them as crystallizations in language arising in experiences typical to the human psyche, what he called "mental dispositions."[53] Jolles did not explore the instability of these forms through mutation, parody, irony, and cultural diversification as Wolfgang Mieder has done in perceiving such practices as the "anti-proverb," but dwelt on the mentalities that defined generic practice.[54] Central to the present investigation is precisely the stability of the form in relation to the stable cognitive operations on which its decoding depends. When a different configuration of neural excitation is called for, it is because the form itself has altered. By entering the simple forms into a contest between socially contrasted players, however, something like the parody proverb comes into view, calling on extended computational processes even while the simple forms persist within the new binary structure. At the same time that proverbs become turns, social gestures, speeches from the mouths of caricatures, pieces of the world views of the contestants, rallying points, privy nips, and by extension even positions representing modern social and political agendas (hegemonic Solomon, proletariat Marcolphus), the metaphoric mode of mind remains engaged.

With that we leave the world of simple forms for hermeneutics of a different kind, and a closing glimpse of *The Dialogue of Solomon and Marcolphus*. "He that feedeth well his cow eateth often of the milk" is sound folk wisdom in a generic sense: perform your duties and reap the profits; care for animals and they will supply your needs. It is a perfectly apt "situation label." A rather different question is whether wisdom expressed in terms of cows is subversive by nature. To reason out in appositional terms just why we might be led to think so implies a great deal of presumed wisdom about the nature of society and the nature of discourse, some of which may be misleading or anachronistic. Such readings depend largely on a predisposed political view of the world, namely that persons as species are always persons as genus. Marcolphus lives at a subsistence level of economy. His wisdom is tailored to reflect those demanding realities. That much is attributed to him simply in terms of who he is represented to be – an epitome of the lowly rustic. But he is otherwise the creation of those who assigned him a role in the contest. Marcolphus performs like himself, of course, but evinces no motives behind his participation. He speaks in rustic proverbs, anticipating thereby, perhaps, his later role as a comic trickster, but there can be no certainty by more than loose association that he speaks for his "class," understands collective values and pretensions, has the slightest of political instincts, or a sense that his performance defies the "class" represented by his opponent. For a start, his proverbs are not consistently oppositional; sometimes he merely supplements the Solomonic dicta with examples of his own in kind. Thus, how much damage he does to the Solomonic proverb or world view in form or substance along the way is by no means clear, although many have seen his performance as an assault on high culture and mockery of its stick-in-the-mud values.

Moreover, if Marcolphus is the creation of clerics, by maintaining his scatological perspective as they do, they conspire to frame him as low, crude, offensive, and precisely what the *rustici* were taken to be in the eyes of the elite. He is by no means their champion or hero. Moreover, they assign victory to him, the underdog, on the basis of volume rather than content. The contest, meanwhile, is a creation of the schools, a kind of *altercatio* or *disputatio* that was a central part of the curriculum, making the matching of wits a parody of a standard pedagogical practice, while at the same time teaching Latin through the translation of folk culture into the language of the Church. Yet the *Dialogue* is decidedly ludic in its brokering of high and low cultures, although it was no doubt confined initially to pedagogical environments. More generally, the work forms an anthology of current proverbial lore, reveals the medieval mind in a recreational mode, testifies to the recurrence of forms by which we may investigate the stability of mental categories across time, and makes estimates about the common denominators that bind us together as a species.

But if the proverb frames the generic as specific, does it not also do the inverse at the level of the group, insofar as our stereotyping habits compel us to see all persons as representatives of their social classes and hence counters in the conflict of interests that divide those classes from one another? Is "class" not itself a human universal? Arguably it is not, because it emerged too late in evolutionary history as a concept to have built up values in the genome. Nevertheless, far more recent notions of class identity most certainly arouse emotions around the beliefs attendant upon them. Marcolphus may or may not manifest class stereotypes in his depiction of individuals, or in his own rustic demeanour, but the innovative format opens the potential for political interpretation. It is the first collection of its kind to do so in the Latin West. Ironically, that dimension may be its salvation in the canon of literary survivors. The politics of literature has become a relevance marker, but the making of meaning in first instances remains a primary consideration and the more demanding assignment once the operations of the decoding instrument, with all of its systemic particularities and procedures, are taken into consideration. Therein lies a new beginning.

CHAPTER SEVEN

Romance and the Universality of Human Nature

Heliodorus, *Aethiopica* and Robert Greene, *Menaphon*

Élie Bergougnan, after translating *Les Éthiopiques* into French, concluded that the Greek novel never achieved a real masterpiece. "L'analyse de l'oeuvre d'Héliodore montre bien, à côté de qualités réelles, le défaut capital du roman grec et son impuissance à dégager le principe de vérité, source de vigueur vitale qui lui a manqué."[1] (An analysis of the work of Heliodorus clearly reveals, compared to the nature of reality, the fundamental flaw of Greek romance: its incapacity to reveal principles of truth – the vital force of which it is completely lacking.)[2] Heliodorus's context is the whole of the Greek cultural world, by then in perceived decline from the great age of classical philosophy and civic institutions. By the third century of the new era, the Roman Empire into which Greek civilization had been absorbed was itself in decline. Writers were allegedly unable to impose a new synthesis of knowledge and being based on representations drawn from observed life. Their fanciful imaginations had been reshaped by an age of nearly anarchic regionalism, lawlessness, and religious superstition, whether in the name of Jehovah, Isis, or Thoth; they could not rise above popular sources and materials. There was nothing based on "l'observation de la réalité" (the observation of reality).[3]

Bergougnan was not alone. The anxiety over the realism of Greek romance has been central to critical considerations of the genre throughout the past century, beginning in 1912 with S.L. Wolff's *The Greek Romances in Elizabethan Prose Fiction*. The issue figures prominently in the work of Tomas Hägg and B.P. Reardon and is alluded to directly in such titles as J.R. Morgan's "History, Romance and Realism in the *Aithiopika* of Heliodorus," or E.L. Bowie's "The Novels and the Real World," or "Apollonius of Tyana: Tradition and Reality."[4] The "reality" with which these critics were concerned, as it turns out, pertained principally to the historical and material world, to the conditions of travel, governance, outlawry, religious practices, and civic life, in the second and third centuries; secondarily, it involved the accuracy of the represented social values and mores that emerged with a civilization in decline. In relation to the documentary unreliability of the portrait of the times, they took stock of the narrative improbability arising with the hyperbolical exploitation of select historical features in the creation of extreme contingencies and probationary conditions for the characters. If piracy and bands of brigands were in operation, then fortune might arrange to have the lovers come up against them incessantly and without reprieve, beyond all circumstantial probability.

But those matters are best relegated to contextual byways and matters of convention. If there is truth in these stories, it must be found in the ritualized and emblematic representation of human passions arising from the universal preoccupations that pertain to human nature. The challenge is how to discuss "the passions," the "universal," and "human nature." Of paramount interest is the psychology of romance, with its trajectories of desire and volition embodied in the preoccupations of the characters. The truth we are seeking is in the patterns of commitment and the lived obsessions of the lovers, and in the hedonic evaluations of experience that constitute our own empathic investment in their destinies. The question is how we may establish the bedrock realism of these subjective tropisms of human nature beneath the characteristic hyperbolical representations of romance and how, in turn, they may be labelled as truths.

Heliodorus's heroine, Charikleia, might have been seriously damaged by the traumas of her youth, but while she is a victim of her history, her character is not framed by it.[5] That she bears no scars would disqualify her as a *Bildungsroman* character of interest. Nevertheless, she is a highly motivated creature to whom we readily grant the fullness of personhood in order to believe in her actions as welling up from the drives, beliefs, and volition that continue to signal her quintessentially human energies, now as a generic representative of her gender and her race in search of love, safety, and her originating family. Her life is driven by self-advancement agendas, none of them ideological. She is propelled by a desire for refuge, by the need for paternal protection, by her tenacious and self-defining love for Theagenes, by the wish to rediscover her birth parents, by an unwavering need to preserve her chastity, and by a desire to survive at all cost, as long as it includes her beloved and her own purity. These are the conventional values and motors of romance, but the full account of their trans-mythological truth remains open to investigation.

In the pursuit of her desires, Charikleia must deal with hope and despair; faith and doubt in her beloved; fear of violation, incarceration, torture, and separation, and do so in the limited ways available to Heliodorus for orchestrating her attendant moods and emotional states. All of that is perfectly well known. The lovers desire each other, rightfully fear the world, and progress through it, one contingency at a time, until they achieve union and slip into quiet domestic anonymity after their story is of no further interest to us. But what of their compulsive, suicidal, purist, self-denying, fatalist moods in rapid succession, their clinging, testing, jealousy-tormented psychological fixations evoked by inner drives and incessant environmental calamity? Romance is ultimately about the making of a bond between eligible, self-directed, idealizing lovers in a hostile, alienating world, a bond the preservation of which is their entire motive for being. Moreover, it is about maturation and self-understanding and about their quest for stability, security, and community. What are these to the human condition, and are their quests more than mere adolescent fantasies?

Arguably, in their story there are patterns of human behaviour, excessive on the surface, that nevertheless represent values of an ethnographic kind, behaviour that is "true" according to the drives and conditions of being human. Dare we go further, to say that these are belaboured stories of bonding in adversity and the formation of a fought-for mutuality between lovers as prospective replicators acting in the great drama of the genes, according to their psychological programming? We have not hesitated in the past to impose the behavioural models of Freud, or Jung, or Lacan on the idiosyncrasies of characters in our search for unifying themes, eager to explain their conduct according to the urges and instincts arising in the depths of the psyche. A post-Darwinian exegesis may simply be another explanatory model tossed into the ring of so-so metaphors about the origins of the human; that remains to be tested. Yet such enterprises will continue in an effort to find what is true about the literary representation of the quintessentially human. That truth, in turn, needs a standard, and that standard requires an analytical system that connects literary representation to bedrock human nature, including the universals that are transmitted from story to story in the ways that constitute and define genre. In the simplest of terms, those abiding preoccupations urged by a brain designed to seek its primal interests in social and geographical environments will have a shaping hand in the kinds of stories we tell about human experience.

Such an investigation is at once banal for those who build their social sciences on Darwinian principles, but heretical for those who do not. That challenge must be taken up in due course, but it is only fair to forewarn readers that a demonstration of the universals that arguably inform romance must somehow involve the genetic basis through which, alone, the common legacy of the species – what it means to be human – can be accounted for. There is no other imaginable explanation, but it is hardly, as yet, a received idea. Northrop Frye detected the archetypal traits profiled in romance, but he was powerless to provide more than a mythological rationale for their pervasive and iterative presence; the genes may be our last hope. For if the genre is to become a "scripture," as it was for Frye, its truth factor must pertain to the triumph of human wishes in actualizing courtship grounded in desire and loyalty, themselves the emotionalized belief conditions that sustain human reproduction. Romance themes are adaptive themes. The question is whether that is answer enough to those who disparage the genre, beginning with the critique of Heliodorus's work for its lack of truth.[6]

As a model for humanist translation and adaptation, Greek romance, with its embedded formulae, was transported to the Renaissance literary world, resulting in a conflation of the Hellenist ethos and values with the conventions and themes of chivalric romance, pastoral, and the Petrarchan lyric. Yet the Renaissance romance remained vitally concerned with such Hellenistic themes as cruelty and contingency, lost identities, tests of chastity, faith, loyalty, erotic trauma, and dangerous adventures. Heliodorian psychology is thereby brought to the bar of our assessment through the works of Renaissance imitators, ready to espouse the same generic truths about the human. Now we have a common repertory of the psychological urges that define romance plots carried across centuries. Those urges remain valid, of course, only if they are universal and remain pertinent in such diverse periods of social history. Only then is there hope that we, today, can successfully develop a theory of mind for these otherwise culturally distanced representations of persons.

But in actual practice, matters are quite different: we read their "antique" minds without let or hindrance because we understand our own thoughts and beliefs, and on that basis we have imaginative access to theirs. Such understanding is made possible because nature, as it were, through generalized genetic design, made certain that our brains share in the concerns across time that pertain to our species. And now the thesis: the more synchronized the mental design systems among brains, the more similar their emergent hedonic and behavioural properties will be. Thus, however individualized by cultural or personal predilection, the more behaviour conforms to the basic expressions of biogenetic design, the greater will be its quotient of literary as well as anthropological truth. The hedonics of love are felt and experienced in personal, contextual, cultural, and temperamental ways, to be sure, and yet every love story is our love story, because the hedonics of love are generic. It is a paradox to ponder.[7] Moreover, to the extent that Heliodorus's protagonists arouse well-wishing and concern in their struggle to find happiness and repose, they confirm the "truths" of our own natures, which form the basis of our empathy.[8]

Before proceeding with the thesis, however, we need to settle on a representative Heliodorian romance in Renaissance England to place under comparative investigation. One of the earliest is Barnabe Riche's "Sappho Duke of Mantona," the first tale of his *Farewell to Military Profession*. There is also Sidney's accomplished and celebrated *Arcadia*, which is patently indebted to Heliodorus's "sugared invention of that picture of love in Theagenes and Chariclea."[9] But I have settled on Robert Greene's *Menaphon* for the typicality of its style and episodes.[10] These works span the 1580s and no doubt take their inspiration from the elaborate paraphrastic translation of *The Ethiopian Tale* published in 1567 by Thomas Underdowne. (Longus would not appear in English before 1588 and Achilles Tatius a full decade later, although these authors may have been known through their French translations.) As stated, the conventions and materials of Heliodorian romance had come to London as part of the humanist recovery of the ancients, including tales of shipwrecks and pirates, lovers separated from one another, lost identities, longing quests, passing generations, enigmatic oracles, narrow escapes, sexual predation upon the innocent and the vulnerable, abductions and attempted rapes, illusions of death, sexual rivals, the bonds of mutual love under duress, obsession with chastity, long-delayed episodic plotting over wide geographical areas, and final deliverance, reunion, and return to something like home and sanctuary. There would be modifications and variations, but the Alexandrine ethos was behind them all, constructing a vision of the human condition through its repertory of situations and emotions.

Greene's *Menaphon*, written in 1588 and published in 1589, is arguably among the most Heliodorian of all the English Renaissance romances. To be sure, there are the conventions of pastoral, while the suffering lovers are now married parents driven apart by dynastic interdictions. But it remains a story of long-term separation and longing, accompanied by the fear and danger of betrayal brought to members of a nuclear family. The accidents of fortune still produce the tell-tale shipwreck, leading to a loss of original identities, which allows, in turn, for forbidden sexual encounters between a son and his mother, and a father and his daughter. The underlying values of ancient romance are entirely replicated and predictable. The disposition of the world with its manifold threats, initially tempered by pastoral restraints and decorum, ultimately delivers the same density of malice and accident. Sephestia, the mother, wife, and daughter in question, after being exiled by her tyrannical father, becomes Samela upon landing on the shores of Arcadia with her babe in arms following shipwreck. Despite the title of the work, named after the hopeless shepherd who protects and pursues her, Samela remains the central selfhood in the work. This young woman, for her beauty, soon becomes the universal *desirée* of the Arcadian world, now separated from a beloved husband, deemed lost at sea. Her story entails seventeen years of privation, careful negotiation for hospitality and sustenance, the nurturing and education of a child soon to be abducted, devotion to the memory of her lost husband, and the preservation of her chastity in a society that seeks to invade her privacy with importunate addresses of love, or through admiration of her as their goddess – the nonpareil of feminine beauty. If she cannot be seduced, she must at least be worshipped as their homecoming queen. She, like Charikleia, is blessed and cursed by a beauty that attracts all eyes and exudes a mystique that incites erotic desire. The emotional and erotic economies of the Alexandrine world remain manifest, and in being so, align this story with the promised biogenetic themes that inform the genre through their embodied representations.

Here is an opening example. Charikleia and Samela must endure life as beautiful women in social circumstances where erotic arousal in relation to that beauty is nearly out of control; it is a large factor in their respective stories. That may be a comment on the mores of individuals and communities, or the lack thereof, but it is concomitantly an assertion of something close to a human universal. Steven Pinker believes that it is; with regard to sexual attractiveness, "people outside a culture usually agree with the people inside about who is beautiful and who is not, and people everywhere want good-looking partners."[11] The point would be merely anecdotal if it had not been tested. But why the cult of feminine beauty? Would friendship and companionability not be far greater assets in the founding of long-term relationships? Whence the power of beauty alone to polarize desire, fire dreams and longings, elicit fantasies, or bring men to despair, suicide, insane adventures, or mortal combat?[12] Was it not the stunning beauty of a woman's face that launched the Trojan War? If desire and the universal standards of beauty are not taught to every new generation by cultural transmission, how are they to be explained? Behind that entire economy is an evolutionary backstory, having to do with the genetic programming of each male of the species to secure the best and most fertile partner he can hope to attract – programming which gives emotional priority to women who convey to the sight all the physiological markers of health and fertility. What appear to be merely the shallow conventions of romance may well have their adaptive rationales, and thus their truths, through selective design. Beauty in women is linked to the imperatives of desire and in romance this emotional economy is routinely translated to regions beyond the restraints of socialization.

How would that story work? Both women are under constant solicitation or threat of violence by desiring males. Both have made their commitments, and yet Charikleia will not announce her relationship to Theagenes, and Samela will give hope to suitors, each woman strategizing her beauty in relation to her best perceived interests, playing for safety or promotion through temporizing, escape, or a restraining control of male erotic energy. Thus, romance, at one of its bedrock levels, is about the economy of feminine beauty and its unprecedented power over the hormone-driven imaginations of a profoundly gendered species, coupled with the ability of women to employ that beauty in their campaigns for the best quality males while controlling them through enchantment, or through the exploitation of rivals. Meanwhile, the male gaze will be conditioned by the inherited values that have prioritized the traits that constitute the feminine ideal. In general, she will be symmetrical, robust, pure, with shining and healthy hair and strong nails, but otherwise perfectly average, not too tall or fat or masculine, having all the quintessential female traits and proportions, small chin and nose, imperceptible eye ridges, a delicate jaw line, high soft cheeks, all signifying an estrogen rich, nubile, virginal female. Such a prize will be young for a longer reproductive career, having the large eyes, red lips, and moist, tight skin of adolescence, and in her chastity she will show promise that her future children will belong to her designated partner.

We may all be smiling at what appears to be a mere stereotype, until we recognize these features in blazons and the profiles of heroines, and that both genders negotiate their social worlds in these visual terms. Beauty, by its very rarity, commodifies the female, inevitably, but women, in their instinctive partiality to strong, protective males are themselves the makers of stereotypes through eons of mate selection. After all, they themselves are single-handedly responsible for the physical and temperamental traits that characterize the most coveted males (thanks to the genetic legacy of their mothers and grandmothers who engineered them through selection and reproduction).[13] The values of biogenetic history are nowhere more evident than in mate selection processes, and those are nowhere more evident than in generic romances. There will always be debate over the role of culture, but culture must have the lesser place, because the interests of the genes are ultimately heard loudly in the stories of the most handsome and winsome specimens available and their quests for each other. There is little new in this appraisal of the human economy for students of "how the mind works," except to make it the foundation for a quality of literary truth in a genre that features it in such salient and emblematic ways. It forms a biogrammar for the interpretation of stories.

Gender feminists may find every element of this economy a matter of disgust, leading to efforts to devise a cerebral blank slate for the species (a total genetic amnesia) upon which new values of their designing might be written through cultural conditioning. But that will prove discouragingly revisionist. The desire to overwrite the imperatives of the genes that designed male and female bodies and brains according to their adaptive advantages in ancestral mating environments will not be easily accomplished. Whether a practised denial on moral grounds is desirable for the future is another matter; one may resist the truths of the genre in the name of modernity, fashioning, and choice.

Romance, meanwhile, has endorsed a pervasive, if sometimes formulaic, representation of the basic strategies of gendered sexuality from the delicate advances of handsome, tender, athletic, devoted, idealistic, and vulnerable males in stark opposition to the raw drives of sex-starved loners, outlaws, and opportunist rapists so vigorously resisted by heroines. The genre also calls for the imposition of absolute standards of chastity and sexual continence unto death for both the heroes (against their wills and temperaments) and heroines as a precondition to successful bonding. These conflicting energies make for the extreme negotiations and unalloyed levels of hedonic support that regulate sexual relations from gendered perspectives in an eroto-phrenetic world. Charikleia regretted her beauty on occasion, but she did not shrink from playing the goddess in Delphi and making herself eye bait to the city, although it qualified her for sacrifice in her own country, while Samela, too, met compromise in allowing herself to be as fetching a shepherdess as she had been a princess. Such are the risks in exposing their winning faces as come-hither instruments for attracting the highest quality mates within their compasses. A life of perpetual peril in the gender wars carries overtones of social critique, but civility and restraint through law and order or moral self-invigilation in a society of reliable guardians and closed social structures are antithetical to the equally vital fantasy of love bonding under extreme duress. Chaperones are few in romance, and that brings the raw negotiations of love, most emblematically, back to the principals in the face off, and thus to the raw truths of their respective natures. The ultimate formula of romance is to build the foundations of erotic conflict on reliable observations of human nature pertaining to lovers, rivals, and enemies, and to set that struggle in motion in a uniquely dangerous social and geographical world.

Pulchritude is but one dimension of the romance complex, however. Subsequent to the initial mutual gaze, nothing is more elemental than the desire of paired lovers to enjoy the exclusivity of their entente. Hence, the anxieties that attend upon their insecurities form a significant part of the emotional energy of the genre. To add that such torments contribute to the time-engineered goals of "selfish genes" (to draw on the thesis of Richard Dawkins) merely supplies a more elaborate rationale for the obsession of lovers with their sexual and social insularity. Why, we may well ask, is courtship under fishbowl investigation the heart of romance plotting, and why does the testing of the lovers' emotional commitments include imagined suicide pacts in the event of sexual violation or the death of one of the partners? Why should the extreme misgivings and forebodings of Charikleia and Theagenes concerning sexual purity become the template to a host of literary successors? Presumably because their promises of reciprocal suicide are, to their mutual bond, something like a doomsday scenario, making breach of promise too costly to contemplate; couples are in it for the long haul when the cost of parting is self-administered death. In this way, the interests of the genes are carefully looked after by brains programmed to deliver the hedonic instructions that will fortify relationships at a cost level that guarantees stability for the production and protection of children.[14] Otherwise, these lovers might have joined the shallows of the 1960s and indulged themselves in casual and attachment-free sex.

Once the magic of mutual desire has drawn the lovers into a common emotional sphere, intimations of uncertainty and doubt, of wavering devotion or straying attention, fantastic or real, inevitably follow. Romance is not only about beginnings but also about the emotional craving for every possible assurance of the exclusive enjoyment of the beloved over the long term, a primal sentiment most positively expressed through pledge and promise, but most poignantly through fear and jealousy. This hedonic crisis does not end with marriage, but incorporates a lifelong jealousy, felt not only by humans but also by many birds and mammals. Fear of cuckoldry was a Renaissance leitmotif. That it is experienced emotionally has both social and biogenetic explanations, because the psyche is most apparently hardwired to favour investment in one's own genetic offspring and to avoid being tricked into expending resources on alien genes. Those replicators who felt jealousy in ages gone by were clearly more successful and thus passed the jealousy mechanism along to their offspring. It is embedded in the genome, almost as though children, before the fact, instruct their parents in the emotional grounding that looks after their needs, especially during a period of prolonged childhood and training. On the score of cuckoldry, the species can be murderous, not by training but by instinct, and while one may deplore the potential results arising from this architectural feature of the brain, the hedonic promptings that instill such pathological anger can never be erased. The male fears infidelity due to his anxiety over paternity, while the female fears philandering less than she fears the alienation of affection provoked by a more appealing woman, which could cost her protection and essential resources for her children. In anticipation of these possible eventualities, lovers begin their search for guarantees in the displays of militant chastity or cost endurance in risky quests or unflinching service during courtship. Jealousy is a biogenetic truth and a big piece of the romance formula.

The conflict between generations, particularly between imperious fathers and headstrong daughters, is also deeply rooted. *Menaphon's* Sephestia (Samela) is exiled by an angry father because she has deigned to marry the man of her choice against her father's will. The plot is as old as the hills and arguably as true.[15] The assumed autonomy of children motivated by powerful desire notoriously pits love imperatives against obedience and duty and often leads to execration and exile before time or political leverage brings truce or reconciliation. Romances are heavily invested in this overarching plot. If there is truth in this crisis, it pertains to the profoundly interested perspectives of opposing psyches. Fathers, and especially those in positions of political or economic power, have believed in the almost sacred right to exploit the sexuality of their daughters to their own ends in extending political and financial ties beneficial not only to the family but to the tribe or nation. Daughters, by contrast, have an equally sacred sense of their right to autodetermination in the selection of their sexual partners. Rarely do we side with tyrannical fathers obstructing the imperatives of love, yet their positions reflect a biogenetic disposition intolerant of disobedience because it deflates their political prestige and cripples a survival strategy which seeks to combine additional resources and powerful associates through marriage alliances. Female defection is treason to the community whose benefits she might secure through a strategic political union. Typically, in romance, the betrayal of an inter-tribal marriage leads not only to hatred and alienation but to foreign invasions and collective blame in the breaking of promises. Tyrants become really ugly when their children disempower them, and they are notoriously slow to forgive.

In explanation of this psycho-history, Lionel Tiger and Robin Fox have profiled this father-daughter conflict as a universal arising with the bartering habits among males in hunter societies.[16] Small bands formed alliances with other bands leading to vital cooperation in resource management through the exchange of gifts. Where commodities were scarce (and they usually were), they traded daughters, building the kinship bonds through "in-law" arrangements still visible in the negotiations among royal houses almost down to the present. Motivating the fathers was the power and prestige associated with gift giving and hence personal political prowess. (The quest for status is also a human universal.)[17] The sexual tastes and mating proclivities of the daughters were largely discounted and conflict emerged, particularly when "falling in love" (also a biogenetic phenomenon) became the daughter's emotional imperative to disobey.[18] Many a plot has been made of this: fathers with power instincts bred by ancestral success; daughters with love instincts bred by reproductive success. Romance centres itself in yet another set of conflicting biogenetic imperatives.

Moreover, that generic conflict is easy to construct when one considers how much a woman's quality of life as a nurturing mother depends on a compatible, healthy, and loyal mate equal to the task of providing high quality sperm, protective prowess, and material sustenance over the long term, and the degree to which she is wired to rely on her own instruction in making that choice. The subversive empathy elicited for Sephestia through her exile and suffering is indicative of an artistic campaign on behalf of this intuitive "right" of the prospective mother to resist choices made on her behalf.

This campaign was apparent in ancient Greece and continued throughout the Middle Ages and Renaissance down to the nineteenth century through variations on the romance quest whereby adolescents were released into challenging worlds to discover themselves through the trials of pair bonding. J.Q. Wilson in *The Moral Sense* went so far as to assert that this conflict of genetic predilections between father and daughter contributed to the philosophy of the Enlightenment and participated in the abolition of feudalism and slavery.[19] The point is rhetorically compelling if problematic in terms of demonstrating the analogous reasoning by which mate choice was actually equated with escape from feudal oppression and slavery, but clearly all three were long-term struggles for the freedom of the individual from diverse oppressions. It has been a hard-fought entitlement for women now inscribed in law in Western nations. As a result, power relations within the extended family are somewhat attenuated, particularly in situations where the tempering perspective of a mother is present, although parents continue to express their reservations over mate choices violating their sense of ethnic, religious, intellectual, or class standards. Romance as a genre (as well as erudite comedy) led the way in eliciting empathy for the pretensions of the younger generation against the dictates of fathers, causing Northrop Frye to remark on the ritual containment of paternal power through the rites of passage conducted by such plots, and of the frequent transfer of that power to a new generation as part of the new social contract at the end of the action. Heliodorus and Greene concern themselves, thus, not merely with the settlement of affairs between lovers in isolation, but with the conflicting demands that mark the divided interests of the generations – matters particularly poignant when they involve political power and the prestige of rulership.

That the behavioural motifs characterizing the race may have their universality accounted for by the consolidated emergent properties of an adapted brain is both an explanatory argument and a hermeneutic opportunity whereby the leitmotifs of literature may be aligned with the leitmotifs of a species designed by natural selection. John Updike, in a strangely nuanced statement, declared that "a writer of fiction, a professional liar, is paradoxically obsessed with what is true," and "the unit of truth, at least for a fiction writer, is the human animal, belonging to the species Homo sapiens, unchanged for at least 100,000 years."[20] That the "truth" of literature pertains to homo sapiens is only to be expected; we have a fixated interest in the members of our own species – those with whom we calibrate most of our social interactions. But that literary truth is aligned with those elements of our being that have not changed over millennia is rather innovative, for it is an implicit recognition of the stability brought to humankind through the imperceptibly slow processes of genetic adaptation and mutation affecting the entire human family.[21] It would be interesting to have Updike's list of those ancient and unchanging characteristics preserved by an engineering system predating all cultural developments and all languages. Genes, he would no doubt agree, do not shape specific beliefs, nor do they determine the cultural solutions adopted among members of groups for the regulation and coordination of their societies, but they repeat in systemic ways the mental architectures that define our computational and emotional capacities, retaining in those modular designs the emergent mental states that serve the survival interests of the species. Those states are not the specific choices of the hour, but the mechanisms which condition the making of those choices in setting priorities, projecting feelings, fixing attention, conditioning memory, and invigilating consciousness in nearly all matters of our survival – mechanisms too vital to be left up to word-of-mouth communication. Actual knowledge must be acquired, as Pinker points out, but the ways of knowing are innate and closely calibrated to the adaptive features of neurological design.[22] Romance is specifically preoccupied with a subset of those conditions.

To be sure, genetic design narrows the choices for dealing with environmental challenges. Choice is reductionist by definition, including those prompts that abet the decision-making process. Falling in love is reductionist insofar as it delimits the processes of mate selection in the interests of getting them right, as well as in managing rivals, investing in one's own progeny, staving off kin interference, and protecting the collective family from predators, starvation, theft, loss of life, and related perils, largely by the way our brains make us feel about these issues. These are not behaviours which nature has left up to chance or local instruction. Human nature comes loaded with behavioural predilections looking after the destiny of the genes. That the heroes and heroines of romance should reflect these same biases, even in exaggerated form, by building them into the rituals of courtship, testifies to the presence and operative power of biogenetic universals shining through a plethora of cultural and individual variations. Thus Pinker could write, "I think we have reason to believe that the mind is equipped with a battery of emotions, drives, and faculties for reasoning and communicating, and that they have a common logic across cultures, are difficult to erase or redesign from scratch, were shaped by natural selection acting over the course of human evolution, and owe some of their basic design (and some of their variations) to information in the genome."[23] Pinker is restating a set of convictions now held by a growing community of witnesses, several of whom appear in the elaborations to follow. These elaborations aim to solidify the hermeneutic bridge between reasons of evolutionary psychology and the functions of literary genres. More on the science of bio-universals will follow, but let us swing back to *Menaphon* and its generic actions distilled from the plot.

The closing episodes of Greene's romance consist of interlocking, high-tension conflicts involving the males of a single family over the sexual conquest of a woman whose identity as daughter, mother, and wife is unknown to them. These are not the circumstances of a son's secret Freudian longings for his mother and hatred for his father, or of a father for an incestuous relationship with his daughter. These are accidents arising from identities lost through time and distance, resulting in a triumvirate of alpha males, each seeking to mate, two of them by force, with the celebrated Arcadian beauty queen. Their behaviours are not those of deviants or the sexually jealous, but of men in combat mode for the most attractive sexual partner, men who are subsequently appalled by the circumstances that induced them into such error. There is little doubt that the plot has been manipulated to the edges of the plausible or even possible, but the emotions presented in succession reflect mind states hedonically driven and coordinated to extreme circumstances. This plot is a rich configuration of crises keyed to the deepest instincts regarding the most primal matters of survival and reproduction. When matters were made clear, "Democles, seeing his daughter revived whom so cruelly for the love of Maximius he had banished out of his confines … leaped from his seat and embraced them all with tears craving pardon of Maximius and Sephestia" (174). This is not the response of a father with a fixation, for he had no knowledge of her identity, but of a father mellowed by time, a father whose advancing years brought him at last to confess the right of his daughter to choose her mate, particularly in realizing that their child was "a matchless paragon of approved chivalry" and fit to be crowned on the spot as his successor (174). Once again he is a father and grandfather, at long last content that Sephestia, on her own, had done her part in securing the future of the dynasty. She had, in fact, through her choice of a husband, done far better than her father in securing a high-quality heir to the throne. The more shocking feature, implicit by circumstance, but never discussed, was the horror of de facto, situational incest that would have ensued had either father or son realized his ambition with regard to this unidentified woman. The taboo passes through the reader's mind as a catastrophe narrowly averted, and there is presumption that it must have passed likewise through the minds of its potential perpetrators. That Greene could make such wholesale drama out of this provisional horror depended, to be sure, on what such a perceived taboo means to the human psyche, and why its interdiction is so widely and emotionally endorsed. That question is not entirely resolved in explanatory terms, whether the taboo is a violation of nature or of ancient custom. But whatever its origins, it is, in the context of romance, continually represented as a universal human interdiction of the most absolute kind, generic to our most horrific thoughts and distasteful feelings. That such feelings are biogenetic I would not venture to claim, but it is a motif that runs deep in the human psyche all the same.

Let us return now to another of romance's leading motifs: the dramatization of the *coup de foudre*, the lightening flash of falling in love, more weakly expressed in English as "love at first sight." It has been celebrated throughout literary history as a mystery, a blessing, a curse when it is unrequited, a destiny, a momentous and compelling occasion worthy of narrative record, and an imperative to action. But less thought has been given to the brain mechanisms responsible for its genesis, its biological raison d'être, and its value to the organism. Quite simply, if it is both a genetically grounded property of the species and a consistent feature of romance, we have another potential match, a literary motif reflecting a psychobiological universal – an archetype. Inversely, that truth is established by achieving the status of a literary convention pertaining to a genre. This paradigmatic feature was already evident in Heliodorus, who tells of two adolescents impervious, even hostile, to love's enticements, but who, moments after first seeing each other, find themselves emotionally shaken and quickened by a strange and enthralling desire. Awed by their own sudden feelings, they struggle to account for their captivated states of mind. Two physiognomies among thousands of potential choices have established a mutual and exclusive destiny; no further information about the instant beloved is required on either side: beautiful Charikleia; handsome Theagenes. What follows is massive innuendo after minimal inflection (as in the poetry of Petrarch or Cavalcanti), a fixation of the imagination, erotic desire, feelings of helplessness, sensations of physical illness, confusion, a sudden self-consciousness and reticence, even a death wish should the enterprise lead to failure, because some element in the design of the brain has dictated from the depths that this is the only he or she acceptable as a life partner.[24] Brains, in effect, must be retro-engineered in a way that explains this response mechanism; there must be an adaptive principle that accounts for this intense arousal.

Skeptics will demur, but this response mechanism occurs in all cultures and it takes each of its subjects by surprise at a precise yet unpredictable moment when a stimulus is embraced by exceptionally excited receptivity.[25] Romance writers specialize in monologues expressing this transition. Were the *coup de foudre* phenomenon not adaptive, the mechanisms for its triggering would not have been universally embedded in the human genome. And because its occurrence in life is common to members of all societies yet so infrequent in individual lives, like the unique blossoming of a rare flower, this moment of awakening holds eternal fascination, accompanied by a nostalgic regret that it can rarely be repeated, for it is specifically designed to make only one individual so precious and emotionally costly to the beholder that all other contenders cease to hold any interest – nature tricking us to look after its own agendas by so emotionalizing the choice that a bond can begin to grow that excludes rivals and philandering to the ultimate benefit of the offspring. This trait became fixed in the genome because those who experienced it over evolutionary history were more reproductively successful and thus passed it along as part of the brain design of their offspring, and ultimately of the race.[26] The falling-in-love mechanism is a stock feature of romance psychology grounded in the ways our brains do business with the social world.[27] It is a representational truth.

In *An Ethiopian Tale*, Charikleia rides out in a carriage drawn by two white bullocks, the very measure of feminine pulchritude in her purple gown, gold bands of intertwining serpents, her hair cascading and curling and crowned with a bay wreath, holding a bow and torch, its light surpassed only by her shining eyes. Theagenes, who "had never felt anything but contempt for their whole sex" (422), was in the crowd. His matchless masculinity was nevertheless the perfect complement to her unequalled beauty; we intuit the inevitability of the match long before they do. Heliodorus describes the moment in detail, for "the young pair fell in love, as if the soul recognized its kin at the very first encounter and sped to meet that which was worthily its own" (414). Fear and confusion immediately follow as the emotions take over the direction of the erotic theatre. Kalasiris the priest is about to become mentor to both, the medium through whom their passions are explained to their faltering reason. Theagenes takes to his room to deal with the hedonic kidnapping of his thoughts. Both lovers are made hostages to minds doing the heavy chores of match-making, during which time the emotions fortify the exclusivity of the choice and inaugurate a period of courtship which further separates them from the "promiscuity" that led them to each other (am I explaining away the mystery of the rainbow?). Further candidates are now excluded, even though lovers' doubts keep them in states of agitation, prompting them to set up tests, entertain fantasies, and exact pledges. This is the emotional work, so pleasing and so painful, that underpins the success of a one-time mate selection bolstered by the confabulatory praise and estimation that rarifies the beloved even more. To be sure, it is the necessary prelude to their loyalty to each other through the untold tribulations to follow (or the ability to pronounce their wedding vows). It is the innate thinking behind long and chaste engagements, beneficial to the reproductive success of a species characterized by physical vulnerability and the long dependency periods of their offspring. Emotionalized bonding is nature's ploy for increasing reproductive success, and romance dwells on the transaction as the felt qualities of love whereby couples are lured into doing the work of the race – nature's little rite of passage that polarizes the emotional life into an investment quest unto death. Adaptive pressures put it all in place.

This same transformation recurs in *Menaphon*, when the shepherd for whom the book is named first pronounces his indifference to foolish love, only to be smitten by the sorrowful Sephestia newly landed on his shores. No sooner does he lay eyes on her glorious face from afar than he begins his mental blazon to her tresses, her brows, her eyes, her neck, and southward, turning his apostasy to amazement. His relish for her every feature is disguised as pity for her plight. The misalignment of their minds, the mistaken intentions, the misplaced hopes come to a crisis when she is compelled to rebuff his, at first, delicate advances (106–8). Greene provides his own account of the transformations wrought by erotic longing, which Menaphon openly frames as his desire for a child. But it is Samela's beauty that becomes the imperative of his emotional life, as though in punishment for his former recalcitrance. These emotions and hidden states of mind set in motion the action chains around which the plot is initially organized: first the lingering hope and then the dawning awareness that Samela will never be persuaded to love. Even the delicate and respectful negotiations that characterize desire contained by pastoral restraint are manifestations of the emotional intelligences of both players.

Meanwhile, we are treated to a most delicate invention, that in which true husband and wife encounter one another in complete anonymity, yet find themselves secretly and emblematically reconfirming their love through the spontaneous mutual attraction inspired by resemblances. The substance is a chaste and sentimental friendship as each lives in memory of the true beloved in the presence of his or her simulacrum. This loss of identity turns Melicertus, superior among the shepherds, and Samela, unequalled among the shepherdesses, into epitomes of desire for those who look on, as well as for each other, while the bonds of remembered passion from their former lives impede all but polite and innocent exchanges simultaneously shot through with erotic energy. It is a study in the social emotions configured according to the altered circumstances prevailing in their respective minds. Even here, behaviour traces its origins to the logic of the emotions.[28]

In accordance with these principles, Donald Symons brings the assessment even closer to the materials and preoccupations of romance. "Courting is thus a … series of strategies and counter-strategies, with mutual love as the goal, in which sex and commitment are manipulated, each partner attempting to maximize gains and minimize losses."[29] Plots are made of these negotiations, directed by generic goals and emotional prompts. The lovers of romance must do the work of their species in protecting chastity, making promises of devotion, holding rivals and gatecrashers at bay, bonding under hardship, fixing memories, seeking the well-being of the other, negotiating with family members, and working toward reintegration into the communities from which they have been abducted or exiled. Wherever and whenever these profiles find alignment in the common causes of the race, directed by the same repertory of biogenetic emanations, there is an invitation to gloss such literary creations in terms of their fidelity to human universals. The result, once again, is a renewed expectation that vital literatures must endorse our biogenetic substructures – but I must not over-protest.

Another element of romance narratives, often drawn out (to the evident delectation of readers), is the courting sequence itself. There are many variables, but the ritual follows an order that begins with sight before moving on to social gestures, acknowledged glances, a few words, then to socializing, playful exchanges of information, more elaborate shows of interest, and eventually to touching, then caresses, eroticized verbal expressions, and finally to public displays and private proclamations of their love.[30] This sequence is pursued according to their respective interests in the formation of the couple and continues for as long as the female encourages the male to ward off the attentions of other males. The uncertainty of that continued encouragement is part of their negotiation and a source of underlying anxiety. There are, in fact, versions of romance which dwell on the trials of communication between the prospective partners, the anxieties over virginity, the flights of excessive and threatening feelings, the risks of over-aestheticizing the initial glance, of fixing on phantasms of beauty rather than on the flesh and blood girl, or fetishizing her body, or becoming obsessed by touch, or of attempting to leap over the polite verbal and social niceties and move to the desired coupling – many of these misunderstandings arising from the conflicting and immature emotions shaping the courtship. Charikleia has fixations of her own about such matters and delivers her ultimatums with lockstep precision. In living through their emotions, the nascent lovers of romance shape their destinies along predictable profiles because their emotions have been designed by a common genome; they are performing the rites of the species, which are nearly synonymous with the challenges of dealing with the intentional states of gendered minds. These are foundational matters of cognition much as they are foundational to the conventions of romance.

Tiger and Fox add, moreover, that the themes of romance, as with the conditions for breeding in general, are political, for "politics involves the possibility of changing the distribution of resources in a society – one of which is the control over the future that breeding allows."[31] This provides a bridge for resuming the analysis already begun. Romance is about future reproduction and breeding choices are about making winners and losers in accordance with the strategies of the candidates in competition. *Menaphon* is, in part, the touching story of one of those losers who gives his all to win the woman of his heart's desire, only to discover that he remains, for want of status, the farthest behind in the race to reproductive success. Sometimes heartlessly, romance is also about who is denied access to the best females in accordance with hierarchy and the rarity of the female as a marriageable "commodity." Once a choice in partners was allowed or demanded, "the species has been irretrievably concerned with who can marry whom and with the relationship between position, property, and productive copulation."[32] In *Menaphon*, matters play out according to expectation as Samela gently dissuades the déclassé males while fighting off the ineligible alpha males who falsely assume her eligibility and her inferiority to them respectively as king and warrior (unwittingly her own father and son). Samela maintains her place at the head of a female dominance hierarchy, asserting her right to choose, while, through her natural superiority, she commands the "attention-structure" due to her class, for all eyes are turned upon her in veneration. Yet such dramatic ironies and extraordinary circumstantial anomalies mean nothing without the most predictable strategies of the most predictable game of the species. By ancient instincts, it is a fight for a greater share, for control and privilege, while, paradoxically, success also entails cooperation, self-restraint, even self-sacrifice, as in the case of her lawful husband, Melicertus. Fairy tales repeatedly express this behavioural paradox. Successful candidates must be able to drive off rivals and retain the females who are won. Melicertus had done this work, twice, in courting Sephestia and in serving Samela with perfect circumspection and self-control, without recognizing they were one and the same. These are mere variations on the bonding that pertains to romantic love.

It stands to reason that these negotiations are shaped by the biogenetic impulses that have led to the most successful negotiations of the past. Among them, females, in keeping with the subcortical information they simply "know," will seek out partners likely to indulge and supervise their offspring and not abandon mother and child to fortune. Little does it matter that the modern world provides alternatives, because the hedonic prompts framing these transactions between potential mating pairs will have been formed in ancestral environments when reproductive success depended on such cooperation. The genetic futures of women who chose disloyal men were repeatedly jeopardized, whereas those who found loyalty found the trait genetically fortified in their offspring. Their only tactics were delayed sexual rewards and putting their lovers up to high risk feats, while searching for signs of an emotional investment they could actually believe in.[33] Charikleia is obsessed by these games. Realizing, temperamentally and instinctively, that loss of virginity destroyed her bargaining position, she resorted to subterfuges of many kinds. As Northorp Frye observed of her, "Chariclea's dedication to virginity is not part of a general commitment to moral integrity. It certainly does not imply that she is also truthful or straightforward; in fact a more devious little twister would be hard to find among heroines of romance."[34] But in accordance with the raw negotiating positions of the two sexes, this is precisely what we should expect. Samela is a true daughter of her prototype, but in a gentler key because her world is, for a time, constrained by the conventions of pastoral; the exchanges are entirely poetic, conventionalized, and polite. But in managing poor Menaphon, she is equally devious and equally resolved. Nothing takes place at random, and much of that limited randomness reflects the design propensities of gendered, as well as socialized and class-defined, brains.

Within structured and public social contexts, the interests of the mother and child become a concern of the group, for while prospective mates may initially prefer each other above all other contenders, that emotional entente is subject not only to the waxing but also to the waning of high feelings. Hence, because the conditions of life provoke mistrust generally, collectivities, suspicious of the fancies of adolescents, impose their own strictures in the form of rites, public vows, ceremonies, and material symbols of those emotional bonds in a climate of provisions and sanctions. The mother-child bond, to paraphrase Tiger and Fox yet again, is a precious resource meriting collective protection. "Most of the societies that use mated pairs as the basic units do not dare to leave the necessary stability of the family to the vagaries of adolescent emotion."[35] But this is precisely the juncture at which romance imposes its variations upon the social laboratory; romance also comes to its end in the collective interests of the homecoming. For the genre recognizes that adolescents at the bonding age are often separated from the regulating group, even driven out by the elders for having broken other injunctions, as in the case of Sephestia/Samela, her husband, and child. This is now the nascent family driven into exile, separated by accidents, the mother-child unit isolated and in desperate need of substitute protection. Partners such as Sephestia and Maximius may find eternal loyalty in their natures alone, but many another tale takes the lovers through a final phase of public declaration and solemn vows. Public concerns make their own demands, and many romances conclude with the assumption of political power and responsibility.

These are closural matters, however, for it is the predilection of romance to allow the lovers to reach their mutuality and maturity as autonomous persons independently of clan intervention. Feelings are therefore given priority over contracts in the constitution of the couple. The lovers in *Menaphon* are atypical in being older and already married. But in turning back to Heliodorus, we find the partners as adolescents just at that awkward moment when girls become vulnerable and boys become emotionally unstable, when they are no longer boys under the direction of older men, yet not accepted by them as equals. At this age, adolescent boys find themselves attracted to females when they least expect it, do silly heroic things to attract attention, and inaugurate the activities that proceed not only toward mate selection but toward recovering a place in male society. In *Menaphon*, Pleusidippus, Samela's abducted son, must make his way at court, avoid an imposed match, establish his autonomy as a warrior, exile himself, pursue the beautiful Samela in a misguided bride quest, while all along fulfilling the directives of his own emergent adolescent drives, supplied to him by subcortical urgings, before returning home to accept the devoted but scorned girl initially assigned to him. No special case need be made here to explain male adolescent volatility, hormones, emotional exile from the adult males of his society, the wanderlust mode, and the sudden awakening of sexuality. His life episodes form a little *Bildungsroman* in precisely the contours we expect, not only from the genre but from the genome. As with Theagenes in the *Aethiopica*, his emotions become the only instruments of instruction, and thus the stories that unfold are closely aligned with the archetypal. Pleusidippus will eventually find himself not only reintegrated into the world of male power but made the head of that world because his mother had, during the mother-child bonding period, given him the security and the values that made him a bold explorer of his world, a brave and intrepid youth, as well as a natural prince, for she had inculcated in him the values of his class. He embodies these two instincts: to find his breeding partner and to reintegrate himself into the male society of power and influence. He does both. Romance performs this double enterprise again and again, the wandering outward and the return to a home, whether his own or that of his love partner who shares with him her own right to rule.

Hence, there is a telling degree to which romance juxtaposes with the outbound quest for the beloved a nearly equal insistence upon returning to a homeland, whether to find refuge, family, origins, reconciliation, and the contextualizing of identity among one's own peoples, or to complete a dynastic adventure that includes the assumption of power and the displacement of an older generation. There is a sense in which the genre of romance is a vehicle of memory, even in its structural design, involving a place of origin to which the hero or heroine must return and there assert the legitimacy that comes through a consolidation of the family as an emerging generation. Such memories and desires become tantamount to nostalgia insofar as "home thoughts occupy dominant positions in memory and are accessed first at times of distress."[36]

There is, then, the sense of a second set of imperatives that contribute to the making of a genre, the values of mind associated with "nostos" which are, likewise, part of the ethology of the race – a biogram of human consciousness.[37] Often, during the questing portion of the tale, this awareness of a place of refuge and sanctuary is displaced by the preoccupations of the lovers with each other, while the design of their story holds a consciousness of home in abeyance as an order of closure. Home, for Frye, was a component of "deliverance" from eternal wandering and endless cruelty; it represents the recovery of a lost self by anchoring it in a severed past which constitutes the continuity between origins and the present.[38] Place becomes not only geography but also a condition of the self after exile and initiation. Stated otherwise, just as the psyche actualizes a programmed destiny in aligning the self with the beloved, it takes guidance from the emotionalized platforms of thought prioritizing the homeland and all that we cherish of "safety [and] habitat selection."[39] The adaptive value of knowing how to favour and find the way home need hardly be stressed; home is a place of reciprocal support and the sheltering of offspring, and thus nostalgia is a maladaptive trait only if it compromises the performance of the hero during his quest. The homing instinct, to the degree it represents eons of adaptive successes, becomes another salient mode for making environments intelligible.[40] Moreover, as a hedonic state linked to the natural world, through artistic representation nostalgia becomes a part of the conceptual universe of this literary genre.[41] Homecoming thus may be understood to express one of "the fundamental structural relations within the author's own world-picture or cognitive order."[42] If our reclaiming of places within communal structures is juxtaposed with the imperatives of mating, arguably as tropes embedded in the genome which designs the brain platforms responsible for these emergent states of consciousness, then biological truth is genre and mimetic genre is literary truth – a piece of nature generic to the species and of certain hermeneutic importance.

Literature, to be sure, tells stories of individuals with all the satisfying particulars of their lives and circumstances, but it is illogical to think that they are not equally bound to the inherited traits of the species, performing variations of the universals that define us. It is only to be expected that romance performs the values of our evolutionary history in its representations of courtship and the trials of constructing the bonds between partners, the search for the integrated self, and the longing for home, community, and sanctuary. Females see sex as a service granted to males under their right of control. Male sexual jealousy is more violent because males are less certain of the paternity of their children. Men are more ready to engage in sex as opportunity allows, while women are programmed to modify and control those urges. Older men are acceptable to women because their sperm quality is undiminished and the powers to provide enhanced. Younger women are preferred because their reproductive years are longer. Prospective mothers are the rarer commodity and therefore make greater demands in mate selection and, in commodifying their beauty, they are able to inspire aggressive competition, which they can use to their advantage. The greater the differences in investment in the care of children, the greater the dimorphism between the two sexes.

We are ultimately a gregarious species, accustomed to surviving through negotiations within small groups, attached as they will be to extended families, cities, and even states. Not surprisingly, we are emotionally driven to mating, now associated with exile and with the homecoming that brings the anxiety of alienation to a close, accompanied by a quality of quest, self-knowledge, and the passage into maturity and responsibility. This is a random list of tendencies and probabilities, justified according to the logic of a mate-selecting species. They achieve their universality in the mental dispositions which define the species, dispositions which, in turn, rely on the tilted predilections of our cerebral architecture. Tying literature to these insights is a matter of hierarchical reductionism to show how behavioural genetics can explain significant areas of cultural behaviour, actions, and impulses not traceable to learning but to cortical and subcortical productions.[43]

Because we have placed romance at the centre of this enquiry, attention is turned to those elements of a narrative type concerned with the pairing of lovers, the external and internal obstacles to their bonding, their geographical and emotional isolation in the world, and their return to social stability and deliverance from contingency. The present study asks whether the designed brain as profiled by evolutionary psychologists has any rivals in accounting not only for the universals of human nature but for their reflection through the continuity of literary genres – genres now inviting classification by something other than mythology and seasonal associations, deep as those motifs run in the literary consciousness. Insofar as fundamental erotic desire, imposed self-control, jealousy, fear, the preservation of chastity, feelings of honour, the attractions of social mutuality and sexual pleasure are now understood to arise from deep-seated information centres closely associated with the thought properties of vulnerability, paternity, exclusivity, and nurturing, we can begin to read the conventions and archetypes of literature according to a more scientifically honed sense of human nature and to formulate our hermeneutic insights around those values. To be sure, individuals and cultures provide the customs and temperamental variations of these *récits*, but the drives behind them originate in the innate design of a species.

Returning to the opening salvo about romance, realism, and truth, the emergent properties of a designed brain assume a "truth" quotient, not pertaining specifically to the mimetic depiction of local mores and customs, historical times and settings, or literary conventions, but to the representation of those universals that drive the action and reveal the generic psychologies of the characters in the pursuit of self-actualization in challenging environments. There can be no more compelling or veridical a story than that.

CHAPTER EIGHT

Suspense ....

Suspense is one of those workaday terms so integrated into the discussion of literature that definition would hardly seem necessary. It receives pro forma entries in most literary handbooks, but never provokes more than a statement of the self-evident: that it is a "state of uncertainty, anticipation and curiosity as to the outcome of a story or play, or any kind of narrative in verse or prose,"[1] that such anticipations arise "particularly as they affect a character for whom one has sympathy," and that plot types vary in ways that affect the ethos of suspense: those situations in which the outcome is uncertain and readers are concerned with how they will be resolved, and those in which the outcome is inevitable and readers, in their fear, concentrate merely on knowing when the catastrophe will be complete.[2] Indicatively, Roger Fowler's *A Dictionary of Modern Critical Terms* gives it a pass altogether.[3] But even these generic representations of the concept must venture such quizzical terms as "state" and "sympathy," both of which are seen to inhere not in texts and narratives, but in spectators and readers. Suspense, then, must have two sides: that which is invested in the design of the story as an emotion prompt and that which is a feature of mind.

This critical divide can be stated in many ways. One could say that evaluating the emotionality embedded in a text is an act of literary criticism, while the study of the emotionality aroused by literature belongs to cognitive psychology. These are the opposing points of critical departure familiar to students of aesthetic questions in general – in this particular case, that which authors know about organizing narratives to produce suspense, and that which readers "know" through the constitution of their brains about situations of alarm or disorientation involving themselves or others, and the compelling limbic reinforcement that impels them toward ends that will release them from incertitude or danger. The critical challenge is to decide whether a study of suspense should begin in narratology or psychology, for without the *récit* of events in time there are no stimuli, even though suspense as experience does not reside in texts but in brains.

Wolfgang Iser calls for a balanced approach to literary study in general that "lays full stress on the idea that, in considering a literary work, one must take into account not only the actual text, but also, in equal measure, the actions involved in responding to that text."[4] But that does not help in determining whether suspense should be defined according to a minimum prescription of narrative procedures and characteristics, or as a literary subset of a human, genetically conditioned response mechanism selectively engineered to meet a range of stressors in the environment. That same challenge originates in literary criticism as early as Aristotle's *Poetics*, where he discusses tragedy both as a controlled pattern of character and narrative plotting apt for arousing a predictable set of emotional responses, and as those emotions themselves generating a collectively felt response within audience communities: a structural formula and a mental state.

The first question in the creation of a rapprochement between these domains must deal with the very capacity for mind states to be emotionally moved by literary configurations, insofar as literary suspense borrows from the vocabulary of limbic arousal responses. The second question pertains to what readers are brought to feel in experiential terms about make-believe persons in make-believe situations of danger. Are these also make-believe emotions, for then it must be determined what an exclusively literary emotion might be. Subsequent questions must deal with the range of suspense arousing situations, whether they involve only protagonists, their desires and prospects, or whether suspense applies to any motivated pursuit of information deemed vital to mental composure. These distinctions and their relationships will prove critical.

Almost universally, in the limited number of critical studies that exist, literary suspense entails liked characters under duress whose futures are perilous and uncertain – futures about which readers hold strong preferences. Such a concern is clearly vital to the genesis of empathic-based suspense, but is it a subset among all the computational jags evoked by literature that include curiosity, problem-solving sequences, or pattern completion? Or do they represent two incompatible forms of mental absorption? Structural resolutions held in abeyance by incomplete data may be of a different order of cognition from the felt anxiety for an endangered person. This would pertain to the information shortfalls associated with jokes and riddles, the skill- or chance-determined outcomes of games, or the outcome of tricks targeting others leading to exposure or ridicule. Even the anticipation aroused by murder mysteries, which are principally about reading for clues and finding culprits, "with the cleaver, in the dining-room," would no longer constitute true suspense.

Yet, for reasons of parsimony, there are no grounds for imagining that the limbic system has distinct categories of attention arousal for games with undetermined outcomes concerning which spectators hold strong preferences and stories involving persons for whom the same criteria apply – namely, unspecified outcomes and preferences. Suspense involving characters involves incomplete social patterns, thereby conflating expressions of anxiety. Or not? What is clear is that characters in distress tend to epitomize literary suspense, but that literature also produces patterns craving completion, problem-solving situations, situational riddles, structural ambiguities, and much else that places the mind in a state of epistemic quest and anticipation. Rendering the matter even more perplexing is the prospect that suspense mechanisms are entirely hardwired, which is to say, the products of genetic coding and mental architecture whereby the mind is predisposed to respond only to those circumstances determined by the biases of inclusive fitness to constitute alarm states. By which criteria, then, should the term be defined? It would seem that if brains produce emergent mental states in accordance with design, then correlates will have to be drawn between how minds work and what literature means.

Suspense studies per se have emerged only in the last two decades, and largely among film specialists.[5] For them, not surprisingly, the most challenging questions pertain to the many visual as well as narrative devices whereby spectator anxiety can be incited to the maximum and sustained over long sequences. Shark attacks, car chases, the detected presence of aliens, helpless babysitters in monstrous houses threatened by menacing phone calls are not the only kinds of scenarios they have in mind, but they are, by and large, the kinds of hyperbolical expressions that epitomize the phenomenon and make for the best examples. These situations offer, in common, a protagonist representing a sympathetic point of view, or at least a morally sustainable point of view in the person of a tough guy upholding right against evil. Such protagonists are then placed in circumstances of diminishing chances for victory, survival, or escape. Plot, in these instances, amounts to little more than representing those diminishing options in a climate of supercharged emotion – principally fear. In this economy of events, the operative features are qualities of empathy, categories of social emotions such as fear, hope, or despair, but also, of necessity, mental computations concerning risks and prospects, and something like a moral arrangement of things approved and things disapproved that are tested in the action and selected by the outcome. This list circumscribes the principal features.

Noël Carroll made moral considerations an integral factor in the suspense formula. By "moral" he means simply that readers hold clear preferences for the things they desire in relation to the evils they fear. These values, likewise, may relate to our mental dispositions concerning survival and the binaries of thought that relegate options into clearly defined choices attached to felt approbation and disapprobation. Typically, those moral choices are tantamount to the potential benefits or losses facing sympathetic characters – those we deem to be most like ourselves, who most closely share our collective values, or who are the most reliable and trustworthy.[6] Nevertheless, the means and perimeters of empathy remain problematic. Although many cinematic suspense sequences are long, single episodes, allowing the attendant mind states no relief along the way, Dolf Zillman underlines the fact that suspense plots are not necessarily so, but often roller-coaster through episodes that raise and lower prospects, answering to one contingency while encountering another.[7]

Peter Vorderer builds on Zillmann's conception of suspense based on "empathic distress" felt for characters in relation to their perceived degrees of danger.[8] He goes on to examine the differences among texts regarding strategies for provoking arousal, the variations among readers, the nature of suspense situations, the emotions involved, and the "perspective taking" that accompanies these emotions. Suspense must also involve "problem-solving" to some degree, and thus some computational component synchronized with the events provoking the arousal, although Keith Oatley reveals the potential conflict between critical thinking and the felt qualities of fictional experience, as outlined below. That problem will persist. Equally troubling to the classic definition of empathetic protagonists facing good and bad alternatives is Vorderer's concession that empathy is not the only criterion for suspense, insofar as we can treat characters virtually as moral categories – agents within a struggle for values – or generate feelings around them through relived memories of our own seen in parallel to the fictive situation. That is to say, suspense has to do with resolving injustices for the societies affected by human agency and with the degree to which we interject our own sense of well-being into the outcomes for fictive worlds. In all instances, it comes down to aligning representational stimuli with the response mechanisms to environmental changes embedded in human nature – what happens in social simulations that arouse the limbic system. This, in turn, is related to our access to information in relation to unknown outcomes, thereby setting up a new crux, the so-called "suspense paradox." If this limbic response normally requires uncertainty – unknown outcomes – nevertheless there are stories as stimuli that can produce a complete state of suspense after multiple readings. You can know the ending and still feel concern over the well-being of the protagonist. The anomaly matters if, as some have thought, it touches on the essence of the entire phenomenon. Answers to this puzzle have been ingenious but not entirely convincing – a problem to which we will return.

Some of these critics write with an awareness of the cognitive dimensions: that what happens in art happens in minds (which also happens in brains); and that human emotions run on their own genetically prepared programs built into the phenotype. Few can imagine the limbic production of this mental state without persons, an empathic bond, an uncertain future, and something like a relatively simple alternative between options deemed "good" and options deemed "bad." Because feelings run in apparent binaries of good and bad, they must correlate to clearly defined circumstances, like the good guys and the bad guys, the cops and the robbers, or their moral equivalents. Stories built around ambiguous motives and moralities raise uncomfortable problems of their own; are they antithetical to suspense? The collective critical project has been to arrive at a consensus concerning suspense as a feature of the dramatic involving human agency and social conflict, or human agency in a threatening natural world, and the means whereby spectators and readers come to "feel with" these characters. There is a need to dispel the challenge launched by Aristotle in the *Poetics* pertaining to a split response often translated as "pity" and "fear," the former because we care about characters' well-being insofar as they are like us, and the latter because we also identify our interests with theirs and thus internalize their dilemmas. Only infrequently are these critics troubled by the quality of suspense that arises independently of some form of empathy in relation to a "moral" order of desire. If other qualities of stimuli leading to suspense were to come into play, the classic definition based on empathy for likeable persons would require serious modification.

Keith Oatley pressed toward a more comprehensive analysis by outlining a hierarchy of attentions apt to arouse suspense. He divides them between those outside the narrative – aesthetic and structural evaluations thought about independently of the experienced representation – and those inside the narrative, directly pertaining to the events and characters. This is a "have your cake and eat it too" arrangement that allows for suspensefulness based on anticipations relating to structures and their completions, but now separated in kind from the suspense that arises in emotionally saturated narratives about persons in dilemmas. His ultimate purpose is to make a nearly complete identification with endangered persons the quintessential precondition for arousal. In his survey of the emotions evoked by reading, he passes through what he calls "emotions experienced from outside the membrane of the text," by which he means all the ways in which readers feel tension over the strategies and designs of literary works. "A writer can invoke a schema and appeal to the reader's curiosity," for such schemata provide a state of incompleteness that, in turn, incites a need for completion and relief.[9] In this discussion he is concerned with how authors foreground art, pattern, strangeness, and distancing, "dishabituation," discrepancy, and purely aesthetic challenges. Such an approach begins to point once again to social emotions and artificial emotions, those we feel for people and those that pertain to design and artifice.

Again, Oatley's concern is with those "inside the membrane of the narrative world" through which readers relive emotional memories, empathize with characters, and ultimately become one with them in a common experience. The "membrane" metaphor, derived from Erving Goffman's book *Encounters: Two Studies in the Sociology of Interaction*, extends his inside-outside analysis of social relations to strategies of reading, separating the computational operations associated with incomplete structures from the felt qualities of experience arising with imperiled protagonists. This resolves several kinds of problems, all of them pertaining to the conflict between critical and computational thinking and direct, absorbed emotional involvement with characters and their plights. The successful identification with characters through the adoption of their goals and emotions as our own is tantamount to the complete abandonment of all those meta-conscious evaluative activities that cannot participate in the experiences of the transmuted self.

In this article, Oately is, to be sure, aware of the polemic concerning identification with characters, but downplays the distinction to be made between identification and empathy, the latter requiring that we have feelings of "well-wishing" for persons in dire situations without completely blending our own beliefs, desires, and assessments with theirs. Moreover, what readers know and what characters know and feel is rarely synonymous, which raises problems of a different order. But the crux to be resolved creeps back into consideration because there can be no meaningful understanding of suspenseful situations without those fundamental computations that determine risks and probabilities, often in terms of patterns and forms. The suspense felt for characters, after all, involves situational computation and a logical calibration of risks. Even the simple fact of knowing the imminent dangers to characters before they do means that the mind states of protagonists and readers are in differentiated problem-solving modes.[10] This factor, alone, works against the notion that readers do not indulge in abstruse reflections while maintaining an affective involvement in the story. Readers cannot appreciate the levels and qualities of danger, calibrate the future in relation to the past, or estimate probabilities concerning that future in light of the status quo without a sophisticated degree of computational thinking – within the fictional envelope. Granting even a little activity of this kind, however, tends to dissolve the "membrane." To state the issue more directly, there is less certainty than imagined concerning the roles of empathy or identification in suspense as pure emotional states versus suspense as a state of intense, quasi-subliminal calibration of odds and probabilities around social – or non-social – situations. Oatley's identification theory barricades itself from broader theories that equate suspense with a generic limbic response that measures the distance, through computation, between the status quo of nearly any dynamic structure and its undetermined outcome. Suspense at first seems to be merely a felt quality of experience, but may prove to be, even principally, a complex form of strategic computation.

Turning from the textual to the psychological, an additional set of issues pertains to the nature of suspense as a mind state, namely, why we experience it, and whether literary suspense is itself the representation of a make-believe emotion in a make-believe setting, a real emotion in a make-believe setting, or a real emotion in an equivalent-to-reality setting. Only by making a speculative detour concerning the nature of imagined-representations-as-real can anything be resolved about the phenomenon of suspense as a limbic reading of mentally reconstructed environments both perceptual and imagined. As a premise, if not an axiom, it would appear that literature creates images of sufficient experiential reality that the human systems of attention and emotion both accept and treat them as real, even though the rational faculties recognize them as fictions. Will it prove that we actually have two brains that perceive reality in different terms, that what we invoke in the imagination can be submitted to the feeling brain as urgent and pressing? That interface may be one of our most defining characteristics as a species. But if it is so, it has perplexed our theories of mind.

One of the most challenging studies to this approach is Kendall Walton's *Mimesis as Make-Believe: On the Foundations of the Representational Arts*, precisely because it is predicated on the ostensibly commonsense belief that fiction is always fiction, no matter how it moves its readers. In a very vital sense he must be right, for any profound confusion between the two – taking fiction for reality – is tantamount to treating hallucination as perception.[11] If ever there had been a gene mutation allowing for that confusion, there is little chance that its bearers could have long survived. But the argument leads to the concomitant belief that if fiction is make-believe in all of its components, then the responses to fiction must be part of that same make-belief. Even the attendant emotions would be merely simulated for the sake of the fiction, or, in Walton's words, "a person's actual moods are simply carried over into the fictional world."[12] They are then proven fictional because the story has power to change them without reference to real moods.

These arguments are of particular value to Walton in providing answers to aesthetic mind-teasers such as the "paradox of tragedy": that in art readers may find pleasure in sorrows that would otherwise prove intolerable to them. The answer lies in the fact that those sorrows are themselves merely fictional and aesthetic.[13] But can there be a parallel limbic system that produces real tears on a purely fictional basis, and what would have been the pressures in ancestral environments to have produced such a system? In the same vein he resolves the "paradox of suspense": that readers can find the same levels of tension in rereading works with known endings, again because the emotions are fictive or aesthetic. The non-reality status of the story extends to a non-empirical knowledge of endings, so that stories can be resolved again and again. But the question to be answered is whether the nature of fiction deprioritizes certain information in a way that allows us to know it, yet search for that same knowledge on multiple occasions with the full limbic reinforcement of a first-time cognitive lack. That would seem entirely doubtful. We read in relation to knowledge, including knowledge from the real world. Is it that readers become so immersed in stories that they forget what they know outside of the story regarding where knowledge of its ending has been deposited? Is this a form of aesthetic amnesia? Do stories re-establish such doubts as part of their "world?" Or is it that the knowledge of endings is not outside the fiction but a part of the fiction? Is fiction like the historical imagination that allows us to place ourselves in the past on a contingency and uncertainty basis as if the future were unknown? Richard Gerrig's solution lies within this configuration of ideas.[14]

But if the cracks in the theory have not already become apparent, they should become so now, for can memory also become fictional? Can we also merely make-believe remember that both Romeo and Juliet perish in Shakespeare's play, thereby allowing the ending of the play to remain eternally in question? Is the representation of fiction so categorically different in all respects, apart from the fact that it belongs to the imagination rather than to the perceptual world, that it requires an independent set of computational and emotional mind features? Does learning become permanently ephemeral because eternally fictional? This cannot be right, for so constituted, the human imagination would have no inclusive fitness value, no capacity to have its representations tested by the perceptual faculties through access to the computations that locate us in the physical and social world, or to the fundamental emotions that give urgency and direction to consciousness. There would have been no adaptational reason to confirm the imaginative faculty among the phylogenetic traits of human nature. No theory of selective amnesia or of fictional emotions can account for the suspenseful pleasure of second readings.

Richard Gerrig examines this problem under the heading of the "toggle theory of fiction."[15] By this he means the switch that readers throw when entering the fictive world mode. It invokes, out of context, Coleridge's now famous phrase concerning "the willing suspension of disbelief," taking it for another version of the toggle switch. But in fact readers do not consciously suspend; they actively believe by treating the equivalent-to-reality of fictive representations as reality, through which all the faculties of mind pertaining to the world of percept and extension are engaged. Gerrig's arguments against the "toggle" fallacy are based on the simple notion that humans cannot will themselves into an alternate reality. They can merely subject themselves to the contents of the imagination, inwardly directed, or textually directed, and allow the diverse faculties of the brain to perform their essential interpretations based on intuitions, subliminal knowledge, and emotions that are virtually untouchable by aesthetic predilections or cultural agendas. By reasons of parsimony as well as demonstration, these cognitive-limbic modes must be the same as those that orient persons in the real world.

Among them is the capacity for suspense, one that must have had a parallel function for our primitive forebears, presumably as an attention enhancement mechanism in the presence of danger. This excitatory mind state performs in conjunction with a limited access consciousness, capable of remarkable feats of concentration, but on a competitive one-thought-at-a-time basis. Suspense arises within a particular configuration of perceptions in consciousness still to be defined in its broadest sense, and prevails for as long as that configuration is a volatile sequence of events in time, uncertain in its issue, and of great importance to the observer. To be sure, one such quality of importance is concern for the well-being of others, likewise on a basis yet to be determined, but by no means the only concern. In short, brains wired to care for others have been adaptive because our survival has been determined for eons of evolutionary time by cooperation and group dynamics. The modality of suspense appears to quicken two kinds of responses: limbic fear and anxiety over the uncertain destinies of those deemed group members, as determined by chance or by planning; and rapid, essentially subliminal computations regarding options, risks, and probabilities. Regarding literary suspense, these affects and computations experienced in real time are performed in precisely the same way, but now in relation to the equivalent-to-reality representations produced in the imagination. The superimposition of brain parts with their contrasting emergent properties has given us our world in these terms.

Imaginative representations may be textually driven – they are the kind we are ultimately concerned with – but not necessarily. The imaginative faculty also has access to memories, nightmares, daydreams, hallucinations, and, in a far more pertinent sense, all forms of provisional scenarios of the future from which we can select subsequent courses of action, subjecting them in advance to the emotions as well as to reasoning – intuition as well as logic – allowing us the better to follow our "hunches" on an equivalent-to-reality basis. This feature of mind is perhaps the most important adaptive and strategic trait of the species. Suspense, in a sense, is the emotional component of survival-related provisional thinking. But fundamentally we are concerned here, not with memories replayed in narrative fashion with all their limbic baggage attached, but with scripted signs and their capacity to evoke social worlds, together with their attendant emotions.

Through this evolutionary-psychological approach to suspense, a few new axioms begin to appear. They pertain to issues worthy of book-length investigation that here must be offered with hypothetical brevity. Mind architecture, it can be said, has made the human organism survival-efficient because it provides an adequate response and arousal system for the exigencies to be met in the environment. Authors provide representations of environments in which those same response mechanisms can be called upon, and in the same degree to which they are aroused by perceptual dangers and challenges. These recognitions take place at relatively subliminal or subconscious levels. The mechanisms whereby humans build up representations of persons and places, of motivations and intentions, of dangers and options, lack all access to ideological and cultural input; they are reactions according to the phylogenetic properties of mind. For Peter Ohler and Gerhild Nieding, "the set of cognitive operations which is responsible for the construction and modification of mental models remains quite constant," despite differences in emphasis and detail.[16] Or to state the concept in the words of John Tooby and Leda Cosmides, "the design of the human psychological architecture structures the nature of the social interactions humans can enter into, as well as the selectively contagious transmission of representations between individuals" so that much of what we know is what the species knows in the form of an "encompassing functional superstructure of virtually universal, complexly articulated, adaptively organized developmental, physiological, and psychological mechanisms, resting on a universally shared genetic basis."[17] In this theatre of subliminally constructed representations, suspense becomes an automatic barometer of attention priority in relation to perceived stressors, accompanied by the many, mostly subliminal, calculations of risks and probabilities that arise in the presence of danger – from loss of reputation to an attacking tiger. At this point we are nearly drawn into the tautology that suspense is what suspense does, that it defines the world by its autonomic activation.

To turn a narrative into a representation entails making a running mental model of as many of the relations and descriptions within the text as the mind can seize upon. The final product is the passage through consciousness, on a selective and sequential basis, of an equivalent-to-reality representation that takes advantage of the same memory analogies and fill-in processes that pertain to the representations of perceptions in real time. Humans are constituted to make their own imaginary representations as much like the real world as possible. Only then do they attain the critical mass that makes them useful as social models. The mind participates in these constructions both computationally and emotionally. This process requires a vast amount of subliminal "reasoning," the processes of which can be intuited only in the nature of the finished product. Narrative representations emerge in "filled-out" fashion in relation to memories, what is known by analogy with the perceptual world, what is minimally necessary to the category of personhood, along with evaluations conducted to establish liking and disliking – in short, the many fusions of narrative signs with categorical imperatives that result in an adequate setting and a working social environment constituting the ontology of the fictive world.

With regard to narrative, each constituent moment is a status quo that is reduced to a memory context for all succeeding events. Subsequent events in turn become part of the memory schemata in which the past is constantly epitomized, thereafter serving as incrementally emerging contexts for each new present. With regard to this stream of episodes, the mind is teleologically oriented because it is equipped to meet present challenges through the projection of multiple provisional futures. Stories that incite such projections on behalf of characters, in relation to structures or to information gaps, generate suspense, which is the emotional quotient of future prospects calibrated against the current and evolving status quo. Suspense is therefore always partly emotional as a mind state, and partly computational, whether subliminally or consciously. Because of this flow of events in relation to a future that is computed according to desired and feared outcomes, the complete representation is perceived in quasi-structural terms. There is an emerging shape to events that is marked by the rise and fall of suspense in sympathetic and parasympathetic terms. When one shoe drops in the room above, we wait in expectation for the second because humans, generically, have two feet. Pattern awareness and pattern completion are primordial.

Eric Rabkin broaches many of these topics in *Narrative Suspense*, for he too detects a need for "the possibility of multiple operative fields working in the text while the reader is consciously aware of but one," touching on the complex contribution of meta-conscious operations in the construction of "fictional" representations. Elsewhere he observes that "subliminal suspense is integral to the experience of any narrative which we read voluntarily."[18] In this latter statement he implies that suspense is at work in far more contexts than in the classic twist of a doorknob by an undisclosed person on the other side of the door. The challenge remains to determine how many different kinds of narrative prompts play on the suspense-arousal mechanism.

Before turning again to the matter of suspense in relation to liked characters and the difficult question of empathy, there are two preliminary matters requiring amplification – the computation of probability in relation to endings, and the projected completion of forms. Both are apparent components of suspense as an affect-computational phenomenon. They are closely linked. Noël Carroll builds a generic model of the suspense narrative in terms of questions asked situationally or episodically, followed by their answering episodes. His concern is that mere anticipation is not enough; suspense entails options, justifying his claim that an "appreciation of relative probabilities is at the heart of suspense."[19] True story narratives make us ask, will the stepmother be contained, or will the robbers manage to break into the vault before the police arrive? Both anticipate future resolution; both invite second-guessing on the part of readers in relation to what they know of the circumstances, characters' aptitudes, human nature in general, or of the genre of the work in hand. These calculations may be entirely subconscious and not clearly formulated. But their necessity is certain, given that suspense has no phenomenological basis before some feature of mind has calibrated both the nature of the danger and the probabilities of escaping it. If the prospects of deliverance are high, the suspense is low and vice versa. If the dog between the protagonist and the pathway is just a little one, standing still and not barking, we sense only slight volatility in the situation and low stakes. Yet this measurement must be taken in some computational sense before the level of suspense can be established.

This is a form of cognitive mapping, which is a basic drive. It is the source of deep attention in situations of instability in search of homeostasis in future events. Suspense measures and establishes the intensity of that attention. Moreover, it is a feature of mind not only to look for best explanations of events and compute probable effects but also to achieve this in the form of provisional models of the future. Readers attempt to know endings before they arrive, just as in the perceptual world potential victims read over in the imagination as many drafts of the future as possible. With stories, readers may be passive, perhaps, and trust entirely to the writer to lead them to the conclusion held in secret. But to a minimal degree, they must be proactive in an orientational sense – constructing the fictive "world" and keeping track of characters and their intentions – and they must sense options as a condition for anticipation. Without this level of participation, there can be no sense of story. Moreover, insofar as suspense establishes a mental state pertaining to a present situation evaluated according to optional outcomes, there is a distinct sense of sequence as a temporal design seeking its completion. This is basic to discussions of narrative, but the logic of completion-design must be equally hardwired in the mind.

There is a particular quality of emotional involvement with characters to which we now return, one that allows readers to feel suspense on their behalves when events conspire to bring characters unhappiness or worse. Gerrig confirms that "to a large extent, a theory of suspense must include within it a theory of empathy."[20] The word stands for the transfer of emotional interest in the self to a similar interest in others deemed worthy. We do not feel it for villains, fools, and malefactors, even though they may be made to suffer horribly for the enormities they perpetrate. But how and why humans have the capacity to experience emotions pertaining to situations in which they are not directly involved remains a tease for behavioural and cognitive psychologists. Theories abound.

The nature versus nurture claims re-emerge. Is empathy learned by example or is it one of the predispositions instilled through inherited mental architecture? Cultural constructionists hold tenaciously to the former, but their theories have been badly damaged by the reasoning of evolutionary psychologists. Our genes may dictate to us a concern for others as one of our most effective self-enhancement strategies achieved through the dynamics of groups. The case has been made for an actual "altruism gene," given the strength and universality of the disposition to cooperate with others on a small-cost-to-self basis in anticipation of reciprocal support in the future should circumstances necessitate. Robert Trivers launched the debate in his 1971 article "The Evolution of Reciprocal Altruism," in which he shows how this ostensibly self-defeating adaptation actually serves the best interests of all the members living in groups.[21] But altruism remains self-oriented even though it benefits others in the process, while empathy appears to be more generous in the absence of all expectation of reciprocity or reward, particularly in the contexts of literature.

Dennis Krebs attempted to clarify their relationship: shared genes are the biological part of altruism, while empathy is the psychological part.[22] Perhaps. Another approach pertains to the mimicry response. Dolf Zillman allows that "primitive motor mimicry ... may be thought of as an evolutionary residue that respondents bring to the theatre and that dramatists can exploit for the creation of involvement and affect."[23]

But the paradox of mimicry is that we need not like the character whose physical pain makes us jerk back our head or pull back our hand in parallel response. Even seeing the hand of a notorious malefactor chopped off may provoke the reaction. Empathy, inversely, requires appraisal and approbation across a potentially large spectrum of traits. Is it then likely that empathy is an extension of the mimicry instinct? The space between is large.

The "science" of empathy is difficult to demonstrate because it is difficult to imagine what empathy must have been in its most primitive phases, and why it was selectively affirmed in the genome. Arguably, not only humans but primates of all stamps (the categories could be extended) make use of their limbic prompts to know who to fight for and who to fight against. That alone constitutes a form of "feeling for," which might provide the basis for socialized expansion based on the mutual interests among family members, extended outward to community members who merit trust, support, and protection according to the principle of the advancement of members of a common gene-community, or the principle of reciprocity in a more general sense. Patricia Churchland endorses this principle in *Braintrust*. Her basis for a theory of empathy is not in simulation (mirror mechanisms), largely because what we feel does not depend on what the other is feeling, but in who that person is to us, and why their pain or anxiety is associated with us in terms of who we are. Empathy is interested in its orientation. Her entire approach to the economies of the brain are predicated on the hierarchies of caring.[24]

But with final answers still pending, an account of this escape from the solipsistic envelope may have to fall back on a "folk" approach: that empathy exists by dint of our first-hand knowledge of our own immediate experiences. That humans sometimes cry over the losses of others, feel pity for tragic protagonists, and worry on behalf of close kin in trouble is incontrovertible. Adam Smith recognized this need in humans as the basis for any form of moral society, namely, a deeply engrained interest in the fortune of others, which makes their happiness necessary to us.[25] How deeply engrained, specialists are only beginning to assess. Martin Hoffmann opines that "humans may be built in such a way that what happens to others is at times as motivationally significant as what happens to themselves."[26] Joseph Carroll makes a case for cooperation over combative individualism, given that, "as Darwin understood, it is the very nature of social animals to have sympathy for their fellows. Among human beings, the union of sympathy with the higher mental faculties results in the capacity to exercise imaginative sympathy for experience that is not absolutely identical with their own."[27] There is no shortage of these definitions by fiat. According to identity theory, we do so because we make our goals and emotions synonymous with others. According to empathy theory, we do so because we evaluate their circumstances and feel concern over their prospects. It is an emotion based on reciprocity to the extent we are anxious to reveal it to others as proof of our loyalty and sensitivity to the group. Steven Pinker offers an engaging meditation on the problem of altruism as it relates to empathy. "Many people still resist the idea that the moral emotions are designed by natural selection to further the long-term interests of individuals and ultimately their genes." In this vein, he speculates on the sincerity of past acts of generosity, and the degree to which they were selflessly group motivated and egotistically projected on the group in ways that would ultimately benefit the self. As he states, the first who expressed generosity may have prospered because it made it worthwhile for their neighbours to cooperate with them.[28] But again, empathy wells up where it cannot be seen by others. Perhaps it is a form of self-projection, given our reliance on our own experiences for estimating those of others. That is different from the notion that we feel empathy for those who are most like ourselves, for how, then, could empathy be felt for the heroine of a thirteenth-century Japanese novel who is distanced from the reader by time, culture, and perhaps gender, who in fact never existed (equivalent-to-reality representation aside), and who has no reciprocal capacities? For the creature of the "selfish genes," empathy remains an anomaly. Yet it is recognized as a strong social trait, liberated even more by the distancing of fictional representations, where there are fewer grounds for rivalry and jealousy. Empathy in fiction comes easily, not only because we are anxious to hope for and pity others, but because we are anxious to cooperate with authors in favouring those they seek to establish in our favour through the management of the point of view.

How are levels of "feeling-with" to be sociologically determined, and can this determination be employed as a guide to the increase of empathy that accompanies the amplification of suspense?[29] One direction out of the empathy conundrum is to relinquish somewhat the idea of intense "feeling-for" as a glut of undirected hope and return to the computational dimensions of emotions. Empathy, like suspense, entails a calibration of circumstances. The protagonist must first be understood in terms of goals, for readers are "happy for" or "sorry for" characters not only as their prospects for achieving those goals increase or decrease but also by how much they believe these characters merit rewards and the avoidance of loss.[30] Empathy may not be so much the search for kin, or an alter ego, as it is the expression of an economy of equitability and merit. This was touched on before. Accordingly, the "prospect-based emotions" of hope and fear grant limbic support to the pursuit of approved goals in a climate of risk. Empathy remains a category of emotion aroused on behalf of others, including those presented in imaginative representations. But it functions "situationally" in a rich economy of related emotions pegged to dilemmas and their outcomes: not only hope and fear, but relief, jubilation, or disappointment.

The empathic bond, moreover, means a preliminary evaluation of persons, for in narrative relations, there are not only those who find approbation but those who are disliked and blamed. Empathy thereby represents a moralized point of view, a separation of the sympathetic from the antipathetic. If suspense depends on the degree of liking as well as the perceived levels of vulnerability and danger, together with the strategic diminution of prospects by authorial design, then the preliminary computations pertaining to suspense grow ever more complex. There must be a centre of emotional interest vested in a person, whether for being deemed most like us, pathetically helpless, admirable in cunning, morally upright and trustworthy, relatively less sinning than sinned against, the only person in the action with an interesting ambition, or the only human in a hostile landscape – a seemingly infinite variety of interests and attachments that offer to constitute the point of view for the reader.

Nevertheless, we are undeterred, for in these matters we take ourselves for experts. The evaluation of agents is an activity of daily occurrence. In Shakespeare's *All's Well That Ends Well*, two characters in a macroplot designed on the romance model are destined to mate. But one sets impossible conditions and attempts to flee while the other plays devious tricks to meet those conditions in order to gain a man who doesn't want her. Thus no character enjoys the empathic centre of the play, although readers feel compelled to establish that centre in one character or the other before the action is complete in order to know how to feel about the ending. The process is a form of scorekeeping in matters of trustworthiness, transparency, loyalty, devotion, rights, merits, class, autonomy, and freedom of choice in an effort to resolve the compelling need to wish well to one party and to blame the other. Humans are intuitively skilled and adept at making these computations, and can register merits and demerits to a refined degree. These capacities play a key role in determining persons worthy of empathy in their plights and enterprises.

The former discussion has accepted at face value the notion that where there is literary suspense there are characters who serve as the centre of concern through empathy, without which readers could have no emotionalized concern for their destinies. Readers are desperate to find that centre because they are eager to locate their own moral vision regarding the action. Because they are not in the story, yet live it as a representation of the real, empathy is a means for establishing a reading perspective through a person who most represents their values. It is part of the fictional experience, much as choosing a team is part of the experience in spectator sports. The analogy is incomplete, because sports teams may be chosen arbitrarily if both deserve to win, whereas, by their qualities of conduct, characters establish hierarchies of merit. Perhaps we settle into a character-centred point of view in hoping for that person what we hope for ourselves: prestige, reputation, self-esteem, health, success in finding a sexual partner, friendship, protection for offspring, fair-play in all things according to a personal understanding of that economy, giving and receiving aid through mutual interest groups, building and realizing personal goals. By the same token, we may wish many things for others in accordance with their worthiness: avoidance of pain or untimely death, prosperity and recognition, successful careers, happy families, or inversely the defeat of rivals, social cheaters, scrooges, and hypocrites. All these suppositions accept that without some centre of empathy, the story would lose its powers of absorption.

But the antithesis can be broached in a number of ways. Marlowe puts the empathic bond up for negotiation in *The Jew of Malta* by inviting readers and spectators to see in Barabas initially a man more victimized than victimizing, but ultimately a man whose inhumanity is too patent to allow for an ounce of sympathy. Without the erstwhile protagonist, however, the play has no alternate human centres of empathy. (Much of the play's experience may lie in this strategic degradation of the hero.) Yet, insofar as the outcome of the play remains in doubt till the very end, there is an inevitable degree of suspense. Marlowe's play is a reminder that dramatic narratives may retain a high level of attention, even where empathy has been displaced or eliminated altogether.

It is time, then, to give more weight to statements like the following: "the most frequent problem-solving structure for suspense [involves] circumstances in which the readers' initial state is ignorance and their goal state is enlightenment."[31] The formula hints at cognitive jags and knowledge lacunae of many kinds apt to generate suspense; they may be marshalled under the heading "the anxiety of ignorance." Eric Rabkin recognized that, in a more generic sense, suspense pertains to artificial structures in search of their endings.[32] Within this category, he drops back as far as metaphor and sentence completion. There is a vested attention involved even in such entry-level formulations as "He is a … " We expect a metaphor to fit the context, such as mouse, or lion, or gem. Not to accept this imperative "is to lose the ability to read."[33] Inset stories develop structural and thematic intentionalities that must be met by the reader as problem-solving situations in the alignment of epitomes and meanings among narratives. Even metaphors, mapped so quickly as to seem instantaneous, nevertheless require the structuring of little plots somewhere in the mind that map the two elements of the figure and draw them into meaning. This is to start rather far back where suspense is concerned, but this approach is a reminder that strategizing, computational orientation, and problem-solving may be the larger part of the suspense response, and that the anxiety over probabilities, survival plans, and alternative futures are quintessential to narrative suspense. There would seem to be no categorical difference between the modest anxiety displayed in awaiting the completion of literary *formes simples*, the joke, the riddle, the trick, and the potentially greater and more socialized anxiety of awaiting expectantly the outcome of situations threatened by imminent disaster. Both are forms of epistemic or information lacks, both have access to limbic responses, and both establish levels of emotionalized attention, if not complete absorption. There is the particular divide between the stress of self-orientation in incomplete forms, and the stress over concern for others within the narrative forms to which their desires and decisions make a contribution. Moreover, for all the reasons set out above, this displacement of the emotional centre from self to other relies on a common set of emotional and computational mechanisms. Most of the critics cited have confined the meaning of the term to persons granted empathy, and moral choices, but there is also the suspense that attaches itself to forms, situations, and non-moral criteria. The balance of this essay is concerned with identifying a few of the literary creations that explore the liminal areas between the two.

Let me start with the "banana-gun game," a make-believe sequence at the threshold of what may be adapted into a simple literary model. My two children are at the breakfast table, back when they are just little kids. I stuff a banana in my pocket and chant in as suspenseful a voice as I can the "da-duh, da-duh, da-duh-da-duh-da-duh" that invariably precedes the draw-and-shoot routine. Do not ask me why they loved this so much, and over and over again, with little diminution of interest or of nervous response and release. (Is this another example of the "paradox of suspense," because the routine was unchanged from episode to episode, and yet anticipation levels ran unabatedly high? Or is it a case of emotional contagion spreading from episode to episode?) The banana-gun game manifests all the ingredients of suspense as a psychomotor response in a fictionalized setting of stress and release. They are not the good guys and me the bad, but it is a game of aggression. The choice of pulling or not pulling the banana remained an option, even though the routine was fixed. Thus it is a game of putative threat – like tickling or peek-a-boo – and escape. The plot moves through a stress field from the strange to the reassuring in laughter or shouts, which can be "read" for as many times as this stress-release sequence brings pleasure or interest. That duration seems to be measured by the thresholds of the limbic system: attention heightened in the presence of danger offering virtually no escape except in the leap from menace to play form. I'm good old dad; this is just a banana. Or, this is the same routine, and we're not dead yet. Or "bah-da" is the invitation to get scared. This is play, but again the emotions are real. Dare it be asked what kind of equivalent-to-reality children can present to themselves in these circumstances in order to hike up the limbic response? As a little plot in search of an ending, this game suggests the generic link that exists between all interest-compelling, incomplete narratives and their capacity to trick an architecturally determined attention response system to come out and play for real.

Literature strategizes in many ways to gain access to emotionalized attention, not only by posing dangers to favoured characters but by disorienting readers, withholding information, most of which become sequences in relation to unknown outcomes. Tragedies allow for empathy and hope, although the general tenors of the endings are known. (Why we should read to our empathic misery is a question all its own.) Intrigue plots have unknown endings, but often no morally preferable characters or positions. Yet these and other forms have in common not only a capacity to hijack consciousness for the time required by the narrative to complete itself in reading or on stage but to rivet attention emotionally to the degree the terms of that story can be brought to matter to the well-being of the reader. The banana-gun game identifies and activates the generic response system involved. The question is how broadly that system is engineered to read reality.

There is little reason to linger over the suspense associated with jokes, riddles, and games. These are efficient instruments of suspense, particularly where stakes are placed on combative skills or where intellectual and decoding abilities become the measure of personal prowess and esteem. The trick, however, falls into the gray zone in which suspense and the social narrative are once again clearly linked, but with revealing departures from the classical, generic, formula of the film critics. The trick is a suspense structure in its own right, typically consisting of an elaborate provisional idea that will, through its successful execution, alter the social circumstances for the planner and perpetrator, or for others on whose behalves the trick is deployed, as well as the victims. Let those "others" be the helpless lovers facing eternal separation if they do not manage to contain the wishes of an opposing parent, and let the trickster be a witty servant willing to risk his job security by devising an elaborate scheme whereby the blocking character can be compromised, embarrassed, or reconciled to the match through deception. Several things happen when this tactic is employed. The provisional thinking on behalf of the lovers is transferred to a secondary agent within the action who invariably proves better at the job of strategic scheming than either the lovers or we, the spectators, would have been. In a larger sense, attention is transferred in general from the macro-plot of the lovers and their play-long desire to overcome obstacles to their union, to a series of micro-plots constituted of the machinations set in motion by the trickster-as-agent. Empathy drops out of the formula, although it may linger, on hold, as it were. Suspense is now displaced to a plot-within-the-plot involving strategies of craft and expediency to achieve anticipated ends.

Such plots become entirely computational in relation to a correct analysis of the target victims, their propensities and vulnerabilities, and the probability of achieving the expected ends in light of the potential punitive repercussions. Needless to say, the effective writer of these trickster routines will toss in a number of contingencies along the way, testing the intriguer-agent's ready versatility, emotional self-management, and insights into human nature. Such tricks are teleologically oriented, defined by estimates and scenarios created around the future, emotionally calibrated in terms of risks, and empathic to the degree that we wish the trickster well if only on behalf of those he serves, and provided that the targeted persons deserve what is likely to come to them. Nevertheless, such tricks are often conducted in an amoral climate of ends justifying means, wherein suspense belongs only to the completion of forms with their morally indifferent social repercussions. They appeal by their capacity to make play into a form of social reality. Only a pale form of suspense can be resurrected over the uncertain plight of the lovers and their formulaic deliverance into happiness at the end, one that has been placed in the background throughout most of the action.

The Jonsonian world order likewise proposes characters seeking a variety of goals, few of whom have claims on our empathy and approbation on moral or likeable grounds. One may see the action from the perspective of the knaves, themselves tricksters of the confidence game variety, let us say, with clever goals and strategies for gaining illegal money. Or one may follow the action from one victim to the next – albeit none merits compassion or well wishing. Quite simply, empathy for either rogues or gulls will prove misplaced. Without moral protagonists as the objects for well-wishing, suspense must inhere in other features. Micro- and macro-trickery are the game structures that determine fortunes. Jonsonian suspense is about information management and its gradual dissemination. Security in social situations depends on an up-to-date critical mass of comprehension concerning customs, power echelons, protocol, the identity of persons, their functions and states of mind. Even though readers are made party to the true identity of characters in disguise, they feel stress on behalf of those within the drama who labour in error. Much of the expectation horizon for these plays is vested in just when and how the *anagnorisis* or discovery scene will emerge, and what the effects will be on the company gathered when the proper identity of those in disguise is made manifest. *Peripeteia* sometimes names the salient climactic moment when frauds and dissemblers are exposed, punished, or ridiculed, with all the suspenseful interest otherwise invested in liked individuals. *Measure for Measure* is such another work. Just who holds the empathic centre of this play is open to debate, but there is situational suspense in awaiting the Duke's return, given all he knows about the society he is reticent to judge or to excuse.

As equivalent-to-reality representations, trickster plots hover at the edge of the probable. Yet they create the conditions of their world, and readers are disposed to lend all that they know of reality, of human agency, of dissembled intentional states, of gullibility, of the criminal mind, of material and hedonist appetites to the representation of those worlds. These plots are coherent and causally rich, contingency-laden transactions in consciousness, and as such incite curiosity, the forecasting of futures, and perhaps something like well-wishing for the clever on a temporary basis, or simply a distanced interest in human motivation and its consequences, while taking in the shape of the narrative in prospect of a choice of endings. The computations they arouse also belong to suspense. Tricksters are characters and functional agents who plan and execute by calculated design; the tricks are schemata projected into the future as social events. But given the absence of empathy, these plots challenge all theories of suspense excluding computational components, and blur the lines between the structural thinking outside the story and the strategic planning within. Thus, the definition of suspense must incorporate plots in which human agents vie in social terms for advantages within their social and economic environments, against a background of odds and probabilities actively calibrated by the spectators, for there is self-evidently a high limbic investment in the outcome of such actions. Jonson, in short, is a master of suspense by dint of the attention aroused by his brilliant social designs and their drive toward closures in which the alteration of human destinies according to an economy of rewards and punishments transpires in the absence of empathetic bonding or the creation of a sympathetic point of view.

This then seems increasingly certain, that computational reasoning cannot be separated from the suspense response and therefore must be brought "inside" the fiction. That readers can think and feel at the same time poses no challenges to a "folk" approach to human nature, even though the mind-brain procedures involved remain deeply challenging questions. Mind states are mysterious because they are colorations to cognition. Emotions capture consciousness and demand rational analysis, even as thoughts and memories provoke feelings. There is little question of the human capacity to assess because they feel, and to feel because they assess. The interface between these two systems provides a running double reading of both percepts and images, accounting for the richness of experience in either mode of consciousness. Attention turns to suspense in social negotiations when the intentional states of others become mysterious, teasing, or threatening and call for deliberate response, or in the natural world where safety depends on memory, tactics, contextual calculations, and estimates of ability. Trickster plots incorporate both dimensions of suspense into a common endeavour: that which pertains to the completion of designs, information jags, and the strategic planning of ludic manoeuvres in game-like sequences, and that which still searches for moral preferences through character evaluation and the destinies of persons. Empathy and moralized options will always form a part of the profile of suspense as a limbic investment in narrative, but not all suspense plots offer that option.

Ultimately, then, if the "themes" designed into phylogenetic human nature are taken for the common denominators of anxiety and attention arousal, then suspense is to be defined by the alignment between narrative circumstances and the mental reading of those circumstances in conjunction with arousal thresholds. With the mind so conceived, suspense is either the emotional component of absorption in incomplete events presented to the imagination as real, or the reader's emotional and computational investment in the prospects of characters in beleaguered circumstances represented as real. Suspense pertains equally to representations originating in percepts and to those originating as images driven by memory, scenario spinning, or scripts. It is not a social emotion that interprets the environment and defines courses of action, but the attentional component of those emotions that keeps the mind excited about its options. It may in fact be defined as the arousal dimension of any emotion, such as hope or fear, or the cognitive urgency associated with vital information gaps or disorientation. In that regard, where it arises as a mind state, it exists phenomenologically and defines itself in terms of all its efficient causes.

CHAPTER NINE

Laughter's Shortfall

The Aesthetics of Renaissance Tragicomedy, *The Witch of Edmonton* and *The History of James the Fourth*

Regarding tragedy, Aristotle left a critical legacy that, for all its potential ambiguity, is at least a complete statement, one that places openly before us such notions as *peripeteia*, *anagnorisis*, and, although Aristotle used the term only once, *catharsis* – a compound emotional quality arising from the felt conditions of closure. In a tauntingly brief section of the *Poetics* he also began a definition of comedy, one which calls for persons of relatively low social status who invoke our laughter or sense of the ludicrous by donning comic masks, replicating ugliness, or engaging in distorted antics. Little in all this would appear to be overtly injurious to the characters or profoundly moving for the spectators, just as Aristotle prescribed (5.1449 a6).[1] Implicit in his brief account is that laughter – that strange stimulus-specific human response mechanism – enjoys bedrock status as the functional marker of the successful comic formula, thereby aligning an art form with a specific psychomotor response system. For him, the act of laughter itself is the cathartic effect; its achievement is the function of comedy.[2] That tactic was a brilliant one on Aristotle's part for slipping past the need to anatomize the genre by its social contents (which has nevertheless remained a temptation over the centuries, beginning with Aristotle himself); just let the design of the human genome decide what comedy is through what the limbic system is willing to entertain as an adequate stimulus for the production of laughter.

In the analysis of comedy to follow, then, in the spirit of Aristotle, laughter is taken as the first or foundational measure of all things comedic – not only what causes laughter, but what laughter is as a characteristic feature of human mental production, although the final analysis will also move beyond this limbic response. We are, in a primary sense at least, a comedy-generating species because what constitutes comedy is an extension of what we find in our verbal and situational environments that incites this unique psychophysiological reaction. Moreover, Aristotle places the emotions of comedy in binary opposition to tragedy, suggesting that both emotional expressions are, in some real or metaphorical sense, symmetrical, therapeutic, purgative, counteractive, and, by implication, adaptive. Hence, the meanings of tragedy and comedy are to be read in relation to what it means to feel these emotions as reflections of our own conditions and prospects in changing environments – for that is what our emotions tell us. That is a start. But making a single and unified theory of laughter out of the many explanatory positions regarding the comedic art forms and the kinds of comedic laughter, with cursory nods to the key thinkers along the way, remains a challenge worthy of a trite metaphor. And to compound the challenge, the ultimate riddle involves not only the nature and meaning of laughter but those potentially bona fide comedic art forms in which laughter fails to define the closural mood.

The study hence sets up a series of inaugural questions, among the most challenging in the study of aesthetics. What does laughter mean as an ancient stimulus-response mechanism, and is that ancestral meaning carried over into all manifestations of the stimulus-to-laughter endeavours of comedy? Is laughter the *sine qua non* of all successful comic designs, as Aristotle prescribes? Do artists invariably target it as the ultimate confirmation of their comic skills? Or if not, what other hard-wired limbic responses might be named as the biogenetic markers of plays with happy endings – plays which do not indulge primarily in the masks, ugliness, and antics traditionally cited as motors to laughter?

Given Aristotle's authority, his sketchy critical reflection on the comic in the first book of the *Poetics* received detailed examination and amplification in sixteenth-century Italy as part of the humanist enterprise.[3] Academicians were concerned that contemporary plays meet all the criteria of the ancients, however innovative their intentions. Their primary method of elaborating Aristotle's definition of comedy was to extend the symmetry between the two genres, making comedy conform structurally to the "laws" of tragedy.[4] Thus Della Porta, in the prologue to *La sorella*, defends the formal correctness of his tragicomedy by assuring readers that it "follows the same principles of peripety and agnition that Sophocles used in his *Oedipus tyrannos* – a play greatly praised by Aristotle and one used as a model for tragedy."[5] Antonio Riccoboni concurred that "comedy is an imitation of a base action in that genus of vice which causes laughter," a genre that "through introducing a purgation of minds by the pleasure derived from the ridiculous element" proposes a comic as opposed to a tragic catharsis.[6] By implication, laughter simultaneously alters mind states to social ends within group structures. Moreover, the capacity to laugh creates a behavioural target at which cultural configurations may direct themselves; art itself may be calibrated in accordance with the mood potentials of the human psyche. But I shouldn't put too many words into Riccoboni's mouth. In explaining a similar theory, Antonio Sebastiano Minturno emphasized the "unexpected event after which some remarkable mutation follows, against all our expectations and with the greatest pleasure," whether in bringing discomfort to troublemakers or relief to those who merit it.[7] In that definition, he allows for two kinds of comic reaction, the second of which suggests pleasure as opposed to laughter following a radical reversal of the protagonists' fortunes. The least meaning to be taken from these statements is that comedy involves plots characterized by sudden reversals or "mutations" capable of producing pleasure and laughter in accordance with Aristotle's use of the words hēdonē (pleasure) and gelōs (laughter).

From mid-century onward, however, Italian theorists and playwrights would experiment with a *tertium quid*, namely a hybrid of the two classical genres. They would speculate on theatrical forms in which persons of serious demeanour and high social standing confront the terrors of tragic peril and suffering before finding escape through the mechanisms of comedy. Often for good measure, and in anticipation of those comic endings, the "entourage" characters might include frivolous or blustering stage types whose comic routines punctuate the serious action. Sir Philip Sidney recognized the incongruity in censuring the English for thrusting in "clowns by head and shoulders to play a part in majestical matters" thereby devaluing both genres.[8] Whether Shakespeare, for example, considered the "low comedy" antics of his buffoon characters from Lance to Dogberry as a warm-up for a final comedic emotional stance, or a smokescreen attempt to arouse the *de rigueur* laughter in an alien story will remain forever a moot point. Abetting or complicating these matters are the efforts of critics in the shadow of Mikhail Bakhtin and Michel Foucault to read the hijinks of the under-lifers as cyphers of issues of the highest and most sober concern.[9]

But the real conflict is between the laughter of quick, local escapades, and the slow movement through suffering to a stasis in deliverance. More broadly, such plays advance from adversity to prosperity through the juxtaposing of aesthetic forms insofar as the former state is accompanied by the ethos of tragic expectations while the latter displaces the former through the means and mechanisms of comedy. These plays, in effect, pass through contrasting cathartic millieux in the general direction of happiness, typically through the emotions attached to romance prompted by our instinctual hopes for lovers: their escape from danger, betrayal, or death; their longed-for union; and their potential for creating offspring. Despite the complementary festive slapstick, the justified containment or social punishment of the blocking characters, or the banishment of excessive household folly redolent of the old comedy, the new story was of a different kind based on the emotions of goal achievement.[10] Now, arguably, it was attached to the most urgently Darwinian plot of them all, the struggle for survival and the self-actualization of good and attractive persons capable of love and devotion in a world beset by potentially catastrophic disasters or redefined by near fatal error. That they escape peril through sudden reversals from sorrow to joy and secure a reproductive chance for themselves should thereby form the stuff of cosmic laughter. It is the better of the two plots that surround the human condition: the falling expectations that lead to demise and death, and the rising expectations that secure the future of the race. Should the limbic system not be in perfect alignment with the environmental pressures of this great survival story, which designed the human brain around the emotional responses to success and failure? And should Aristotle not have been right that laughter is the universal marker of the comic condition through its power to communicate a sense of well-being within this great drama?

But I put thoughts into his mind, for in truth, Aristotle did not actually say so, and that is the problem under investigation. With this *tertium quid*, we have a genre that is either essentially tragic in ethos and theme, but with a sudden almost incongruously happy ending, or a gradual and systematic movement from *trompe-l'oeil* tragedy to fully anticipated comedy, thereby making problematic the limbic response which seals the genre. Through this third, conflated emotional stance, laughter itself comes in for reassessment, not for what it responds to, but for what it fails to register. Should the success of lovers inevitably make us laugh?

A first issue is the matter of catharsis. It would seem both critically novel and yet self-evident to say that Aristotle's view of the tragic experience specifies a sequence of events established through mimetic or representational activity that builds toward an emotional response. Stated otherwise, the purpose of the tragic narrative is to produce not only knowledge, moral reflection, and inter-referential awareness but ultimately something felt along the nerves. Aristotle urged that the effect was not amazement or spectacle alone, but something of its "own peculiar kind," something apt to "make anyone who hears the story shudder and feel pity even without seeing the play" (1453 b26–7). Aristotle accepted as axiomatic that feigned actions have the capacity to incite real emotions, to make us experience dangers and fears not our own, to which we react in full psychosomatic fashion. By implication, everything that pertains to the order of fable, the condition of the protagonists, and the mechanisms of reversal is designed to initiate targeted kinds of excitement through exchanges of information between the cerebral cortex and the midbrain, where feelings and emotions arise. This theatrical instrument of hedonic production in consequence imposes its felt reading of the changes in social environment according to the binary pertaining to improved or diminished prospects for protagonists. Conventions, decorum, the unities, noble characters, high poetry, strategic design are all deployed to the end of achieving the desired negative or tragic response, awkwardly deemed "cathartic."[11] Thoughts, thereafter, are free with regard to the cultural and social purposes of art forms designed according to phylogenetic human response systems. What matters is that the Aristotelian formula for tragedy is prescriptive as well as descriptive, for those incapable of joining in the communal expressions of regret and loss bring suspicion upon themselves and are emotionally excluded from the community united around those feelings.

Much the same may be said of those who fail to bond with groups over common values through laughter, or worse, confirm alien values by laughing when others see nothing funny. These group dynamics may ultimately confound the word "catharsis," but in themselves convey a profound regulatory power attained through the theatre; art, through the emotions, is a litmus test of socialized humanity and therefore defines groups. That could matter, hugely, but it means modifying the Aristotelian medical metaphor that links art to homeostasis. If the meaning of art is invested in the categories of emotions it arouses, then art must be understood in terms of the meanings of the emotions – a matter of subconscious instruction, hedonic drives, readings of altered conditions in the world, and group communication. Only then can we parallel the arousal states of tragedy and comedy and ask whether they belong to a biogenetic binary arising in the design of the midbrain. The question, extended, is whether tragedy has its exact counterpart in comedy as a narrative form designed to target a specific phylogenetic substrate of human emotional production. On this last notion, some clarification is in order to make further sense of the Aristotelian formula.

Laughter may be defined as a virtually uncontrollable but usually pleasant reaction to a wide variety of stimuli from tickling to off-coloured jokes which, tickling aside, are culturally constructed to target a limbic response of ancient evolutionary origin. Axiomatic to all that follows is that comedic constructions must perform a version of that which produced laughter in ancestral environments, or we must abandon that backstory as having any explanatory value. With an increasing conviction that the "logic" of laughter has remained constant throughout recent millennia, it follows that the laughter-producing brain draws "structural" and "thematic" analogies between efficient present stimuli and efficient ancestral stimuli.[12] But it also means that laughter today means what ancestral laughter meant as a primal form of reaction and communication. Any real science of laughter must be based on these premises, involving along the way a cultural appropriation of a single limbic response mechanism through the underlying analogous associations by which the brain, in responding to the new, still thought it was seeing the same old thing. There are questions remaining about what this expression of the mind can mean as a form of ancient communication, or wordless speech, and these questions must, in a sense, be aligned with the laughter-production capacities of the many mental flips and sudden surprises built into the codified verbal and situational patterns first examined in the commentaries on the famously missing second book of the *Poetics*: the *Tractatus coislinianus* and the twelfth-century *Iambi de commaedia* of John Tzetze. These formulae, based on ambiguation, joke routines, false expectations, and paradigm shifts, or mock injuries to desensitized and marginalized persons, are the earliest attempts to anatomize the varieties of sudden mental readjustments in vogue in the ancient world that provoked laughter. Largely, they have to do with wrongness in the expression of conformity-bound social customs and verbal formulae. Such shortfalls may arise in slips, accidents, stupidity, or ignorance all the way up to inexcusable moral and social deficiencies and value impairments, each condition having its appropriate form of laughter. Or, paradoxically, they may begin with confusion or dissonance and find resolution in reframed modes of discovery and resolution.

Now for a preliminary detour into gelotology – the science of laughter. For the record, as efficient causes, the formulae just outlined increase neuron activity in the thalamus, the hypothalamus, the mammillary bodies, and the cingulate gyrus – regions of the brain whose collaborations in the phenomenon of laughter have been confirmed by neurobiologists through magnetic resonance imaging or magnetoencephalography. V.S. Ramachandran, in *Phantoms of the Brain*, recounts the case of the Philadelphian librarian who suffered an aneurysm in the thalamus that provoked such convulsions of laughter that she died of suffocation.[13] These triggering areas are at the centre of the mammalian brain, suggesting that for whatever reasons this psychomotor reaction evolved, it has been there for a long time, indeed long enough to permit biogenetic association with similar reactions in cognate mammals such as chimpanzees.[14] Ramachandran describes this nucleus as "a relatively small cluster of brain structures ... a sort of 'laughter circuit'" barely larger than a fingernail.[15] It is this circuitry, constituting part of the midbrain's mechanisms for the production of hedonic states, which comic playwrights and actors seek to activate, especially if they are stand-up comedians, clowns, or storytellers intent on unifying audiences in mockery and derision. This is accomplished indirectly through the cerebral cortex's capacity to become completely absorbed in the verbal and gestural signals required to generate the desired level of emotional excitement. As Aristotle set out to explain, certain kinds of stories must be told to elicit the desired kinds of emotions, and artistic practitioners have shown remarkable proficiency in their crafts to these ends. Nevertheless, we remain unsure of the essence of laughter in these stories, unsure of the kinds of social signals laughter implies, how many kinds of stories are able to bring this reflex to pass, and whether some kinds of stories thought to qualify will inevitably fall short of the mark.

Helen Mayberg of the Rotman Research Institute has concentrated on questions pertaining to brain activities relating to depression. One experiment, however, involved healthy, happy, "normal" people to whom were read accounts of former traumatic events in their own lives. The results were traced through magnetic resonance imaging, and thereby provided an activation profile of what may be called the sorrowing segments of the brain. The process created a reversal of mood through the effects of narrative, one admittedly with which the subject had an intimate involvement. What the experiment revealed applies equally to the imaginatively generated world of drama; the social data makes its appeal to the midbrain and the cortex, the former reading hedonically through feelings, the latter seeking to control feeling through planning and reasoning.[16] It is a simplified but still useful model to describe this hermeneutic indecision: the emotional evaluation of events pitted against aesthetic distancing and rational analysis. My guess is that Mayberg's sorrowing segments of the brain as revealed through MRI scanning will remain open to scrutiny because imaging and specific brain activity are not perfectly aligned, but that brains produce sorrow as an emergent state in relation to precise social data is beyond contest.[17] It happens somewhere in the brain, and it happens only when there is a modification of material activity. Such modifications have been deemed the work of both tragedy and comedy, and must hence entail a quiescence of the rational in a way that allows emotions to convey their own sensations and felt meanings.

The traditions of art are, in a sense, cumulative in passing on, through practice and prescription, the trial-and-error progress through which story types and their responses are coordinated. Stories are open to endless inferential interpretation, but the activation of an emotion similarly named or manifested by a community of witnesses narrows the interpretational range. Reversals in fortune then become critical, for they are measured by emotional transformations, and those transformations may, in themselves, become thematic, and particularly so when the body expresses them to public view with "doomsday" urgency through emotional displays.[18] What, then, can that unity of laughter mean, and did it mean the same in ancestral environments when many laughed together? The choices are far more limited in relation to the lived circumstances of survival environments. Was it a gesture of group reassurance after threat, a group expression of glee following conquest or a narrow escape? Or was it a gesture of trust and détente among would-be aggressors or competitors – the signal for trust and cooperation? Each form of awareness implies reversal or a change of mental frames. And why did laughter become the biogenetic marker of these occasions? Probably because laughter, as a limbic or subcortical response, does not arise through volition, but through the emotions now made additionally valuable precisely because they are remarkably difficult to fake. It was a genuine indicator of emotional involvement.[19]

That retro-story becomes harder to write from a more recent perspective. Either we have appropriated this spontaneous evaluative system and imposed it on everything that is able to prompt laughter in modern contexts (jokes, mock aggression, comedic narratives, nervous uncertainty, puzzles resolved, self-denigration, message deflation), or these many social transactions are, to our analogue-hungry brains, precise equivalents to those ancestral stimuli, including all the plots of successful comedy. Either the brain is good at colonizing response systems engineered by ancestral environments to do the work required by recent cultural novelty and invention, or it is good at deciphering the essences that unite entire categories diversified only on the surface.

There is heavy work here for cognitive philosophers, but the brain is both exceedingly plastic in diversifying the applications of its limited repertory of embodied emotions and exceedingly adept at building its model of the world out of templates grounded in essential samenesses and differences.[20] Moreover, in keeping with the necessary means of selectivity and adaptation, the brain systems were all established in relation to their roles in enhanced reproductivity and survival. Those with the greatest complement of adaptive traits reproduced themselves the most efficiently and passed those traits down to posterity. But along the way we also learned to imagine, to play, and to target reactions. We learned how to construct jokes, riddles, and the ruder forms of comedy, cementing the laughter response to these minimalist expressions of danger and escape, disorientation and resolution, producing a sudden micro-surge of superiority and dominance, or the solidarity felt in groups united in excluding cheaters, cripples, or idiots, or united in their delight with the antics of self-appointed clowns. This sense of pre-eminence may be a direct counterpart to the perceived "moral deficiency of comic characters" in whose worlds we have placed ourselves.[21] From a different perspective, the evolutionary backstory which links laughter to deliverance might be illustrated in such cinematic creations as *Jaws* in which the two survivors, upon discovering their mutual escape after a horrendous and prolonged struggle with a monster of the deep, break into a long sequence of uncontrolled laughter – a sequence which seems entirely right. This laughter comes from stories around campfires dealing with narrow escapes from predators.[22] Comedy can be the art form of false alarms, of danger that is made benign, with the audience looking on from the safety of dramatic irony. Laughter may also arise spontaneously from the moment of eureka that follows a dangerous depravation of knowledge in a hostile environment. Or, insofar as emotions are sometimes group infectious, comedies may be incremental in their laughter production as buffoonery begets more buffoonery or ridicule is bolstered by the anonymity fostered by mobs, even to the point of cruelty without censure.

Kant thought of laughter largely in terms of a sudden cognitive realization that circumstances constructed as dangerous were now deemed trivial.[23] He proposed that laughter was produced by "the sudden transformation of a strained expectation into nothing."[24] This is a Darwinian tale of self-advancement through escape, even if the pattern is as diminutive as the punch line of a joke. Others have concentrated on the laughter of cruelty in keeping with Hobbes's theory that laughter was an expression of superiority, of conquering the weaker, as if laughter were the emotional prerogative of the alpha male or of those who successfully ganged up against him. That too may be another Darwinian stance, socially packaged, group defined, and based on a sense of delivery from insecurity, inferiority, or dominance. Laughter, like crying, is a single psychosomatic mechanism that was nevertheless adapted to a number of stimuli, the context of each establishing the meaning of the response: laughing to celebrate, laughing to scorn.

So what, then, of the potential link between comedy and tickling? Tickling is a form of aggression that is understood as non-threatening as long as a familiar or trustworthy person administers it; the laughter makes it known that the mock attack is not real. Laughter may begin and end as the response to paradigm confusion between fear and reassurance, much as Kant predicted. Tickling by strangers is too frightening to produce laughter, while self-administered tickling is not frightening enough. In this context, Ramachandran sees laughter as an "all's okay" signal, which, as a ritual that turns aggression into play, has generic implications.[25] One may ask whether, in this, we are not coming close to the common denominator of the comedic. In an extended sense, our mammalian ancestors' acquisition of the ability to recognize aggressive activity that merely masqueraded as real, but that was conducted only as a form of make-believe, may be the defining factor.[26] This is what tickling seems to be all about. Pascal Boyer called the phenomenon "decoupling," but he was not the first to spot this vital distinction.[27] Most immediately it pertains to roughhouse play among cooperators, but pertains to all forms of communal sharing of information and attention and what can be achieved through the building of group solidarity. Decoupling means that we can distinguish between the fictional and the non-fictional, yet experience the fictional in a fully ontological and emotional way. It was the beginning of play, and the precondition to provisional group planning. Now narratives that frighten, amuse, sadden, arouse incongruence or suspense could be shared in communal settings in order to build communities through the language of the emotions. All mammalian mentation was once concerned only with the real until the mind learned to deal in counterfactual projections, and a case may be made for the natural selection that confirmed and enhanced this trait among replicators.

Laughter thereby becomes associated with several aspects of the decoupled experience: first it signals the state of playful intentions, of make-believe; then the cautious response of groups to ambiguous and potentially compromising stimuli; and finally a spontaneously discovered consensus. J.A.R.A.M. van Hooff sees the response extended from social aggression to verbal or intellectual aggression. Laughter begins to attach itself to "decoupled" stimuli that nevertheless express the same social values, as when jokes are mastered through paradigm shifts, friendships are confirmed through ribbing and joshing, mock insults are turned to compliments, or when self-mockery brings cooperation from those over whom the mocker might assert power and control.[28] This too is danger made trivial for purposes of socializing and cooperation. Yet laughter also seems to arise with groups of the superior bonding over their conquest of fools, superstition, and entropy of all kinds (the default beliefs which make others vulnerable) – all of which may be relayed through tales in which the audience is included among the victorious.[29] How may this be linked to the "agonistic encounters" in which laughter signals the downgrading to play or the release from fear through escape or deliverance from danger?[30] These are all among the little plots that fall within the compass of the risible, themselves a form of mental play that will, in due course, tempt a playful and curious species to extend the cultural prompts apt to trigger laughter. Such experiments would be conducted through trial-and-error approaches to comic storytelling, each new story type seeking its confirmation of membership through the laughter it elicits.

What constitutes the cultural forms of laughter production has engaged some of the finest philosophical minds in the Western tradition from Descartes to Bergson, and from Juan Luis Vives to Freud. Ralph Piddington in *The Psychology of Laughter* provides a resumé of forty-three of the most noteworthy explanations, including those by Kant, Hobbes, David Hartley, Schopenhauer, Herbert Spencer, and J.C. Gregory, who "begins his treatment of laughter by adopting the view that 'there are many laughters: laughters of triumph, of scorn, of contempt, of superiority, of self-congratulation, of play, of greeting, and of amusement.'" Even so, like so many others, he recognizes the need for an explanatory fuse that can "'identify some features of human laughter and ... connect them consistently.'"[31] Gregory, in this, evokes a major challenge: if laughter expresses many modes of the self in contrasting stances to the world, how, through an act of parsimony, has a single physiological response system been harnessed to express them all? The question is already familiar. Is there a common value pertaining to laughter that has been socially diversified? Is laughter, subdivided into several kinds, thus imprecise and ambiguous, leading to misinterpretation and animus? Is this irrepressible response now so compound and ambiguous a signal that it requires sophisticated contextual interpretation? Moreover, the potential for laughter remains in a metaconscious hair-trigger state, a state constantly invigilating the world for just those patterns prerequisite to its activation. This in turn defines one of the phylogenetic traits through which we negotiate our general environment. We thrive on what it means to laugh.

Among the more successful of recent thinkers in matching social stimuli to laughter are John Morreall and Norman Holland.[32] The simplest answer, of course, is that whatever produces laughter is the cause of laughter, some of the characteristics of which have been recognized for centuries and compiled in lists.[33] One need simply attend crowd events featuring narratives of all kinds and take careful notes. Juan Luis Vives observed pertinently that "the incongruity of some saying or fact causes in us a pleasure ideally suited to make us laugh,"[34] but in keeping with the traditional notions of laughter stimuli, he does not move beyond word games, absurd interpretations, and gestures. Yet he has touched on an implicit principle; humour and mirth require celerity in the disambiguation leading to bursts of laughter resolving the absurd, the counterintuitive, or the baffling. We recognize this as the cognitive jag theory based on misdirected anticipation or the inexplicable which is then reframed in a way that grants sudden release from the the anxiety of unknowing. In the notion of incongruity, Vives anticipated the principle of "paradigm shift" employed by Ramachandran, borrowed in turn from Thomas Kuhn.[35] Ramachandran realized that perceptions are interpreted through frames, and that laughter ensues when the inadequate frame is replaced with another that fits the data in an unexpectedly perfect way.[36] We laugh too when perceived aggression turns into mock fighting accompanied by disarming laughter, and we laugh again when the verbal aggression of Beatrice and Benedict toward each other turns out to be the sparring of courtship. This is another Kantian plot of menace collapsing into nothing. The question is whether hostility suddenly perceived as play according to altered social frames is what the laughter centre is designed to recognize and record in all its many cultural manifestations. How close are we to a science of laughter?

Where now should we turn in this search for the paradigms of laughter? The structure of jokes has been central because of the brevity of its efficient stimulus: a quickly shifting set of perspectives in micro form. Norman Holland, in keeping with an established tradition, describes the sudden reversal of expectations as an "arousal jag," as in the case of the exhibitionist who thought of going into retirement, but on second thought decided to stick it out for another year. Such "jags," in the coincidence of conflicting fields of semantic reference, are often preluded by "did you hear the one about," which further primes expectation.[37] Arthur Koestler likewise looks to joke structures and altered reference frames to explain the punch line and disambiguation effect through what he calls "bisociation."[38] When minds are receptive, this simplest of all sequences is a sufficient *forme simple*, to borrow a term from André Jolles, to produce a limbic effect, followed by spasms of the glottis and the escape of air which, in passing the vocal chords, evokes a wordless form of speech. But what does this laughter mean that brings so much pleasure in its expression? Does it communicate in chorus a ritual of initiated understanding now shared among those who met the same incoherence? Does "getting it" amount to a form of mastery and self-advancement? Is laughter democratic or elitist in this regard? We are back to the kinds of laughter. Different theorists, different answers. The larger question is whether the emotional goal of comedy is a more elaborate version of this simple form, and whether a well-designed theatrical *peripeteia* is tantamount to a paradigm shift, an arousal jag, disambiguation, or communal gloating over escape from the crisis of unknowing through laughter? For if it is not so, then the sixteenth-century theorists who applied the Aristotelian principle of catharsis (now in the sense of limbic confirmation) to all forms of comedy might simply have been wrong.

To elaborate a bit further on the search for the evolutionary origins of the laughing species, the classical binary of the ancient theatre assumes that the work of the playwright is not only cultural, thematic, and stylistic but also ultimately psychological in knowing the kinds and presentations of stories that move the social emotions. We are still in the thought frequencies of Aristotle. If, moreover, as stated above, the emotions pertain to art in the same way they serve as instruction to the organism concerning its environmental circumstances, then tragedy and comedy must epitomize rising and falling destinies through laughing and weeping, all within the grander Darwinian story of survival to which these emotions were exclusively attached. Inversely, it would seem that stories that blend these binary emotions can lead only to confusion or equivocation; so thought Philip Sidney in his *An Apology for Poetry*. Mixing such matters in dual axis plots, tricking audiences along the way, could only result in "problem comedies." For the brain is designed through hardwiring to make radical distinctions between the hedonic circumstances of life, and to signal them in contrasting ways, in order to keep us vitally informed about our prospects. Thus, the emotions of rising and falling fortunes generate the salient binaries that structure all experience. Therefore, our time-confirmed theatrical forms should correspond to the hedonics of survival, and for the sake of unity go in one direction at a time.

So on to the challenge of mixed genres. Many of our experimental art tales are just a hair's breadth from becoming their hedonic opposites. In half the versions of the story about the lovers best known to us as Romeo and Juliet, Romeo gets to the tomb on time, and we can agree that that must make all the difference – a mere matter of minutes.[39] One version is a tragedy, the other a comedy: death or deliverance. Both versions threaten disaster yet maintain hope until all is lost or all is rescued. This play is likewise, in its tragic form, a failed romance; it is a dynastic and reproductive promise cut short by pointless feuding, which ultimately brings the Capulets and Montagues together in an act of reconciliation and contrition. In that reconciliation there is a frail silver lining in the face of devastation, as will be seen in *The Witch of Edmonton.* Such plays perplex the unity of emotions by juxtaposing contrasting motifs and elements, a perplexity which can only be aggravated by plays openly soliciting emotions at opposing ends of the Darwinian scale. Thus we find ourselves confronted by stories so designed in which the consistently foreshadowed tragedy is miraculously converted to a happy ending, whatever the aesthetic effect. Paradoxically, despite the apparent binary between catastrophe and escape, intimations of emotional ambiguity have always been there, and thus tragicomedy may be our most natural form of storytelling by dint of the half suspected, half sprung reversals that ultimately avert the disaster taken for an inevitability. Or we should say, in more historical terms, that it emerged formally in the Renaissance by way of advancing a new kind of story, in a less rigid world order – a story type that paints moods and hopes in chiaroscuro, but at the risk of blending anticipations and their attendant emotions and confounding the patterns requisite to the production of laughter. But does such a mixed genre, if carefully designed to progress toward a final emotional stance through a sudden reversal, actually baffle the formation of laughter, or might it enhance that laughter by designing the fortunes of the protagonists around a spectacular, frame-shifting movement from despair to extreme happiness? Or does it simply tell a kind of story, both in social content and in a slow and dispersed design, in which the laughter-producing brain can register no interest – Darwin and Aristotle to the rear of the line? The interest that arouses laughter may simply function on shorter fuses than those employed in the slower unfolding of tragicomic events.

But we are not finished with Darwin and Aristotle just yet. Tragicomedy did not come into existence without debate. Cinthio began in the 1540s with his *tragedia di lieto fin*, sometimes called *tragedia mista*, and finally *tragicomedia*, in reference to his *Altile* (1543).[40] His tragedy with a happy ending he carved out of Aristotelian thought, where it is allowed that those actions remain tragic in which the intention to carry out a horrible deed is averted through the fortuitous recognition of the true identity of the intended victim.[41] Curiously, this precise form of escape, for Aristotle, failed the test for arousing the comic emotions; the happy ending still carries the weight of nearly malign events. Without fortuitous intervention, the innocent one might have been killed. This lingering perception of a tragic sense of life in such plays as Shakespeare's *Merchant of Venice* or even *Measure for Measure* or *All's Well That Ends Well* has contributed to the notion of the "problem" or "dark" comedy. Sperone Speroni and Lodovico Castelvetro, among many, discussed these options for tragedy at length.[42] Della Porta, shortly thereafter, wrote a series of plays with dark complications, featuring such obstacles to happiness as incest between a brother and sister.[43] Yet for all their resemblance to Cinthio's tragedies with happy endings in keeping with Aristotle, Della Porta saw his plays as serious erudite comedies, heavy in their potential for error and decline in the eyes of the spectators, yet less menacing, perhaps, because redeeming circumstances existed all along which were merely hidden from their (and our) view. The effect is of tragedy turning to comedy, but only through a kind of *trompe l'oeil* tragedy. There is a subtle difference. In the new mixed genre of tragicomedy, one ethos is replaced by another within a single aesthetic design having its own sense of an ending and stasis in a rush of feeling that might include laughter, but that is no longer within the range of actual laughter production. The formula differs, the audience response is diffuse, with the communal bonding, to the extent it can be anatomized, now resembling the good will felt at unchallenged and innocent weddings. But is the new effect due to a change in social content or to a change in design, such as the failure to make the switch from tragedy to comedy "as suddenly as possible?"[44] For, as will be seen, sudden reversal remained essential to the tragicomic formula, and thus a potential trigger to laughter.

No full attempt to formalize the combined genre appeared before Guarini's *Compendia della poesia tragicomedia* of 1601, wherein he prescribed noble protagonists, a feigned crisis affecting the characters as though it were real, and a surprise ending.[45] Not only is there a double *peripeteia* from weal to woe, and again from woe to new prosperity and security, but there is a metamorphosis from genre to genre, a form of aesthetic and affective contrasts wherein each ethos heightens the other. In schematic terms, no formal invention could be more cogently calculated to produce the laughter of sudden escape or deliverance. Only in those plays in which the conditions of tragedy are unresolved, those in which anger maintains a grip on the moral and social imagination beyond the moment of reversal, should the surge of comic release be frustrated.

But resistance to the genre came early. Sidney recognized its "mongrel" nature and could make nothing good of the hyphenation between antithetical feelings. He did not perceive the effect of foil action, the chiaroscuro feature of the design. He did not see the relationship between tears and laughter, or acknowledge the undergirding of reversed fortunes through formal juxtaposition. He saw only the untoward blending of princes with clowns and an abandonment of formal classical rigour in the mixture of "hornpipes and funerals."[46] Comedians, for him, were misguided in seeking out laughter in any case, for delight is the more dignified response, while laughter belongs rather to deformation and "scornful tickling." In "good chances" we take delight, while we laugh only at "mischances," although in the end he allows that we may sometimes laugh with delight. In this he is in agreement with Benjamin Lehmann, who pointed out that most discussions of comedy are really discussions of satire and the laughter of exclusion or correction, the ludicrous and the absurd; if laughter pertains to true comedy it will be far more difficult to explain.[47] Once more, we come full circle to ask not only what laughter is to tragicomedy, but to comedy *tout court*, and whether a genre worthy of our human dignity and ethical being can tolerate much of the limbic production that accompanies mockery, petty triumph, joking, horseplay, tomfoolery, mimicry, and clowning.

Sidney's complaint, without doubt, was a reaction to contemporary practices that were taking the Elizabethan theatre into what he believed to be inferior artistic expressions both aesthetically and ethically. Nevertheless, Tudor makers learned early that the court, in particular, preferred their history lightened with dumb shows and bumpkins, and their comedies tempered with playful courtship and dotted with English denizens in their humours. Audiences, generally, sought variety, including amusements related to manners, foibles, and faux pas, as well as pseudo history, or tales of beleaguered love with happy endings. They were willing to venture vicariously into horrors and nightmares, but only temporarily. The elemental nature of their plays was, in fact, a form of *drame libre*, more often than not tragicomic before the fact. The point has been made many times, as in the words of Ronald de Sousa: "by and large the Greeks separated tragedy from comedy. So did their classical French followers, who, like most of the English eighteenth century, thought Shakespeare vulgar for ignoring the distinction."[48]

According to Herrick, Tudor-age playgoers knew early on the "tragical comedies" of the Christian Terence.[49] Richard Edwards, in *Damon and Pythias* – a play first acted at court in 1564 and again at Oxford four years later – wrote to a formula he described in his prologue as a "new tragical comedy." The play begins with the tragic oppression of a tyrant's rage, which by degrees is brought to leniency through the powers of friendship, while the escape from danger is accompanied by the thrashing of a villain. In 1575, there appears the *New Tragical Comedy of Appius and Virginia*, along with Thomas Preston's *Cambises*, the latter a tragedy with scenes of burlesque and comic characters combined with the motif of the prodigal son. Whetstone's *Promos and Cassandra* (1578) is Christian Terence combined with Italian romance. Greene's six plays are tragicomedies in all but name, in keeping with the tastes of popular audiences. There can be little doubt that so consistent a collective devotion to the play of mixed genres and mixed decorum represents the way they liked it. So, is the meaning of art by catharsis entirely confounded by the mixture of genres, each modified by the other, or is there a sequentiality tantamount to a stimulus pattern recognizable to the limbic brain and correlating to the stamp of a salient emotional interpretation?

Perhaps the epitome of the peculiar and challenging in this regard is *The Witch of Edmonton*, the collaborative venture of William Rowley, Thomas Dekker, and John Ford, completed in 1621 and based in part on the pamphlet by Henry Goodcole entitled *The Wonderful Discovery of Elizabeth Sawyer a Witch, Late of Edmonton* (a village a scant few miles north of London), published that same year.[50] The work is openly styled a "tragi-comedy" on the title page of the 1658 (and only) edition. There are two centres of action, one involving Frank Thorney's marital and inheritance problems, and one involving the abuse of Mother Elizabeth Sawyer, a poor, decrepit, and angry old village biddy who is ultimately driven to witchcraft by those who accuse her falsely of its practices. Young Frank enjoys the first fruits of a winsome and intelligent servant girl, gets her pregnant, and marries her to make her honest, yet asks for her collaboration in keeping their marriage silent in order not to be disinherited by a rigid and intransigent father – a father who insists, meanwhile, that he marry for her dowry Susan, his rich neighbour's daughter, in order to rescue his own estate from debts. In the second action, the villagers drive the old crone into practices with the devil in the form of a dog (who ultimately betrays her), thereby turning her witchcraft-inspired behaviour into incriminating deeds.

An enquiry at this point should be made into the narrative moments at which individual spectators and readers understand that the conditions of both Frank and Mother Sawyer have crossed the line of no return and that both must meet total defeat. Given the explanatory passages which redeem the witch as a victim of mere superstition and stereotyping and confirm the very genuine love which Frank feels toward his first wife, Winnifride, spectators are wont to keep alive the hope that reversals are held in abeyance (and to perfect dramatic effect) that would deliver these protagonists from their progressions toward catastrophe. Even when Frank is arraigned for the murder of Susan, whose stabbing we have seen on stage but fail to credit as final, and when Mother Sawyer is sent before the justice of the peace to stand trial, we are still looking for miraculous reversals, the hidden explanatory frames that will secure true love, bring the dead back to life, and deliver justice to an abused if bitter old woman. We are writing our own melodramatic tragicomedy. But to our amazement, Frank *is* the murderer he seems, Susan *is* dead, and the old trot *has* consigned her soul to the Devil, for which both are sent to the gallows, right in the play. There is no redemption for these persons of concern. Yet the last act devotes itself to silver linings, consolations, and unions arising around secondary characters, even to the point of reuniting the fathers of the murderer and the murdered through contrition over the misunderstanding of their children. It is as though Edmonton itself becomes the collective protagonist – a community mending after malaise, distancing itself from the woes lately seen, and shifting toward reconciliation and renewal in Katherine and Somerton and the comfort promised to Mistress Winnifride and her unborn child by Sir Arthur, the one-time employer who had reviled her as a trollop.

Moving beyond the indictment of so many contributing agents, connected to social values about which we may have strong feelings – tyrannical parents and village bigots, such as those who malign an old lady for gathering sticks – there is the matter of the prevailing emotion as the curtain falls that unifies and defines a strangely kinetic dramatic sequence. This play dares to complete the tragedy before it turns to comedy, juxtaposing the two parts and their attendant emotions. Errors of judgment lead to death, yet the play moves on to a Pollyanna finale. The play is either the realization of an English ideal in the making, or the ultimate mismanagement of a contaminated form. A rationale may be made for the play in sundry terms, but not I think with regard to the clear Aristotelian binaries, unless we attempt to layer both into a single action, passing from a feeble attempt at pity and fear to a feeble attempt at the rejoicing over deliverance simply because life will go on in some form through the luck of others. The challenge of this essay in tragicomedy is particularly marked, for more typically in passing from one frame to the other there is a sense of movement toward unity in finality. Yet, in English practice, especially with those plays that take inspiration from historical sources by definition not designed by the prerequisites of genres driven by the ethos of closure, the problem of mixed ethos and emotions remains.

In relation to this crux, Robert Greene's *Scottish History of James the Fourth, Slain at Flodden*, written for the 1590–91 season, is arguably one of the better early "tragicomedies."[51] In this history, the corruptibility of counsellors and the concupiscence of the ruler are brought to the foreground as political dangers. Like *Gorboduc*, this work illustrates causes that lead to internecine war. Typically there are sub-actions, and even for the main plot Greene arranges to have much of the dirty work carried out by a half-comic French assassin and an opportunist adviser, Ateukin, who is barely disguised as the Vice of the morality plays. Dorothea, daughter of the English king and queen of Scotland, stands in the way of James's infatuation for Ida. An assassination attempt narrowly miscarries. The wounded heroine finds refuge in an aristocratic household, there abiding her time like a patient Griselda or hiding from the world like Rosalind in a male disguise, or Hermione, dead or alive, who finds resurrection through a statue. Just what Greene was thinking with regard to matters of genre and decorum is open to debate, but clearly the action treats of corruption, suffering, and attempted murder before passing on to reconciliation. It is, in fact, potentially both a Cynthian tragedy with a happy ending and a dark comedy with a romance *peripeteia* in the style of Della Porta. For a genre designation we could try tragicomic-dynastic-political-romance.

While Dorothea takes refuge in the household of Lord and Lady Anderson, news of her presumed death spreads far and wide to both the Scottish and the English courts. The king, whose story this is, mostly lingers in the background, more acted upon than acting, but it is his emotional life that ultimately counts. That he was "ravished in conceit" in having Dorothea out of the way is balanced by the fact that he is nevertheless "incensed with grief" over her death and fearful of "sharp revenge" (IV.v.29–34). (His mixed reflections resemble those of young Thorney over his second wife, Susan, in *The Witch*.) That brief reference to grief is the ground upon which reconciliation might be based. Misfortune then sets in: Ida marries Sir Eustace. Ateukin, the wicked counsellor, unexpectedly falls into "galling grief" at the futility of his efforts and withdraws. The English troops arrive in a vengeful mood under the leadership of their king, thereby dooming seven thousand Scots to die. Royal error was taking its toll, while the hope against hope that Dorothea might still be found is the new expectation upon which the final reversal is based. Readers and spectators are, of course, fully apprised of her status and very much expecting a counterturn, keen to participate in the dramatic effect of her "resurrection" as an act of spontaneous and miraculous deliverance. The highest, yet surprisingly most anticlimactic, moment is that at which Dorothea must temper all recriminations against the offending king, telling her royal husband that he was but misled by his youth and that his betrayal of her was "but a little fault" (V.vi.160).

Again, we return to the crux of the question. The play is a skilled construction, an accomplished compendium of favoured Tudor theatrical features, and a work that elicits a modicum of empathetic interest, yet subdivides the attention in many directions. It presents itself as a representative Tudor tragicomedy with all the attendant contours of tragedy passing into comedy through a series of reversals. But how powerful are the play's emotional claims upon us, and are we ever inclined to laugh through vicarious relief in seeing the protagonists escape from the wasteland of death and war to reconfirm the royal nuclear family? Northrop Frye opined that "the ritual pattern behind the catharsis of comedy is the resurrection that follows the death [of the hero]."[52] But that remains a subjective call to be made by our emotional triage centre; the answer is "possibly" at best, and it may or may not reside in laughter. In fact, tragicomedy may rarely generate the nervous laughter of averted horror, or the spontaneous laughter of danger escaped, or even the empathetic laughter of love renewed. The effect seems muted by interferences, even though the design itself resembles the archetypal tale of peril and rescue. Is the effect of Dorothea's staged reappearance dampened by our awareness that she has been alive all along? Is it that we laugh only at quickly sketched and sudden surprises, while the slow passage of dramatic time completely diffuses the effect of the transition from tragedy to comedy? Or is it that we do not, as Sidney objects, find laughter the appropriate response to stories of reunion after suffering, for the play is about the separation of a daughter from her father, as well as from her husband. Are we not as likely to feel a lump in the throat, as we do, perhaps, when Rosalind – more so in Lodge's romance than in Shakespeare's *As You Like It* – is reunited with her father and with her tried and proven Rosader (Orlando)? Is it that the limbic system can, at times, barely choose between our Heraclitus wiring and our Democritus wiring? Is it that romance tragicomedy paralyzes us with regard to sentiment and mild surprise?

The ethos of tragedy may be exchanged for the ethos of comedy, but if the social substance of that story is not of the laughing kind, then the exchanged patterns of genres will not in themselves produce laughter.

Are the threats and sorrows to be escaped sufficiently forceful to make us genuinely afraid? The best we may be able to say about *The History of James the Fourth* is that it does not make us laugh, but that, in the words of G.B. Milner, "something laughs in us."[53] The world is made well, in its way, by the restitution of maligned innocence and goodness. Concerning what laughter is, tragicomedy may simply lie outside the real discussion, and yet as a story of survival against odds, it might have epitomized that form of limbic expression which confirms the feeling of any member of the species suddenly awarded improved chances for self-expression and prolonged life. Ramachandran's definition of laughter was precisely this: a communal sign telling the group that they were out of danger regarding their interests and resources.[54] Art targets the limbic system and its reading of social values, and that system passes judgment in accordance with its ancient design through which our emotional reactions tell their truths about conditions in the environment that impinge on prospects of survival. Such are the truths of our species to which art must conform.

Paul Goodman would claim to have known as much all along, for "in a serious plot, where everything converges to the same meaning, it does not matter if we notice small details of acting, inflection, timing; but it is of course just these things that set off the loudest laughter in comedy, where the tiniest touch deflates the biggest balloon."[55] His point is that tragic design has the power to accumulate materials toward the realization of a single closural effect, whereas laughing comedy is made up of many diminutive forms, word hits, micro-plots, cognitive miniatures, that touch off micro-limbic explosions along the way. It may prove that comic *peripeteia* is not necessarily synonymous with cognitive jags that catch us off guard, disorient, then force a quick mental calculation to recover our mental stasis. Comic design may not be able to save up for a final cathartic effect as design is said to do in the case of tragedy. Hence, the Italian academicians were wrong in filling out the symmetrical relations between comedy and tragedy indeed, section five of Aristotle's *Poetics* is, itself, in part misguided, although, to be fair, his stimuli do not go beyond clowns and slapstick. Goodman goes even farther to deny the validity of an evaluation of literature on the basis of audience emotions. For him, any assessment that reasons in terms of "an apparatus of local motions of waves and particles and of electrical and hormonal discharges most of which have yet to be discovered" does enormous disservice to literary criticism.[56] Yet when he dismisses malicious laughter, he claims not only that it is "roused restrainedly by every reversal" but that it leads to a total collapse or "deflation" of the entire play. He describes that global laughter in Hobbesian terms as "a laughter of superiority," or in Baudelaire's terms as "satanic," and thus unworthy of the high dignity of art, except that of the sage who laughs "in fear and trembling."[57] The final stance is annihilation of a burdensome world, the cathartic effect being the laughter that accompanies the expression of "ordinarily suppressed destructiveness."[58] The fact is, Goodman is concerned primarily with Jonson, and builds his theory of catharsis to fit only the momentum of Jonsonian satire wherein all is brought to absurdity and deflation. In practice, then, Goodman does not give up the search for a unified effect in comedy based on recognition and final reversal. Nor should we give up our own inquiry too quickly.

To make the larger point, let me play devil's advocate for a bit longer. Let us say that *James the Fourth* is to the spectator like tickling, a game of aggression turned to play, or a game of peek-a-boo, simulating death and resurrection, one that creates alternately the denial of the confirming gaze and its sudden restoration, each metamorphosis resulting in laughter until fatigue or distraction sets in. Just as with theatre, the effect of the game depends on the total absorption of the child's conscious attention, a belief in the reality of the situation, feigned though it is, even while the emotional response remains real.[59] One could say that the game itself is an early form of tragicomedy in the affective life of the infant. Negative games with happy endings in laughter (such as "Ring around the Rosie" when everyone falls dead from the plague, but then jumps up to play again amid laughter) are reinforcements of security played out against a very real capacity for fear. In similar ways, the tragicomedy is a formulaic experiment in *drame libre* that is altogether more "primitive" in its bid for attention than the antics of buffoons, foolery, stage madness, slapstick, or satire. The basic plot to all tragicomedy is expected harm, serious and consequential for the characters, which is dissolved in the emotional surge that accompanies escape, a true meeting of minds, or the blessing of sheer good luck.

Still unconvinced, the counter-argument reasserts itself: the responsive range of *James the Fourth* falls short of a cathartic climax; the effect is muted by distance, conflicting emotions, sentiment, insufficient fear, lack of suddenness, as well as a lack of the "wacky" and the "zany,"[60] none of which is necessarily a reproach against the genre. Benjamin Harrison Lehman observed years ago that the same "unitary consciousness" that Aristotle prescribed for tragedy does not pertain to comedy, for "though we laugh at actions and utterances in comedy, we do not laugh at the comedy in its entirety. For the comedy as a whole is a serious work, making an affirmation about life that chimes with our intuitive sense of how things are and with our deep human desire to have the necessary and agreeable prevail."[61] Not all reversals of fortune tickle the funny bone. Goodman would argue that there is insufficient malice in Greene's closure to produce laughter, for laughter is more often a tool in the battle of status and the collective deflation of the arrogant than a mark of empathetic commiseration in sorrow or in joy. By the end of *James the Fourth* all the clowns have departed, the Vice has banished himself, the comic assassin has evaporated, leaving the stage to moralists, royalty, and mute soldiers. Schopenhauer's theory of laughter, based on the incongruity that springs to our attention in perceiving the disparity between two states such as a concept and a reality, might have applied had the reversal been presented with greater speed, shock, or mystery. There is the problem, too, of mental distancing where thought and judgment remain in control, where the illusion is too weak or indirect to pass messages to the centres of emotion. Or is it our fear that something like festive laughter might be mistaken for the laughter of mockery or disbelief that beggars reconciliation as a sham, and that hence, in the present case, a little laughter would be worse than no laughter at all? The gesture itself is an interpretation signalled to groups and is thus subject to approbation or censure.

Ronald de Sousa brings the problem to the foreground in an investigation into the ethics of laughter, for laughter not only communicates – it communicates values. On the one hand, it responds to propositional circumstances that can be paraphrased, while on the other, the design that incites laughter conveys a reading according to that design (and a piece of our animal ancestry we may wish to hide). There is no emotional response without a reading of environmental circumstances, and that environment must be read in relation to what the emotions are systemically designed to say about the human condition. Laughter is never simply trivial, innocent, or meaningless; nature has no incentive to evolve mere frills. Moreover, nature has arranged that when the stimuli are exactly right, the reaction becomes uncontrollable, sometimes leading to the embarrassment of trying to suppress it. This has to be the will of the rational cortex with a dissenting view seeking to suppress the feeling brain. An emergent state, through heightened excitement, comes under self-censure. Some of the more rigorous clerics of the Middle Ages urged the suppression of all forms of laughter as unworthy of a sanctified mind; Umberto Eco had his fun with this spiritual ethic in *The Name of the Rose* in relation to the lost second book of Aristotle's *Poetics*. For the monks, even pleasure, mirth, delight, empathy, or a sociable smile was excessive. De Sousa is no such wet blanket concerning the laughing arts. He is not, as I read him, pronouncing a global interdiction against an ancient component of the limbic system and its role in defining human nature. But he is sensitive to the fact that most laughter is malicious in its endorsement of the mean-spirited and prejudicial premises upon which it is often based: tribal, exclusive, and unforgiving.

To impugn laughter in this way, however, de Sousa places hysterical and pleasant laughter into entirely unrelated categories – which may prove problematic.[62] The question is how we constrain our emotions to ethical ends, and one answer is to abandon satire and run with romance, because in doing so we marginalize laughter altogether. Thus we return to our argument, whether *James the Fourth* is, to our limbic brain, a formal transformation from uncertainty to homeostasis, from adversity to sudden prosperity, in its generic movement toward survival with its attendant emotions, or a play designed to bypass laughter altogether. Is tragicomedy the defective form Sidney believed it to be because it introduces precisely this kind of cathartic ambivalence? Going back even further, if we locate our dissatisfaction with the Aristotelian classification of genres by their cathartic effects, then we must jettison the concepts of tragedy and comedy altogether at the same time that we censure our own biogenetic design for endorsing the malicious through the same semiotic response that signals cognitive jags or trivial amusement. Of course, censure does not alter the constitution of our emotional brain or its reaction thresholds; those emotions continue to define us (and our stories), and continue to call for the backstory to their origins through the conditions that confirmed their design in the genome. We are caught in a vortex of potentially unworthy meanings regarding stories that are, nevertheless, perfectly aligned with the production of laughter. Milton's Eve, at the beginning of Book V of *Paradise Lost*, agonized aloud over the enactment of evil in a dream as an acquiescence of the will, and Adam gave comfort, but de Sousa offers less excuse for the laughter that arises equally sub-volitionally because its activating properties are base. Even the bonding feature of communal laughter is suspect and anti-liberal. We should always think alone to avoid the contagion of mobs; gluts of feeling are discreditable foundations for cohesion in modern societies because we are so easily lured to support community attitudes that thoughtful persons should avoid. Political correctness is born. The theatres should all be closed, for they play to collective hedonics to solicit pre-verbal forms of consensus. Laugh to belong at the cost of your integrity – the subtext to every bad joke, or play. Laughter is connected to the selfish gene, egoism, group think, and inducements to accept wrongful assessments of reality through its distortions. "L'allegro" should be jettisoned by "Il penseroso." And yet laughter is so closely fixed to the life force, to community, to the triumph of the social over the asocial, and comedy, even of the most festive kind, is an invitation to positive co-feeling and cooperation. What are we to do?

Tragicomedy tells us something about the design limitations to the domain of laughter, simply by telling stories that are comedic in spirit but that nevertheless constitute experimental classes of environmental events not correlated to the circuitry of laughter production. But if laughter loses its universalizing role in the definition of the genre of rising human fortunes, what set of hedonic responses with somatic markers exists to signal the communal success of tragicomic narratives? It is an urgent aesthetic question, for despite the classical divide between the two great genres with their opposing moods, in practice the English drama, throughout the Renaissance, moved in the direction of more "natural" orders of storytelling, those beginning patently in despair and rising in hope, thereby blending and sequencing the binary moods (such as we find in nearly every romance novel). Yet such stories, with their compound reversals, still seek emotional attachments to the protagonists and elicit the feelings that accompany their rising and falling prospects because our limbic brains continue to make judgments about changes in environments, both real and literary, so long as the story itself has meaning, conflict, and consequence. Thus if laughter has been dethroned as the essence of comedy, simply because so many upward moving dramatic narratives do not elicit it, the question must be taken back to the top, whether the genre of deliverance, and thus of adaptation and survival, has an embodied response that defines it in biogenetic terms. Comedy and tragedy to these new stories become mere metaphors for upward and downward prospects surrounded by hedonic states in need of a new descriptive vocabulary unattached to embodied emotions.

In Wayne Shumaker's words, literature is "impregnated with *feeling*,"[63] but that feeling, in the case of comedy, is more diffuse and diverse than Aristotle found the case to be for tragedy. T.G.A. Nelson saw the eternal problem with comedy as the incompatibility between laughter and social reconciliation. What is there to laugh about when the dominant feeling is of empathy, love, and forgiveness?[64] The privileged information held by spectators and readers concerning intrigue and incomplete knowledge tends to diffuse both jags and surprises, leaving an overarching ritual of dynastic fulfillment – hardly laughter prone. Comedy is about the life force, about love and the genes, and while we may laugh over the antics of negotiation, love itself is not the subject of laughter. In the words of Robert Storey, "romantic comedy serves ultimately the gene, not the phenotype that laughter preserves."[65] The joke may be on those early critics who attempted to complete Aristotle's work through an ostensible symmetry between comedy and tragedy based on the emotions. There may have been reasons why he never got around to explaining comedy in the second part of the *Poetics*, for the genre in its aggregate simply resists a master narrative defined by the architecture of the limbic brain.

Nevertheless, the genome ensures that a capacity to interpret certain aspects of the social world through laughter will remain a salient feature of the human experience, however ignominious the social bonding in the exchange, the betrayal of undignified egotism or instincts, or however valuable as an instrument of social management through gossip and exclusion. Theatrical practice has demonstrated that the mechanisms of laughter production can readily be attached to a select range of cultural stimuli, many of which may be incorporated into larger and more diffuse comedic designs, beginning with *beffe*, tricks, wordplay, antics, buffoonery, situational incongruities of many kinds, and unexpected reversals. These little hits may become cumulative, incremental, infectious, thereby imposing a mood conducive to repeated laughter. But at some point, laughter becomes sporadic, uncertain, and self-conscious. It may leave behind a smile of approbation or contentment, of the pleasant, the mirthful, or the relieved. By rights, laughter might have united audiences that have experienced together the reversals that deliver liked characters from torment or death or that suddenly reveal the truths about mistaken assumptions.

But, for the last time, it may be said that tragicomedy – as the art form par excellence of the great paradigm shift, of the escape from danger into security, of the cognitive jolt from a state of despair to a state of prosperity – may nevertheless fail to cross the limbic threshold as a stimulus to laughter, thereby removing this marker from the hermeneutic community it creates, or removing itself from the genre of comedy. Della Porta tried to vindicate Aristotle in the new "natural" form of the comedic for mercantile Renaissance sensibilities, which brought romance into collision with social legislation through laughter; read *The Sister* and you will see. Jonsonian laughter remains troubling, arch, sanctimonious, and arrogant for some, and the debate goes on. At the centre of the debate is what laughter is, and what it means as a human signal about environmental change and the spirit that prevails among those who acquiesce to laughter as a social ritual. But with our big brains, we are in a position to see not only how our art forms have arisen in conjunction with categories of information and limbic interpretation but also that our biogenetic responses can be cognitively deconstructed. Playing to laughter is big business, yet while it is taken for benign, some storytellers in their aesthetic times have thought to marginalize the response in relation to rising stories of finer sensibilities. Laughter remains a marker, but one that not only excludes in its judgments, but by judgment may be excluded.

CHAPTER TEN

Cognition, Conversion, and the Patterns of Religious Experience

Francesco Petrarch's *Familiar Letters*, IV.1

One of the best known of Petrarch's *Familiar Letters* is the first of the Fourth Book recounting his ascent of the highest mountain in Provence, Mont Ventoux.[1] Petrarch went on this fresh air outing in the company of his brother and a couple of servants, quite precisely on 26 April 1336 – or so the letter would have it. According to the account, Gherardo made his way directly to the top while Francesco wandered about looking for the easiest routes or stopped to meditate on the state of his soul. This literal and allegorical excursion culminated with a splendid view of the surrounding country as far as Italy, Aigues-Mortes, and the Rhone River, followed by a bit of bibliomancy toward sundown using a copy of Augustine's *Confessions*.[2] It was a pivotal moment, because the text from the tenth book on which Petrarch's eyes first fell echoed in such an uncanny way the activities he had just engaged in that he took it for an omen and fell into silent reflection all the way down the mountain.

Such a letter, with its moments of meditation, its allegorical exploitation of the features in the physical ascent, and its program of classical allusions informing even the geographical descriptions, is much more than a travel narrative. Its studied appearance, quite justifiably, has led several twentieth-century readers to conclude that an actual ascent may never have been made, that the entire letter was the literary creation of a man nearing fifty rather than a same-day report of a young man of thirty-two.[3] Plausibly, at a time almost twenty years later, when Petrarch began to gather and edit his letters, he either composed or significantly edited this piece. That Gherardo had taken up the contemplative life as a Carthusian monk years after the alleged climb would alone account for his direct ascent to the top in the letter, while Petrarch's own errancy would represent his lack of spiritual progress even a decade after expressing his anxieties in the *Secretum*. But whether early or late in its composition, the emblematic narrative structure embodies the paradigmatic order of spiritual conversion. That is the topic of the present study, for arguably a desire for the decisive transformation of mind states that constitutes conversion is one of the leading motifs in Petrarch's thought – an ever-present factor in his examination of the self. The paradoxical significances of conversion extend well beyond this moment in his life. That the options proffered by conversion become a lifelong dilemma for him makes the Ventoux letter the perfect context for an extended look into the psychology of conversion – the dynamics of a particular religious experience – leading to summary reflections on the relationship of conversion to humanist thought in general as a modus for the management of consciousness.

The letter has by no means escaped critical attention. To the contrary, no allusion or literary device has gone unnoticed. At the beginning of a sixty-seven-page article entitled "Petrarca e il Ventoso" published in 1977, Bortolo Martinelli reminds his readers that there already exists "una copiosa bibliografia specifica"[4] on the work, including early articles by Pierre Courcelle, the archival work of Billanovich, and Tripet's *Pétrarque ou la connaissance de soi*.[5] Moreover, Evelyne Luciani has brought to culmination the many preliminary enquiries into the influence of Saint Augustine on the letters and treatises of Petrarch, while Robert Durling has examined the Mont Ventoux letter episode by episode for its allegorical strategies. Others have dealt with the paradoxical life of the humanist who dwells on his moral shortcomings with full Christian intensity, yet who seeks the blessed life through the reading of Seneca, Virgil, and Cicero, and who pines after salvation with minimal reference to fellowship with God. The research of Giles Constable points out the many references to monasticism in Petrarch's writings and the admiration he held for those devoted to the contemplative life, including the occasions on which Petrarch revealed his own attraction to the monastic vocation. By contrast, Francesco Tateo's study of the *Secretum* is concerned with maintaining a clean separation between the author and Francesco the persona of the dialogue who is hard beset by a spiritual guide to bring his fallen will to the sticking point of spiritual decision-making. For him, the *Secretum* is in no literal sense about spiritual conversion beyond the search for beatitude in the world of humanist endeavours.[6] Giuseppe Mazzotta considers the letter as one among the many surrounding letters, each to a different recipient. In each, Petrarch is seen to deal with different ideas partitioned and segmented from each other. These are the "worlds" of Petrarch, which together form a non-integrated record of his preoccupations, easily misunderstood as successive biographical moments.[7] Among these scholars there is little consensus concerning the significance of Petrarch's impassioned exercises in the reorientation of the will toward spiritual self-improvement, although the voices on all sides need to be heard.

The letter itself recounts a spiritual drama that ends in the long innuendoes of silence. On the way up the author meditates on his own moral shortcomings. Then, at the top, he turns geographical space into a stretch of moral time relating to his own progress of the soul. Already evident is the turning, or conversion, of his attention from things deemed outward to the inner life. When the chance passage from the *Confessions* reinforces the urgency of this interiorizing direction, he ceases to read, speaks briefly of the conversions of Saint Augustine and Saint Anthony, and leaves the reader to speculate on the progress he then makes toward the blessed life – that for which he professed to yearn. In this he creates a kind of hermeneutic loop, for without the structuring of Christian conversion, the episode has no spiritualizing direction, yet the uncertain nature of the *beata vita* gives no assurance that the perfect mastery of the will preached to Francesco in the *Secretum* pertains specifically to Christian goals. Nevertheless, the conversion structure, itself, remains, embedded in the sequence of the letter. It is about mind states and their emotional counterparts, about an epiphanic moment that offers to separate the anxieties of time past from the beatified mind of time future through cognitive revelation. Petrarch is reticent – given his passionate commitment to human love, to travel, to the recovery of ancient writings, to the political life, and to a legitimate quest for earthly renown – to assign precise values to this transaction, or to frame it in soteriological terms, the allusions to the conversions of Augustine and Anthony notwithstanding. Yet "a spiritual conversion under the influence of Augustine"[8] remains not only the nucleus of this letter, but the *idée-force* behind others of his meditational and confessional treatises.

This fact, too, has not escaped prior notice, although inevitably investigators turn back to its autobiographical significance. Evelyne Luciani suggested that in his eagerness to imitate the Bishop of Hippo, Petrarch created an episode resembling "le récit de la conversion intérieure,"[9] but that his examination of conscience did not propel him toward a higher spiritual level, a reading in keeping with Pierre Courcelle's evaluation of the *Secretum*.[10] Giles Constable agreed that Petrarch was attracted to the *Confessions* of Augustine because they offered a paradigm for escape from his more habitual frames of mind, but only toward momentary flights as he profiled them in *De otio religioso*. Even his mountain top epiphany did not amount to "a sudden change or conversion in 1336."[11] Carlo Segré goes further in urging that Petrarch's purpose in these creations was never Christian, but rather a quest for the rational life that would bring him peace of mind both morally and vocationally; Enrico Carrara concludes that he never "underwent one of those resolving crises that renew[s] the hope in a human soul and inspire[s] the solid certainty of a new faith."[12] In this he joins the many exegetes who seek to distance the humanist from his own vehicles based on Christian thought.[13] Francesco Tateo would allow that the *Secretum* is a philosophical treatise on "material and spiritual exigencies, nature and religion, science and faith," and that one option among others for the integrated self was the entry into a spiritual life characterized by "will, freedom and unconditional adherence to God," but he would not allow that Petrarch's writings reflect a "turbulent psychological drama."[14] Such a work, for him, could only be a reasoned endorsement of the life of humanist study, with a mild Augustinian caveat that such studies alone will not lead to the happy and blessed life, which is a work of the soul in conjunction with faith and the will.[15] Yet just such words point to the other side of the hermeneutic loop – that of the Christian paradigm. Petrarch professed that he had "learned from Augustine that no one can become what he wishes to be unless he hates himself as he is." In the Mont Ventoux letter he employed such terms as "surging emotions," attachment to "the filth of earth," and the "labour and sweat to raise our bodies a little closer to Heaven."[16] Metaphors all, perhaps, yet they are attached to a profile of intended action beginning in low self-esteem across liminal spaces and time, at an ambiguous pace, toward a more spiritualized state of consciousness. The trajectory conveys its own power as a dynamic idea.

The component parts of the sequence Petrarch presents as transparent Christian allegory: shedding excess baggage; the negative counsel of an old shepherd; the ascent itself as spiritual labour; the miscalculated detours along easier paths; the triple meditations on inner states loosely co-ordinated to the external sights; the conversion of moralized space into moralized time; and the quest for inner peace calibrated against the distracting perspectives of the world. Each feature is a position along the hike toward conversion, beginning in guilt and insufficiency intensified by strenuous physical effort, and ending in a moment of heightened self-perception in relation to stated spiritual goals.[17] Spiritualized minds that scale mountains can barely resist the mental collaboration required to solicit the revelation that comes only at the summit, nor can they resist the inward plunge toward intimations of beatitude. The question comes down to the permanence of the altered state. Yet the shape of the events alone is sufficient to warrant analysis as a generic religious experience.

The insistent presence of the spiritual experience of conversion asks for clarification as the quintessential rite of passage separating initiates from non-initiates, the blessed from the unredeemed. The crux in the methodology of the following excursus into conversion psychology is whether the examination should look backward only, to Saint Paul, the Church fathers, and medieval theologians for the nature of that experience, or forward to nineteenth- and twentieth-century scholars who, for the first time, sought to examine conversion phenomenologically in emotional and cognitive terms. No special case need be made for the retro-reading of past generic cultural practice provided the universalized analysis is not, in fact, polluted by modern biases and ideologies. That is an eternal problem. But the premise here is that modern investigations of the psyche can shed light on traditions of experience, culturally and historically constructed as they are, because they play out in minds "universalized" by dint of their common genetic profiling and mental architecture. The argument follows that Petrarch's conversion trajectory is circumscribed by the nature of mind, so that inversely, an understanding of the nature of mind in the conversion process may speak to the nature of Petrarch's idea of conversion, not only for himself but for the humanist age. That such an enquiry should be carried out is a question that stands in apposition to whether it can be carried out.

Scholars in the twentieth century, beginning with William James and Edwin Starbuck, launched the investigation into conversion by collecting anecdotal accounts not unlike that provided by Petrarch, each one, needless to say, accompanied by its own configuration of circumstances and preliminary states of mind. Among those witnesses were Saint Paul and Saint Augustine, together with the many histories arising in the evangelical millieux of nineteenth- and twentieth-century America. The common denominators became increasingly clear as the methods of inquiry became more objective. Robert Henry Thouless profiled in detail the dynamic state of pre-understanding regarding the redeemed state, an understanding that coexisted with an adherence to old values during the preparation stage, followed by the shock techniques required for collapsing the old self.[18] William James described the process as a "subconscious incubation" of the new order projecting itself in imagined forms as the precondition to choice. On the mountain, these moments are encased in narrative allegory. In the *Secretum* that same dichotomy of mind is presented in dialogic fashion by assigning to the interlocutors the roles of advocate for the projected reorientation of consciousness, and the reticent candidate still overly committed to the values of the active life. Such a debate could have no meaning at all without the prospect of change – a transformation of a kind that could be named and shaped, however vaguely, in the imagination. For Petrarch, conversion entered a circle of reasoning that defined beatitude only as the absence of desire for the things of the world that aroused in him sensations of guilt and torment, and as a joy that is solicited as its own essence without specifically theistic associations.[19] These are important perimeters, for there are many other potential objectives both cognitive and emotional that have been associated with the conversion experience.

Fortunately, Petrarch did not have to say what part of the mind was spiritually redeemed in modern cognitive terms, for so many of the mental operations we invest with purpose, essence, and identity simply carry out their functions on a blind competitive basis producing configurations of thought we are pleased to imagine as properties of the self and its soul. Nevertheless, it was not without significance that he chose consciousness as the target of these redemption-driven exercises, leading us to speculate on Petrarch's own philosophy of mind. Tellingly, the very words concerning the inward turning that appear to him with such oracular power on Mont Ventoux derive from the heart of one of Augustine's most challenging inquiries, namely his examination of the nature of memory as the theatre of all spiritual experience. Memory alone can supply the cognitive resources for discovering inward mental events; hence in memory alone resides the knowledge of God. The study of mind therefore turns around the understanding of spiritual things that are generated without empirical origins. How can we know that which has not been learned? How can the mind itself be regenerated by that which it is able, through self-discipline, to call up from memory? Augustine's examination of the components of consciousness in relation to the directives of the will, in pre-Cartesian fashion, brought him eventually to a state of wonder at the magnitude of the mind in all its diversity and power, and ultimately to a necessary act of faith. By dint of its architectural design, the mind, for Augustine, was an instrument to be shaped in accordance with its spiritual potential, and Petrarch was clearly influenced by that imperative. The discussion served to centre both the act of conversion and its long-term benefits in the conscious mind and its capacities to enjoy its own operations. For how else may the *beata vita* be defined but as an inner state of selective recall of all things conducive to the most agreeable emotions? Thus, for Petrarch, conversion may be defined as the potential for minds to reorient their moral and deliberative states in the interests of limiting the contents of consciousness to approved considerations and memories through the power of the will alone, abetted by the emotions deemed beatific. In these terms, he points to one of the master discourses of his career, and to one that is, at the same time, apt for investigation in relation to the phylogenetic capabilities of the mind. These are, in turn, based on the human capacity for computational analysis of belief structures interconnected to idealized modes of consciousness. These two discourses, the ancient and the modern, can be brought into alignment.

To be sure, the neuro-cognitive systems involved in religious conversion can only be inferred from the accounts given of the experience itself.[20] Clearly, decisive things happen in the brain at the time of conversion, although we need not return here to the level of neuron clusters and the altered firing impedances involved in learning, or to altered modular functions responsible for prioritizing information patterns in their struggle for a brief moment in consciousness. Yet the beatified mind must begin with the neurobiological competence of the brain to produce it. Conversion has already been labelled an *idée-force*, one which has the power to alter the conditions of individuals or of epochs.

In the context of cognition, it is the process of displacing one major "propositional network" with another. Such a network is the brain's tendency to develop habituated syntax around idea clusters, causing the mind to replay them to consciousness in iterative ways that reconfirm the analysis of the self in the environment; they are the substance of identity. These clusters contribute to the predilections of personality and to a sense of the constancy of the self. They are also responsible for association thinking by which an entry idea produces secondary and tertiary ideas in closely associated patterns – thus the challenge in dislodging them. Stated otherwise, the mind, grounded in the socialized values attendant upon inclusive fitness in a survivalist world of instincts and struggle, is not easily convinced to modify these values in sacrificial terms without a significant affective pay-off. It would be supererogation to reassess the whole of the Ventoux letter as a meta-conscious propositional network through which "ascent" inaugurates the associations that complete the spiritual exercise in its classic psychological sequence from guilt to transcendence, which has been deemed the necessary precondition to conversion. But we are decidedly working in that territory. The component still unaccounted for is volition, the agent of mind that Petrarch insists upon repeatedly as alone responsible for refashioning the self through the installation of alternate belief structures and their attendant emotions. This network analysis has been invoked here to explain the conversion scenarios in Petrarch's thought in relation to the default syntax of the mind built up through habituation that must be radically or methodically displaced in any conversion experience.

Some theorists of conversion resort to schemata to explain the phenomenon, a schema being a "template-like representation of a highly complex system of knowledge."[21] These differ in function from association networks in their capacity to organize and interpret incoming information. The schema is, first, a categorizing feature by which diverse manifestations within a group are reduced to prototypes. But the concept has been widely extended to account for social behaviour in the form of scripts that calibrate actions to situations. In this sense, schemata function in more proactive ways, evaluating, censuring, or suppressing in accordance with their own habituated configurations. They are subject to adaptation through the processes of learning, but they are conservative in nature as part of the survival strategy of the organism – an organism loath to part with values so closely related to the operations of selfhood. A conversion experience in spiritual terms entails the systemic displacement of entire belief and value schemata in favour of new patterns calculated to reweight the organism's response to mental activity. Such wording may sound like the jargonizing of the self-evident, but these are the current terms of choice in the literature of religious psychology to account for the phenomenology of spiritually motivated alterations to consciousness.

William James was well on the way a century ago to a cognitive analysis through reverse engineering. For him, conversion was an exchange of belief structures through an intensely emotional process of subtraction and addition, culminating in a "dark night" collapse of the personality that allowed for the mystical sensation of feeling a new person emerge from the old like a second birth. Such emotional turmoil was essential, for it served to denigrate previous values, thereby forcing the convert to rely on the new religious schemata to interpret all subsequent experience.[22] This "emotional occasion" could be generated through the words of an advocate for change, such as Saint Augustine's in the *Secretum*, or through the convert's own meditations on sin and guilt, as in the Ventoux letter. These templates are easily retrofitted onto Petrarch's conversion narratives and require no special exegesis. The only phase not brought under examination is the last in which a new baseline, emerging from emotional crisis, becomes the norm for judging experience.[23]

This is an explanatory beginning, but the vocabulary of schemata and scripts will fall short of a full phenomenological explanation of the Pauline divide between the old and new selves unless it can be mapped onto a complete anatomy of human consciousness. What is it like to experience conversion as the mind reorganizes itself cognitively and emotionally around a new master discourse of the self? This is more than can be fully expressed currently, but there are some suppositions to make. Conversion experiences occur as identifiable phenomena within the phylogenetic human brain. They may be described generically as paradigm shifts in both ideological and hedonic terms. The theatre of action is consciousness itself, which is an unfathomably complex adaptive capacity to reason, to be aware, to remember, to imagine, to experience sensory qualia, to register belonging, and to identify limbic sensations. Conversion transpires as a unique sequence within this faculty, programming many of its capacities in the interest of permanently altering all subsequent events of the stream in relation to an adopted religious belief system. This entire transaction is deemed impossible without some manner of meta-conscious reflection, that is, an ability to think about a future self in contrast to a present condition of the self.[24] But if consciousness is difficult to define, the self is even more so.[25] Seemingly, consciousness is equipped with a capacity to be aware of its own operations and their significance.[26] This feature, too, must have been confirmed by adaptive measures enhancing survival strategies. Or is it merely the accidental byproduct of an ability to strategize in time and hence to be aware of mind operations in time past and time future? The result is a sense of self, beginning in the alterities between body and not body, self-interest and other-interest, present existence and future extinction, seen from a first person point of view. These awarenesses can be converted into strategic schemata whereby the many categories of self-interest can be distinguished in relation to the welter of sensations passing through an otherwise relatively promiscuous and passive stream of thought. The self, despite all that we have granted to it by way of subjectivity and identity, may be nothing more than goal-driven attribution clusters that invigilate the flow of consciously registered information to the advantage of the organism. Of course, when those advantages extend to spiritual, moral, affective, and group goals, not to mention the desire for life after death, the self and its attribution schemata become infinitely complex. Hence the interest in conversion, because conversion represents a major overhaul of the schemata of the self-functions of the mind.[27]

The psyche, so described, provides hints concerning the ways in which the meta-conscious dimensions of the self may be segmented off through projective play and reconstituted around new watch hierarchies that will henceforth regulate the stream of consciousness according to their own pleasure principles. Some such set of operations must be involved in any transactional conversion process, whether it is motivated by Christian principles, or whether it is an idealized mind state projected in Neoplatonic or related philosophical terms.[28] Conversion may be nothing more than a learning process by which the priorities of consciousness are redetermined by altered subconscious habits.

The account of conversion provided here has concentrated on the noetic elements of a cognitive paradigm shift, one that is arrived at by reason and volition, given their prominence in the Ventoux account. But there are few conversion narratives that do not involve the emotions either as part of the preparatory turmoil, as part of the transition process itself, or as part of the new state of mind, often associated with the joy of deliverance or surrender. As Pruyser stated, "the deity claims not only intellectual recognition, but heartfelt and feelingful transactions and loyalty. Piety cannot exist without emotion."[29] Just how the limbic system is enjoined in the process is a subject that, like the cognitive elements, must begin with fundamental definitions concerning the interface between thought and emotions.

Recent thinking is nearly unanimous in abandoning the notion of the cerebral cortex as a governor of the passions embedded in a more primitive part of the brain. The evolutionary sequence still pertains, but there is a remarkable integration between the two systems, so that efficient thought is inconceivable without limbic coaching with regard to priorities, as demonstrated in the work of Antonio Damasio.[30] Steven Pinker provides an accessible account of the liquidity with which thought, at times, initiates emotions through networks passing through the amygdala to the hypothalamus, as well as of the reverse process by which the sensations in the body present themselves to consciousness for interpretation according to the leading attributions in place.[31] Conversion processes involve both forms of emotional "language," whether as feelings resulting from notions of inadequacy and guilt, or as limbic sensations that the mind can interpret as spiritual experience. The second effect is the result of the persistent autonomy of the limbic system and its access to data not always available to consciousness, as described by Paul Griffiths.[32] Recent scholars interested in the conversion process as a religious experience have sought to identify the leading emotions – fear, disgust, sorrow, longing (as in forlorn or nostalgic desire), surprise, elation, relief, or merely a humming sense of well-being – and to decipher how these play out in the narrative of conversion as the attribution systems are altered.[33] Petrarch, during the Ventoux expedition, recorded the sequence of his emotional states: "I was abashed," I was "angry with myself that I should still be admiring earthly things," "I turned my eye upon myself, and from that time not a syllable fell from my lips until we reached the bottom again," "I wondered at the natural nobility of our soul."[34] Religious exercises may, in the first instance, be directed at orders of thought, but rarely without some "cathartic" intent built in. Conversion, particularly of the sudden variety, follows an excitatory course in order to provoke both change and the permanence of the new state, blending both self-loathing and rapture.[35] Petrarch's writings are characterized by his own very particular reading of these antinomies, allowing to himself feelings of anger and despondency at the outset, and perhaps something close to true joy through intimations of the blessed life – a state of approved thoughts experienced as pure emotion – at the close.[36]

The most eloquent philosopher to report on this dimension of the religious mind was Friedrich Schleiermacher who, in his *On Religion*, published in 1799, stated that "religion is not knowledge and science, either of the world or of God. Without being knowledge, it recognizes knowledge and science. In itself it is an affection, a revelation of the Infinite in the finite, God being seen in it and it in God."[37] He went on to provide his own list of the leading affections of the religious mind: longing, piety, humility, compassion, contrition, and desire for progress in the sacred life. These are properly termed "affections," within the control of thought, as opposed to the passions, which rise up from the animal appetites. Petrarch was clearly no enthusiast, no mystic, no subject for revivalist melodrama, but he was intensely concerned with the devotional management of his affections and passions. In his letter to Gasparo Squaro dei Broaspini (between 1363 and 1369), he relates how writing itself was a means for intensifying his feelings, allowing him to grieve through copious weeping.[38] By so expending his tears, he could then return to the letters for consolation – a lesson perhaps in the creation and use of the *Secretum*. There will, in fact, be no agreement concerning the intensity and scope, much less the autobiographical reliability of the emotions he does express. Yet the *Secretum* is formally and by design a struggle with existential anxieties everywhere suffused with the emotions belonging to the phases of conversion, or the lack of progress along that experiential path.

If the beatified mind, for Petrarch, is characterized by the repose that comes after long and arduous endeavour, the fallen mind is that which finds no energy even to begin that pursuit. Just as his goal stops short of *enthusiasmos* or rapture through an identity association with a god, exaltation, mystical visions, or divine frenzy, so failure does not descend into abject despair, but into a state of torpor or acedia, epitomized by the absence of all desire to ascend.[39] What Petrarch ultimately intended by his conversion narratives will continue to stir debate. Whether he wished for himself the solitary life in order to flee the world and all its vanities, or merely to flee the world that distracted him from his humanist pursuits, remains open to debate. But the mind theatre as a place of turmoil is a Petrarchan leitmotif, as profiled in the ninth letter of Book II, wherein the "outer man wars with the inner man" leaving him no rest,[40] and the will of the would-be convert fails to escape its own lassitude. Not even Laura's death could pass without creating in Petrarch's mind an admonition to change. On the first guard leaf of his copy of Virgil he wrote that in having the "most serious" of his temptations removed, he was reminded again of his "vanishing years" and of his desire to "flee from Babylon," which he hoped to do with God's grace.[41] Hegel, somewhere, described this fixation as the resources of the finite mind seeking to know its own nature as absolute mind.

To know the reason why humankind in intellectual and religious communities throughout the world have been drawn to ideas of consciousness-engineering according to discipline-related, guilt-driven, or shock-activated means is not to be asked here, except to say that: with the idea of volition comes the prospect of turning it upon the operations of mind; with the notion of paradise comes the idea of making thought its own golden age; and with the notion of subjective plasticity come the moralized motives for self-improvement. Petrarch dropped himself into this vortex of Western thought, unwillingly perhaps, for clearly large segments of his being resisted it. But once in, he nevertheless made the Augustinian malaise the crisis and crux of his philosophical thinking, leaving in the Ventoux letter what may be an ambiguous testimonial to personal fact, but a compelling representation of the experiential structure that haunted his imagination seemingly for a lifetime.

What may be said, then, about the Petrarchan conversion crisis as a crux of the humanist experience in general? Paul Oscar Kristeller states without qualification that "Petrarch's personal form of religion had no direct influence upon his followers among the humanists, and [that] his emphasis on man, although accepted and developed by many of them, did not retain its original connection with Augustine."[42] Such an assurance could be taken as a challenge to prove otherwise, insofar as Petrarch's letters and Latin treatises circulated widely. But the matter of linear influences is too confining where the conversion paradigm is concerned, simply because the idea is so integral to Western thought.[43] Petrarch's adaptation of the narrative is more accurately viewed as the manifestation of an archetype along a great continuum that originates on the road to Damascus, if not in Diotima's speech in Plato's *Symposium on Love*, and continues to the present time as a religious commission or nostalgic need to organize the life of the mind around invigilating values pre-approved as conducive to the highest forms of the spiritual-intellectual life. The Christian and Platonist *loci* were conflated in the writings of the Neoplatonists, culminating in the writings of Ficino. Arguably, no thinker of the Italian Renaissance was more preoccupied with the psychology of the interiorized mind, to be achieved by methodical adjustments to the watch hierarchies that pilot the soul. Ficino, in that regard, was the most eloquent *porte-parole* for the *vita contemplativa* of his age, an age that, according to Hannah Arendt, did not come to an end before the very pragmatic *vita activa* of the seventeenth century.[44] Modern scholars continue to celebrate the *Secretum* for its breezes of modernity, perceived in its statements of resistance to a spiritual call, forgetting that it is a work of consolation for a man caught in the throes of mental torpor who could imagine escape only as a work of mind. The historical turning point for Arendt was not a denial of truth and knowledge, but the realization that "they could be won only by 'action' and not by contemplation."[45] In this regard, all that pertained to Platonic paradigms pertained to conversion, for the imperfect vision within the famous cave signified mere human doing, while the abandonment of the cave was tantamount to the contemplation of the eternal truths of the heavens. This redemption through altered optics, achieved through right contemplation, "determined to a large extent the thought patterns into which Western philosophy almost automatically fell whenever it was not animated by a great and original philosophical impetus."[46]

The far-off beacon was the life of pure intellect suffused with religious feeling, for through the *contaminatio* of the Christian model, that quest was invariably invested with qualities of religious experience. Ficino's inclusive system vacillates between pure forms, ideal beauty, beatific mentation, and the knowledge of God – all were as one. Common to all such philosophies is the preparation of the "way," the profiling of operations by which the mind is transported from level to level of contemplation. The result is a system of hierarchies linked by correspondences serving as mind echelons in relation to aesthetic pleasures; the modelled mind becomes its own artifact for enjoyment. Through method, the mind overcomes its own lassitude, and through redefinition, that which constitutes the beatified state also brings to actualization all of the cosmological ideas of the humanist syncretists. The product is a regimen for the soul, a program for self-actualization that progresses from a life among the lower senses to a contemplation of divine things. "Ficino ... wrote in his *Argumentum de summo bono* that the supreme good consisted in the contemplation not of any created good but of the highest good, that is, God."[47] The *Theologia platonica* is, needless to say, thoroughly imbued with notions of conversion through ascent, joining metaphor to volition and contemplation in order to spiritualize the operations of consciousness. The Platonic template for such movement is found in Plato's *Symposium on Love* (210e–212a) concerning the rungs of the ladder by which the mind advances from the contemplation of particular and transitory beauty to glimpses of true beauty.

That a line of influence between Petrarch and Ficino may be doubted does not diminish the structural affinities that join their thought. Both were philosophers of the sanctified mind to be enjoyed as a form of the *beata vita*. Each writer created narratives concerning the progress of the soul, with its positive and negative emotions. The Neoplatonists went on to discuss at length whether God was to be known principally through contemplation or enjoyment, through intellectual understanding or through emotion. That debate reappears in the writing of Rudolf Otto in the early twentieth century in *The Idea of the Holy* (1917), no less than in the writings of the earliest Christians who spoke both of the knowledge of God and the joy of His presence. Paradoxically, the Neoplatonists were inclined to make the will itself the receptor of spiritual joy.[48]

Lorenzo Valla, before Ficino, in *De voluptate ac vero bono* (1431–32) made an inquiry into the nature of the highest good, beginning with the moral virtue of the Stoics and passing through the pleasure of the

Epicureans, including their theories on the tranquil mind.[49] But he too makes his way toward a Christian apologetic in following virtue for the sake of future happiness, intimations of which could be had on earth through the ordering of the mind (chapter 9). Kristeller profiles this state as the spiritualized contemplation of the visible world through a combination of faith and imagination.[50] In this treatise, the virtuous life rises to the blessed life incorporated into the Christian frame of time and salvation. Conversion here is a methodical process in which classical and Christian attribution networks are superimposed to the common end of initiating the mind into a state of beatitude on earth as it will be in heaven.

Girolamo Benivieni completes the classical-Christian dyad in a pair of poems, the first his "Amor dalle cui," made famous by a lengthy commentary by Giovanni Pico della Mirandola based on Ficino's *Commentary on Plato's Symposium*. In later years, Benivieni provided his own Christian corrective in "Amor sotto cui" which exchanges the conversion arrived at by climbing the ladder of beauty for the conversion prescribed by Saint Augustine in his *De doctrina* (II.7). Yet each course was eudemonic in nature, tending toward the beatified mind revelling in the perfect harmony of its own transactions. Ficino confessed that perfection would be a rare and fleeting achievement for a select group of seekers, those willing to labour ceaselessly for the joys they would obtain. His affinities would have been with the enlightened one of 2 Corinthians 12:2 who was "caught up to the third heaven" where he saw visions and revelations, rather than with the Saint Paul of Acts 9:3–22 whose eyes were opened on the road to Damascus, who saw light out of heaven, and was struck blind until he was initiated into the company of Christians under Ananias through the laying on of hands. But in this order of contemplation, Ficino joins cause with Petrarch as a man who sought to define his own life course as one of ascent in relation to a contemplative ideal.

Petrarch drew on several, but not all of the varieties of conversion experience. The emotion that is sometimes ratcheted up to catapult the mind into crisis in order to elicit the transformed state is, for him, systemically or by default, modulated by reason, apathy, alternative commitments, or even multiple versions of the beatific state. But the cognitive-emotional procedures associated with that shift reveal the transactional nature of the challenge, as well as the attendant conditions of the emotions. Petrarch avoids the Faustian dilemma by incorporating the schema of conversion into the momentary fluxions of life whereby the best of both worlds might form a kind of dyad between the *vita contemplativa* and the *vita activa*. That was his working solution.

As a prelude to future centuries, it is revealing that Petrarch was so deeply attached to this religious-oriented duty to spiritualize the mind in keeping with Christian tradition in apposition to his deeply held commitment to the worlds of erotic desire, statesmanship, and fame. There had been a break with the age of faith, but not without a lingering nostalgic disposition for the pleasures and agonies of spiritualized mind control. How to incorporate that experience into the new humanist philosophy was to become a major preoccupation among later thinkers. There was the perceived need to redefine and reconstitute an equivalent-to-conversion engagement with the world through a version of existence in time and space viewed as a progress of the soul toward timelessness, unity, and harmony. As for Petrarch, the record is clear that he never relinquished the attraction that the monastic life held for him. It is equally true that he confronted the world of contemplation as a place "which the self longs for and from which the self is also excluded,"[51] much as he looked upon solitude as both beatific and a temptation to sink into the sorrows of love and self-pity.[52] But Petrarch's reticence has meaning only in relation to an intuitive understanding of the elective engineering of cognitive and emotional mind states according to a defined system of the good. That transactional definition is an elucidation of modern times, but was implicit in all that pertained to "conversion philosophy" throughout the Middle Ages and Renaissance, concerning which the Ventoux letter serves as a cameo and a pivotal witness.

CHAPTER ELEVEN

Folk Psychology and Theory of Mind

John Marston's *The Fawn*

To know the people well one must be a prince, and to know princes well one must be, oneself, of the people.

Machiavelli, *The Prince*

John Marston's *Parasitaster* or *The Fawn* is what it is, a competent, entertaining duke-in-disguise plot, formulaic, but not to a fault if a good theatre company were to take on the challenge. The work is convention-laden to be sure, but for that very reason it is the perfect literary laboratory for reinvestigating the mind-teasing topic of "personhood" in dramatic representations because the character of Duke Hercules of Ferrara (The Fawn), not quite self-evidently, is also a representation of his own personhood, however schematic. He is the depiction of a ruler on holiday, anxious to flee the burden of his office with all its constraints in exchange for a life of freedom, passionate spontaneity, and self-actualization at the distant court of Urbino. To adapt the words of Machiavelli, he is a prince who knows his people too well and therefore chooses to become one of the people not only to punish them incognito (albeit not in his own court and city) but to know himself better through these new experiences.

To assess this character is tantamount to examining the design of the play because of the efficiency with which he imposes his will and point of view on the entire action. In making the move to another court, Hercules redesigns his social strategies, turning himself into "Fawn," a flattering courtier who ingratiates himself with all those in his adopted entourage while working his way into the inner circle of Duke Gonzago. As his new name implies, his modus operandi will consist largely of encouraging others by flattery to indulge their respective follies the better to hail them before the court of public opinion and its reproving laughter. Fawn thereby becomes the play's agent satirist, its trickster *animateur*, and hence a role keeper with an ostensibly diminishing interiority befitting the trickster agent. The action moves toward a ceremonial closure as he draws the entire court into a compromising theatrical inset (play-within-a-play) through which all are indicted as fools, before staging his escape from retaliation by claiming the diplomatic immunity furnished by a tactical reclamation of his former ducal identity. It is a variation on the working formula (as in Riche's *Brusanus*) whereby an innocent ruler, accused of treason, might negate the charges by revealing that he is, himself, the very object of his own putative sedition.

This "idea" of the play, the product of several years of experimentation in the Elizabethan theatre with plot-making aristocrats, is tantamount to a composite structure in which the trickster operator obscures his high social station in order to move anonymously throughout the play's society. The formula invests a stock trickster figure with both public and private identities, bouncing the reader's attention between the concerns of a suffering ruler and the machinations of a social prankster, thereby linking political with social issues and doubling the representational perspective of the protagonist as he seeks flight from one draft of the self by inventing an alternate persona through an entirely new and liberating form of social mobility.[1]

The study to follow is not only about the play's protagonist in action but about the cognitive grounds for the attribution of selfhood to this mutant character, and whether, with those mechanisms more clearly in view, something of hermeneutic interest may follow. For precisely how we establish the personhood of literary characters, and to what ontological degree, would appear to be of paramount interest to the reception of literary texts. In this case, we would say that what Hercules-turned-Fawn can be to us as a person is determined by the categories and inferences habitually employed in interpreting the propositional stances of other minds, and this in relation to what we believe about personhood and the natures of alternate selves. Should these operations be established, we might then be able to evaluate more clearly those challenges to the continuity of the "psychological" self that pertain to the case of Hercules: whether his disguise is a means for self-transformation, or whether his trickster persona challenges our bond of belief. In brief, what do we do, cognitively, in representing such transformational personalities to ourselves, both as plenary subjects to whom, as ontological human beings, we grant the entire package of personhood and as functionaries in disguise who, by assuming that alternate identity, may seem to suspend what little belief we had awarded to them initially? This, in turn, harkens back to the critical crux pertaining to character criticism, *tout court*, whether characters are, for us, agents manufactured through words to fulfill motivational sequences within plots or equivalent-to-real entities capable of evoking all that we may be brought to feel and evaluate in real persons.

This is a longstanding epistemic problem, but one that is always engaging to revisit because it connects back to the more generic question of our cognitive habits in cooperating with and formulating other subjects. For arguably, personhood is a completed Gestalt accorded as a default cognitive commission – one which is bestowed on us by those who interact with us, much as we do in bestowing plenary personhood on others as the basis for any psychological understanding of their actions and intentions. The ontological quality we extend to others would seem to derive from an orientation endemic to the architecture of the brain as a precondition to dealing with their intentions in real and consequentialist terms – whether the data for this attribution of personhood is derived from imaginative or perceptual sources (language or immediate social engagement). This has much to do with how we come to care and feel even for literary characters.[2] The challenge of the present study is in determining just when the literary representation of humans in action may no longer inspire the consequentialist concerns resulting from the weakening of ontological personhood by conventions or rules of play.

The dual nature of this protagonist, as ruler and as trickster, was the byproduct of structural developments in the early English theatre. Marston's *The Fawn* appears at the very moment at which the configuration of elements constituting the duke-in-disguise plot reached its apogee. That date was 1604, and coincides with the earliest productions of Shakespeare's *Measure for Measure* and Middleton's *The Phoenix*.[3] But the inability to date these plays more precisely leaves the matter of priorities and directions of influence beyond assured demonstration.[4] Given their differences, and the degree to which the generic idea of the disguised ruler was already established in the Elizabethan theatrical milieu, there is reason to allow that these plays arose independently of one another. Rulers enjoying an incognito status in order to go courting, or to escape the burden of office, or to spy first hand on the affairs of ordinary citizens were already manifest in such plays as *Fair Em*, *A Knack to Know a Knave*, *George-a-Greene*, and the first part of *Sir John Oldcastle*, not to mention the many parallel motifs in Chapman's comedies (although once again the direction of debts is difficult to establish). Perhaps of even greater pertinence is the formerly mentioned *Adventures of Brusanus* (1592), which features a pioneering version of the motif. Barnabe Riche's protagonist prince disguises himself as a merchant in order to examine at first hand the prevailing conditions of his realm, only to find himself falsely accused of treason. That the ruler against whom the alleged treason is committed is himself, of necessity, entails a recognition scene in which Brusanus resumes his true identity before turning on the maligning Gloriosus.[5] Playwrights, through a cumulative tradition of such representations, were exploring the mental plasticity of their spectators in maintaining a psychological investment in those characters who denied their identities through disguises and indulged in motivational strategies arising within double and sometimes incompatible personalities. Nevertheless, to be the trickster, one must begin as another person playing the role, leaping between mental templates, and thereby generating problem comedy around the subjectivity of a self-manufacturing protagonist.

The English plays of that era – of which the duke-in-disguise plays were a subset – would have been greatly impoverished without these and many related experiments with trickster protagonists cast in a variety of guises up and down the power echelons of society.[6] Not only was the character type instrumental in creating efficient episodic plots from within the action, but these crafty intellects were also inserted into a variety of cruel and competitive worlds to confront their own momentary blindnesses and to sometimes fall prey to superior intriguers as in Marlowe's *Jew of Malta*, or Jonson's *Volpone* – two plays that bracket the historical period in which the formula was most experimentally developed. These plays, at the same time, form part of a continuum that originates in the servant slaves and lackeys of new comedy and medieval folk pranksters. The trickster motif then passes through the fore period of *Gammer Gurton's Needle*, re-emerges in Chapman's gentleman knaves and salon intriguers, and comes to its apogee in the Jacobean revengers and usurpers in their respective political environments. Hamlet represents the final transformation of the trickster from tool character to Western literature's epitome of the interiorized hero, the man of anguished deliberation, inner searching, and political disillusionment who nevertheless contemplates the opportunities and liabilities of taking on the trickster mindset – such as in his handling of Rosencrantz and Guildenstern. Overwhelmed by his own vulnerability, this protagonist chooses strategic dissimulation, but finds himself unable to sustain a role as the Machiavellian practitioner as trickster.

Duke Hercules, with his modest show of interiority, belongs to this same equivocal configuration of anxiety and escape through a disguise that requires all the competence and expertise of an alien self. The confirmed and efficient trickster is, after all, one for whom the trick itself is absolute and uncompromised by moral considerations. But with the Jacobean tricksters, the ethical brain is fully engaged with desired ends in conscious awareness of disputable means. If we entertain both Hercules and Hamlet as persons, one might wonder if there can be a qualitative difference between them except that one frames the means of exposure to elicit shame, the other to justify assassination as talion. Yet the portrait of Marston's Duke Hercules seems minimal in its representation of plenary interiority, thereby setting the problem of trickster personhood more clearly before us. If he too attains selfhood as a mental representation, we may well ask if the mind allows for greater and lesser versions of that ontological category. Persons may have to be persons in the fullness of that concept, however sketchy or formulaic in their representations. The habits of our brains in attributing beliefs, desires, and volition to other minds tolerate no halfway measures. There can be no half persons because there are no evolutionary conditions whereby such creatures could have emerged. If trickery emerges from a human psyche, it must be compatible with all the appurtenances of being human, albeit one who is obsessed by a single tactic for managing social relations. Or is it within the power of the imagining mind to devise humanoids self-diminished by obsessions, humours, and mechanized roles in ways that obviate the need to consider their mechanical agency as thematic shortfalls in what it is to be fully human? Duke Hercules makes it difficult to decide. And that takes us continually back to what we are doing with such characters cognitively as imaginary reconstructions of the human ontological Gestalt. Where does the whirligig stop?

Important to our sense of the selfhood of Marston's protagonist are the few details concerning the frustrations with his life as a ruler. (That same problem arises again in *Measure for Measure*.) Hercules was annoyed with courtiers. Back in Ferrara they had been his bane and the reason for his pressing need to get away. He had been contained by their obsequiousness and by his own sense of duty. Office had made him servile and base in his own eyes, while the "appetite of blood" was calling him to fulfill "wild longings" and tasks of "exorbitant affects." Thus, as Fawn, he had scores to settle and a new life to lead. That is pretty human in spirit. The change he sought appears to be a permanent one in promising himself that "these manacles of form" will never regain control over him (I.i.39–45). That is to say, he intended to adopt the artifices of the trickster in order to reshape his ontological self. (That his ultimate choice of motives and actions should continue to be shaped by his pique with courtiers reveals his ironic inability to escape his past, but that is another matter.) One impetus to the forward direction of the play is our desire to know what he had in mind to satisfy the "appetite of the blood." In fact, Duke Hercules's son has been sent to Urbino "to solicit a marriage betwixt his father, the Duke of Ferrara, and our Duke of Urbin's daughter, Dulcimel" (I.ii.45–7). That might have constituted such an appetite were it not his intention that his son should fall for her in his place, and he to use his disguise by way of encouraging the sexual (and dynastic) appetites of his reticent son through the excitement he would feel in surpassing his father. We relish the situational irony in spite of the transparent formula. This is all fairly fetched but still quite human, emotional, archetypal. As a man of three score and five, his pretensions to a "lady of fifteen" had already reaped the disapprobation of the local courtiers as "an enforcement even scandalous to nature" (I.ii.196, 201–2). If his quest for excitement could not be found in young love, it could only be found, on the part of an aging and idle courtier saturated with the vanities of courts, in the pursuit of a playful yet earnest disruption of those vanities. He would find novelty and a measure of revenge by adjusting himself to a gamut of excessive behaviours the better to arraign them before the bar of comic justice. That, by all means, constitutes a very particular intentional frame of mind predicated on a vision of absent norms, kaledoscopic role-playing, and a coordinated defamation of his carefully groomed gulls.

But now, are these transformations and functional adaptations of a theatrical "self" the extensions of a permanent self, the creation of a literary agent, or the construction of a new and ontological alternate? I confess, in a sense, to making philosophy out of circumstances that barely disturb us in the reading of the play, and yet once into the matter, it seems necessary to think it through, if only to understand better what we bestow on characters as the default productions of our folk psychology (its discussion in the making). Hercules is Fawn to this new social set, but to us, an assumed identity owned, animated, and operated by Hercules, and made to appear as a complete and autonomous self until such times as it accumulates its own drives, memories, and history. Or is this even to be imagined in an early modern milieu? What do we accord to this staged representation, which allows him to function on the basis of human competence in his two ambiguously related intentional fields as duke and trickster?

The only alternative to the transformational self is a protagonist whose single and essential selfhood functions on multiple mimetic levels. As Fawn, he is clearly the playwright's "internal maker," while at the same time he is another man's transposed self which is motivated by private causes born of suffering, deliberation, and the anger of a satirist. One question is how complete a man we recreate in our imaginations around a figure who is a projected construct of personhood sufficient only for the performance of tricks while, at the same time, he remains a schematic Duke in frail possession of his initial beliefs and desires, the product of an implicit interiority. Or is the Duke, himself, a framing character theatrically conceived merely to enter into and retreat from the role of the court scourge? This is as much as to say that invested empathy and human concern should be no part of our psychological commitment to his representation.

Yet, presumably all readers and viewers will see him as something beyond caricature. His caper abroad is preceded by a comprehensive cognizance of his current prospects in consideration of public opinion and politic restraint. We know that a romantic fling in his old age had pervaded his fancy, the folly of which he recognized in time, settling instead for a turn as *eiron*, whereby he succeeded in turning the court of

Urbino into a probative playground, while at the same time furthering the romantic interests of his once reticent son, Tiberio. Frail as Hercules seems as a psychological portrait, we nevertheless expect him to conform to a certain range of human behavioural probabilities in light of the propositional and contingency-addressing mind states provided to him by the playwright. Once a character representation achieves such a level of complexity, we attribute – in this case to Hercules – not only the facts of his career, namely his disillusionment as ruler, his paternal concerns, and court-trickster ambitions, but the mental competence to perform in all of these capacities. We grant to him the full status of personhood – so automatically that we are barely aware of the ontological condition accorded, or the cognitive criteria on which it is based. This is about how our minds construct their worlds according to the schemata of being human. It is about how we collaborate with playwrights in the realization of their characters as autonomous human agents, able to perform anything humans could do in the circumstances in which they find themselves, all along carrying somewhere in our potential consciousness the counterfactual dimension of fiction. How we referee between the two is a standing crux of cognition.

The excursus to follow into the phenomenology of mind pertaining to the calibration of personhood is the heart of the argument. It pertains to what we do as spectators in processing characters as persons and has applicability far beyond the play in hand. The status of Hercules as a literary Gestalt remains the case in point to which we will return in illustration of the general premise that how we read characters through our own theories of mind is largely how we read social representations in life generally – and arguably the very foundation of our interest in all literary representations of the social. It is how and why we read. Our minds do business with characters on an equivalent-to-reality basis, even psychologically, no matter how much they conform to literary conventions, because even within those conventional roles, minds are the makers, and human minds come in complete packages because they belong to a species with a generic brain design. (The phylogenetic properties of the human brain must, of necessity, pertain even to imaginary human brains or they have nothing to say to us as humans.) That same design, likewise, determines how readers and spectators process the phenomenon of other minds, and thus the psychological status we grant to them. Apparently we are hardwired to process data in this way as part of the default equipment we have as a species for representing the human environment to our consciousness. The challenge is to determine whether those cognitive processes become prescriptive for the critical interpretation of the literary persons they enable us to experience.[7] Or, do we teach ourselves, through the insistence of conventions, or the legacy of new criticism, or by some other means, to entertain "the character" according to sub-psychological categories and hence to divest them of motivated roles based exclusively on human psychology. You can see the trouble we are in.

For decades now students of literature have interpreted Coleridge's handy phrase about suspending our disbelief to mean that what we do in making that suspension amounts to a relatively self-evident exercise in pretending to make real what we know to be fiction. But recent speculations among cognitive philosophers have reopened the gnarly question of how we perform that reading of beliefs and intentions attributed to other minds by which we make judgments and predictions about their behaviour. In that regard Coleridge's mantra requires a bit of tweaking, for in effect, we do not suspend disbelief in the less than real, but impose full belief in the less than real. Arguably, we implicitly accord to all such minds the material platforms that condition the categories of emergent properties that constitute their identity and subjectivity. In that regard, everyday life entails a suspension of disbelief toward all other human agents. The literature around this debate, in fact, is vast and still growing, potentially leading any discussion of literary characterization into deep and currently debated matters concerning the "problem of other minds."[8] At issue is the degree to which these debates among philosophers of mind impinge upon our reading of literary characters, for arguing against the ease with which we imagine them as people, there is, on the one hand, the angst concerning our solipsism, our remote and imperfect means for knowing anything at all about other minds and, on the other, the fictional status of those characters which, for many critics, disqualifies them for psychological consideration altogether.[9] New criticism, throughout many years, made hay with the proposition that characters are but artifice and agency in relation to actions, and that to treat them as plenary persons is a kind of hallucination that pertains only to the most immature of readers eager to lose themselves through infantile identification. Do these issues belong to a common debate? What do our brains really do, and how vital is it to the reading experience? There is a phenomenology of reading that is brain-based in the construction of events and meanings, just as there is in the interpretation of the sensory world.[10]

In daily life, as in the simulated societies of literary representation, we cannot allow ourselves to doubt that we have the ability to read other minds and to attribute reliable propositional attitudes to them because, like kicking the stone to prove its reality, we simply know intuitively that minds think, believe, feel, and plan, that all cultural and interpersonal exchange is based on that ontological foundation, and that to challenge it in the name of fictional representation or pure make-believe (or indeed new skeptical theories of mind) must be a waste of time.[11] We read the minds of others as best we can in daily life as a first order of interest in the social world. If plays are social worlds, as well, what possible interest could we have in them if we were not invited to assess characters in similar terms, particularly if they are there for us to learn from? If the reading of characters' minds entails "belief," then how we perform cognitively in order to achieve this "belief" cannot be an irrelevant question.

As a premise, the most self-evident form of "belief" regarding literary representation is our willingness to consider characters not only as agents but as persons. The tendency is so strong that seminar discussions of, say, Shakespeare's social comedies, can barely be drawn in directions other than those based on the social viability and prospects of characters granted the full range of all that we presume to know about the minds of others in daily life – as imperfect as this knowledge must be, even to the point of projecting the chances for the happiness of characters long after the play is over. As stated earlier, this driving preoccupation of homo sapiens the gregarious and communal creature is so integrated a feature of our mentalese that a philosophical investigation into its modes may even appear supererogatory. Yet how we know the minds of others remains one of the most demanding of epistemic questions, for although we may doubt that we can perform this with scientific precision, given the nature of other minds, we know that at least our ad hoc observations, estimations, inferences, and analogous attributions serve us with a relatively high degree of analytical efficiency.

The current thinking among cognitive philosophers, developmental psychologists, and primatologists is that we take a "commonsense approach," one that accounts for the behaviour of others in terms of their desires, intentions, hopes, preferences, and phobias, and that, moreover, for many, this procedure constitutes a valid theory of mind. However, this default approach to knowledge has been assigned the term "folk psychology" because it establishes the propositional states attributed to others either on the basis of dubious empathic simulations or dubious norms. There is, in fact, a heated debate between intentional realists like Jerry Fodor and eliminative materialists like Paul Churchland as to whether the mind actually functions in terms of beliefs and desires at all, and whether the tenets of folk psychology will ever be validated by research in neurobiology and the cognitive sciences.[12] In that regard, our best option for the moment will be to join with Daniel Dennett who states in *The Intentional Stance* that probably commonsense psychology as a theory of mind will not stand up to scientific scrutiny, but that it will remain the operative approach to the evaluation of personhood in everyday life, perhaps indefinitely, simply because we have no capacity to imagine what could replace it, apart from trying to reduce all of our mind operations to neurobiological equivalents.[13] Lynne Rudder Baker in *Explaining Attitudes* likewise holds that in spite of recent cognitive and neurobiological investigations, the commonsense approach to the mental attitudes and mind states of others will remain in effect. This is to accept for the discussion to follow that some form of functionalism will prevail, and that a kind of explanatory dualism will allow us to endorse as legitimate phenomena those qualia-like features of propositional states so difficult to imagine in neurobiological terms.[14]

To all of this may be added an implicit backstory, that human brains are meaning-makers in order to orient us cognitively, and in adaptive ways, to our social environments. Just as this brain provides the neural platforms for turning sense stimuli into categories, sensations, and experience in terms of time, space, causation, opposition, and related categories, so it provides the emergent properties we identify as desires, beliefs, and volitions – each of them a modal state of thought, something like a grammar of social analysis, subdivided into the phylogenetic values, by whatever name, through which we understand the defining categories of human consciousness and action. This is all philosophically quite delicate and potentially derailed by the quale argument, yet it remains experiential bedrock. We just know by "common sense" that we all talk about personhood in these terms as we talk about the parts of trees, by whatever sounds and signs, and assume we are talking about the same things, and we just know, moreover, that as trees have bark, roots, and leaves, humans have desires, beliefs, and volitions and all the cerebral equipment required to establish identities and self-definitions in these terms. By adaptive development over vast tracts of evolutionary time, it is the default vocabulary of emergent thought properties through which we deal with other persons. Within this frame of response to our social environments, we are architecturally predetermined; adaptive circumstances built for us these emergent values for dealing with the contingent fact of other volitions. With apologies to the blank slaters, here we are. But the interpretational variables that still pertain are hardly to be underestimated.

For even the "use issue" remains problematic. Precisely what do we do when we attribute attitudes, values, and goals to others in order to calibrate social relations? Because we cannot know the minds of others directly, do we theorize those beliefs based on secondary evidence and deductive skills, or do we simulate the situation of others based on empathy, mirror neurons, or imitation, which is to say, do we arrive at such knowledge by analogy with our own beliefs and desires, or do we employ theory in order to attribute beliefs based on hypotheses about the mind states of others?[15] The entire debate is, ipso facto, relevant to any designation of personhood in literary contexts because the attributions made in relation to dramatic characters cannot differ greatly from what we perform representationally and cognitively concerning the personhood of others in everyday life. This is simply because there were no secondary pressures in ancestral environments which would incite the genes to construct a parallel psychological architecture for the understanding of fictional persons. This is not even a matter of genetic parsimony. In fact, fictional representations could have no subsequent value unless they were acted upon by the same psychological monitors pertaining to everyday life. The hermeneutics are circular. Valuable "practice" experience must come about in precisely the same terms as those pertaining to that experience in real time and spaces. In short, there is no argument to be made for aesthetic or imaginary social responses and emotions.

Paradoxically, being may be deemed fictional, but when it comes to psychological discriminations and emotional involvement with social transactions, the world is always real, including the emotions and anxieties we expend on behalf of fictional characters. That too is adaptive: better to bestow full volitional capacity and agency on inanimate objects than to undervalue those that possess them. So addicted is the human brain to the liberal imposition of its designed psychological habits that it tends to animate its entire environment with human values and even to use the institutionalized forms of that animation to regulate social and religious orders, but that is another story.

The workings of folk psychology can be filled in from many points of view, beginning with the dimensions of folk psychology in very young children; the conditions of social life dictate that at tender ages they learn to make computations in terms of wants and intentions. Whether parents are pleased or angry with them, whether acts entail rewards or punishments, whether certain configurations of events are liabilities or opportunities – these are central components of their mental universe.[16] Even if these selective preoccupations make for poor theory, what could we hope to understand about other minds that does not pertain to states of belief and desire?[17] Evolutionary psychologists such as Steven Pinker and Peter Carruthers will argue that throughout our prehistoric past, humans made progress in linking more and more complex belief states and desires to given ends, barring accident, contingencies, or competitive opposition.[18] Two areas in particular in which we display a certain virtuosity in reading other minds pertain to mate selection and group selection. Pressures in these domains undoubtedly did much to hone our skills, acting as powerful incentives to develop reasoning concerning social norms and the need to comply with them, for "with norms and norm-based motivation added to the human phenotype, the stage would be set for much that is distinctive of human cultures."[19] There is hence a tendency to reverse engineer back into the genome precisely what we do through the conscious manipulation of propositional attitudes, as Carruthers does in stating that "there may be a 'mind-reading' module charged with generating beliefs about other people's mental states."[20] The knowledge structure for doing these things may be modular, that is, composed of special neural clusters that perform these functions, or, more plausibly, coordinate the "all-over" networks for attending to them. In an evolutionary perspective, it was a matter of adapting more basic capacities of the brain to these more specific ends. Whatever the mechanisms, folk psychology is what we are confronted with when it comes to dealing with the mind states of others, for while in solipsistic terms we realize our epistemic shortcomings in reading other minds, we know that our ancestors were of necessity efficient in minimizing liabilities and maximizing opportunities in relation to the agency of others through observation and negotiation, reading causes into events, predicting by norms, and placing themselves by simulation into the circumstances of others in order to calibrate what they would do in those same situations.

Humans, like their primate analogues, have concentrated on an ever-widening range of propositional attitudes in their co-specifics – we are in fact obsessed by them – because they impinge on our own prospects for safety and self-advancement. For this reason, as well, the "personal" configurations of beliefs, desires, and memories in others have become the markers of personhood – and the status itself is constructed according to the terms of folk psychology. But the cognitive competence to manipulate these propositional attitudes in social contexts entails much more than simply being; about persons we assume a great deal of tacit information concerning the *topoi* of consciousness.[21] The list of potential attributions is long, comprising every feature that is a part of our everyday anthropology: that other persons have minds phylogenetically determined to be much like our own, that they possess emotions, social desires, survival instincts, a sense of self, experience, memories, reflexes, habits, cultural interests and acquisitions, agency and volition, powers of reasoning, a level of curiosity, temperaments, pleasures, appetites – the list could be extended. Clearly we do not do a complete topical inventory upon meeting someone new, but each classification exists as a latent frame of reference for evaluating that person's expressed propositions. (You'll never lack for topics of conversation if you work outward from all these attributes!) Whether these *topoi* are uniquely human or shared with the higher primates, mammals, or sentient creatures in general leaves much room for debate. There are many latent conditions of personhood that we may attribute to chimpanzees, yet deny to patients suffering from advanced Alzheimer's disease. Nevertheless, within "normal" ranges, we make personhood with all its attendant properties the basis for intersubjectivity.[22]

In practice, however, in stylized or intrigue-driven literary representations, for example, do we adjust the degrees to which we apply the entire vocabulary of mental properties, doling them out only as the characters require? Consider Ben Jonson's humours characters, dramatically compelling yet reduced and incomplete as humans because their obsessions make them mono-dimensional, purely emblematic, and psychologically stunted to satiric ends. Do we subtract accordingly from their plenary beings in order to see them as shadows of the human and hence worth our disapprobation and laughter, or do we see them as mere cyphers or topical excisions of the human? Once again, in terms of our folk psychology, can there be literary reductions of personhood such as tool characters, supporting or background characters as defined merely by their structurally imposed functions?

Beast fables play directly on the dissonances we experience in fully personifying animals we otherwise know to belong to far more cognitively limited phyla. That we make the genre work with such ease in our imaginations challenges personhood as a rigid ontological category, yet demonstrates how ready we are to add levels of intentionality, belief, desire, and cognitive competence in order to complete the literary Gestalt, even to animals. Arguably, tricksters likewise belong to an "inferior" status as a collection of uni-motivated deception strategies, yet, given their belief-based intentionality and above average mental capacities for projection and the creation of witty drafts of desired future events, we are inclined to fill in the conditions of personhood. To such creatures we attribute rationality, perhaps the "ideal of perfect rationality," and then revise "downwards as circumstances dictate," or do we?[23] Our working premise is that people will live up to preconceived expectations of reliability, honesty, cogency, timeliness, collegiality, and logical planning until proven otherwise. Yet even those who disappoint us in these terms do not lose their human status, despite our belittling metaphors. Literature plays a great deal around these margins. Just such expectations abet the trickster who plays the satirist among the unsuspecting, and who prevails only until his victims make that downward revision in the sense of no longer mistaking a driven ironist and self-appointed scourge as a harmless plain dealer. Even so, this remains but a chosen dimension of a complete human being. We attribute desires to agents for those things we deem them to conceive as good, and we look for the expression of their best means for achieving those ends. Such conditions for personhood may be endlessly rephrased and endlessly challenged by all manner of exceptions that perplex the assumptions of folk psychology, but to grant the basics for self-directed action is to grant a human brain with all its potential. There is no more engaging exercise in philosophical logistics, because the mind has no realities, but only representations – even of the self and its own consciousness. In that "theatre," we reckon with a language that is built around propositional expressions assumed to be about something, and we attribute at least some limited agency to those who, "all things being equal," manage their environments in terms of beliefs and enacted desires. On these grounds, we meet others person to person. The topic is a great mental tease.

There the debate must stand, that folk psychology does not function by laws of a scientific kind, but that it remains our working theory of mind for the purposes of pragmatic assessment and prediction. Because these operations are so clearly experiential, we can only assume our mental and neurological competence to conduct them. In this way, for the moment, folk psychology alone accounts for the computational representation of other minds and furnishes what understanding we have of the experiences of others. By these same operations, literary characters, and Duke Hercules in particular, become persons by dint of the intentional stances we invest in them – stances that include the self-imposed mental horizons of the trickster as satirist.

Selfhood, by contrast, is a convenient inference based on the operations of consciousness and the categories of memory connected to those operations where first person concerns are tabulated. The self, in metaphorical terms, is like many other things, a director, a theatre, a centre of gravity, a spiritual or eternal substance, a file cabinet, a policeman, but all of these are potentially misleading to the extent they localize, intentionalize, or anthropomorphize such operations. More precisely, one might say that the self is a combination of access to unique memories, a continuous internal narrative belonging to one body, a series of goals and beliefs pertaining to a unique organism and its history, or the capacity to train the attention on self-interested priorities, to make provisional responses to environmental and social challenges, and to choose among them. Cognitive philosophers have been at work on the issue for several decades and the literature increases annually. In general, however, the self at work is a set of specialized modular clusters that preserves, in continually latent forms, high-interest concerns relating to the well-being of the organism.

How these function at the material level of the brain is the subject of a variety of theories concerning the interface between the mind and its operational systems. As procedures of consciousness, however, Paul Churchland refers to the self modules as "watch hierarchies" which serve to invigilate all that passes through consciousness in order to maintain consistency with the beliefs which define the self.[24] These interest categories and attribution networks incorporate experiences, memories, drives, and instincts, as well as ethical values and declared desires – all of them emergent thought modes of a designed brain. This meta-conscious system performs iterative operations of such familiarity as to produce notions of identity and continuity reified by emotional colouring and survival urgency. For this reason, Bernard Baars styles the self as a "context that maintains long-term stability in our experiences and actions."[25] Daniel Dennett combines function and continuity in describing the "I" as the operation of consciousness that has access to all other conscious information whether as ideas, memories or passing perceptions."[26] The self has permanence because it can retrieve the same memories of past events and rebuild the life narrative around them.[27] Karl Popper and John Eccles proposed a parallel modular theory of the brain with the self at the centre of the mind's capacity to theorize upon the data of consciousness in order to generate action, making use of the mind's ability to project the future in multiple simulated forms and to establish probability priorities among them.[28] This self-system constitutes the nexus of reflexive thinking and hence the centre of existential awareness. That such a capacity, with all its attendant beliefs about its own operations, emerged in a Darwinian economy as an adaptive measure in a competitive environment is one of the great ideas of evolutionary biology.[29]

Of particular interest to the understanding of literary characters are two directions of analysis regarding the production and perception of selfhood through the applied status of personhood. The first has to do with the mental modes we employ – whether by theory, or simulation, or the projection of norms – to establish our reading of the propositional states of other minds. The second pertains to the wholeness and competence of literary characters eternally caught between the artifice of the medium through which they are presented and the ontological realism necessary for a full investigation according to the categories of our folk psychology. How do we settle the instability that exists between representational design, the elective and purpose-driven options of makers, and the representation processed as the equivalent-to-reality itself? That crux is nowhere more apparent than in those stories in which characters become their own makers, projecting the self into an artifice of its own which deceives only by being taken for a complete, viable, and psychologically operational system. What do we do as spectators who know that Fawn is the trickster projection of Duke Hercules the man, who would himself become a pure and ontological trickster? This, after all, was precisely how the Duke intended to better himself in a new social environment: by making the holiday self a permanent self. Does he become artifice in our eyes and comprehensible to us only because we incorporate the literary-mythological and operational values of the trickster schemata into our folk psychological calibrations?

However we perform these equivocations in computational terms, they represent for us operations so innate as to cause little alarm, challenge, or concern. We follow such recursions of character without hesitation, compensating for the intentions-within-intentions as workaday equivalents. But our competence in these matters, however operationally buried, bespeaks something important about our capacities for experiencing the world in remarkably nuanced ways, including the categorical ambiguities concerning personhood brought about through self-transformation, disguise, obsessions, or stylized character roles.

Such philosophical questioning has served to convert Marston's play into a kind of thought experiment concerning the psychological imperatives of performance reception. All theatrical or literary characters represent persons, and engage our attention on an "as if" basis. "Personhood" is the generic part they play in the action as one of the grand schemas of reality for which we have developed an elaborate analytical vocabulary: brain meets art created in the image of the brain. The thesis implicit in the previous paragraphs is that our reading of other minds arises in the emergent properties of a designed instrument that orients us to others in relation to their beliefs and intentions. We represent persons and their propositional states by empathy, by theory, by norms, or by a direct and efficient computation of liabilities and opportunities. By default, we do the same even with what limited information we have about Duke Hercules. It is the prerequisite to any form of knowledge of other minds. In what sense, then, a thought experiment?

It is the trickster dimensions of Fawn's character, the degree to which he performs according to structural and archetypal formulae, which tease our experiential computations. The entire argument to follow may be a red herring; it may bring us right back to the conclusion just stated. But it captures two leading *topoi* along the way: the theatrical representation of the self in daily life; and the real life reading of theatrical representations. They are very old topics, of course, of which Shakespeare himself was mindful, and perhaps they have been flogged to death, although without final resolution. Instructively, they converge in a schematic way in this play when the Duke as a theatrical creation processed in spectators' minds as real (or as a theatrical simulacrum) divests himself of that self, or personality, in order to play (or transformationally assume) the identity of the trickster scourge. Fawn thereby becomes the exemplar of the structured personality in which we can momentarily lose ourselves in our consideration of the ontology of literary selfhood. In brief, it is the old problem of the disguise plot and the determination of who's who when identities are exchanged, as in the pointless chestnut about the Renaissance anxiety over sexual identity because Aristotle recklessly opined in *The Generation of Animals* that sex changes were somehow anatomically possible by adjusting levels of heat to the genitals, the transformations conveniently made possible by the symmetry of the sexual organs. Not a Renaissance physician worth his salt, except Amboise Paré, believed it, but never mind; they understood perfectly well the functional specificities of the two sexes, and they knew too of the rapid deaths of women suffering from an extrusion of the matrix outside the body. Nevertheless, the cultural constructivists of the sixteenth century would have it that boy actors playing girls might become girls. This prospect has led recent critics to assume a universal anxiety in the collective consciousness of the age concerning the instability of sex and gender whereby all forms of cross-gender disguises become sites for experimentation with essential metamorphoses. They came to disallow that play and projection could ever be the true substance of disguise.

To disguise the self is always an external manifestation of an internal experimentation with identity, orientation, and the self motivated by curiosity, frustration, or desire. Conveniently, regarding sex changes, they have forgotten that there were no provisions for, or case studies of, boys imploding into girls in any medical literature whatsoever. In other terms, however, we do recognize template transfers, ethical reversals, and holiday makeovers of the personality, thereby juxtaposing before and after mentalities against a background of the temporal self at times grasping for identity through the smooth arrangement of a life narrative. This is very modernist thinking conveniently retrofitted onto a world largely essentialist in its outlook.[30] The approach has taken these critics back into the conversation, and helps to maintain the intensity of the dialogue over self-fashioning, the cultural malleability of the self, and the very right to deny human nature itself as a scientific master plot to imprison the species. The point is that Marston stages a metamorphosis of the self, an escape and a return, and we are curious to know what that representation means as fictional fashioning and as an equivalent-to-reality experience according to the bedrock habits of folk psychology.

Fawn performs in the image of the primitive tool trickster, a personage who is born, not emergent, and functions as a "psychologem," to use the Jungian term by which he has been archetypally cast. He embodies a frame of mind seeking entry into society merely to find social contexts for carrying out a penchant for practical joking. He is a creature, human-like, yet so mono-dimensionally intent on writing his entire biography in acts of trickery to mock and deflate others that he appears to have no other self-reflexive interiority. This is important to our discussion, to the degree we cast Fawn into this frame as the archetypal trickster. Jung explains such a mindset as an emblematic depiction of dawning consciousness endemic to eras past when people were uncertain even of the parts of their own anatomies, much less of possessing a full ethically and logically constituted mind.[31] That is to say, Jung imagined the trickster psyche as a phase of development in the emergence of the modern evolutionary human, and that hence the trickster is but a half-formed psyche crusted over by the neoteric self. I doubt this has demonstrable scientific traction, but the notion of the pre-cooperator and loner as a "psychologem" of the race has mythological appeal. The entire life of the trickster is composed of *beffe* based on the inventive opportunism whereby he creates his victims. Thus, the makers of trickster protagonists in indigenous literatures are, in a sense, asking us to imagine the half-formed minds of our ancestors and to live experientially through them in the tricks they perform. But is this a contractual equivalent to the Jonsonian theatrical world of knaves feeding on gulls as the arrested status quo, hovering between imitation and thematic hyperbole? The trickster mindset impels these protagonists, according to their natures, to exploit the fatuousness of their victims, largely through vulnerability to their blandishments, thereby displaying their own sly opportunist reading of other minds. But then the trickster must have a plenary human brain system in order to read the social intentionality of his victims. And to complete the circle, Fawn is not a half-formed consciousness in evolutionary time, but a modern man with a modern self merely imagining and projecting strategic options in therapeutic relation to a humanly plenary awareness. We have taken this "go-round" before and may do so again.

His single mental advantage is his virtuosic employment of a fundamental human survival trait, namely the ability to rehearse in the imagination a number of potential scenarios for future action before choosing the best for achieving his ends. He is a creature of planning to ends, however those ends are to be evaluated in adaptive terms. What do they give him? Pleasure in the actualized Gestalt of the trick? Pleasure in mastering his skills of deception? For paradoxically, he is as ready to make a fool of himself as he is to secure benefits, as when the Winnebago trickster gets the hyperbolical penis of the fertility god he once was whittled down to human size by a chipmunk while employing it to prize the animal out of a log.

What benefits, then, select out the psychological talents of the trickster and give them actualizing priority? In making such choices, he employs his own brand of folk psychology in reading the misplaced beliefs and goals of others whereby he hopes to deceive them. A secondary skill is his capacity to follow events from a safe distance, yet remain proximate enough to control and redirect interim contingencies. (Mosca in Jonson's *Volpone* is the self-epitomizing master of this trade.) Trickster is merely the incarnation of this singular adaptive measure. Those who are experts in such fraudulent skills enjoy distinct survival advantages. If

Jung is correct in this, however, there would be no reason for the human as cooperator ever to have emerged except as a means to contain those adept at amoral deception. Hercules as Fawn may seem to conform to the criteria of the type, and thus we may run with Jung, seeing in Fawn the archetype of the shadow, a latent force for mischief (if not worse) in every person, and a benefactor only by accident. Tricksters have no moral agendas, no empathy, no sense of reciprocal negotiation and fairness; they read other minds only to exploit or baffle them for the joy of the trick and its occasional material benefits. But Fawn, as the satirist and social regulator according to a program of grievances, is fully invested in the psychology of social reciprocity – the economy of cheaters and cooperators. His role is bent upon purging a cooperators' society of parasites, malingerers, and sycophants. His self-imposed primitivism is merely an operative mindset requisite to outing the defectors and deceivers within a social order vitally invested in the maintenance of reciprocal altruism among all its dependent members. To the plot he is a functional agent, but in himself he is a practitioner of justice by playing the scourge. Fawn embodies the trickster, puts on his mode of thinking, and perfects the operations of his psyche. But he is motivated by memories of the past, by social grievances, granting to present activities a causal continuity with all that he had been and still is beneath the disguise. Even so, the tease prevails pertaining to the plasticity of selfhood through an interiority that imagines the actualization of the self in an alternate mode.

So to what extent is the conversion of the self ever possible? Is identity essentially attached to a body, the only one in which it can ever dwell, or is it a configuration of mental states in constant transition and belonging only to a metaphysical or existential condition? Many have assumed the disembodied potential of the soul as the essence of the malleable self. Others have assumed that selves may be sufficiently reconstructed through will and habituation to constitute psychic renewal or rebirth. That is hardly the case in the play, but Hercules does express a desire to start over, to lose himself in a new frame of mind – one which may or may not be construed as existentially and ontologically valid. Is that option denied him? In the course of the play, it is an option both assumed and abandoned. Yet it has become ontologically real in the minds of those who endured reprobation and loss through the trickster's agency – sufficiently real to necessitate Fawn's recovery of his higher status in the hierarchy in order to escape retribution. Only the reality of the ducal self could save him from the reality of the trickster self. Yet, it is just those parallel realities which trouble a convenient essentialism with selves which are more ethereal, less transparent, less embodied, and more than a single point-of-view when the self succumbs to competing desires, value systems, and conduct schemata.

That same ostensible disruption of the protagonist's interior life characterizes the metamorphoses of Doctor Faustus from scholar to mage to village prankster, a man who returns to his former self, as it were, only at the moment of death. His death is the only destiny he can know, a reminder that identity is fixed to a material body, which includes the materiality of his own brain and consciousness. Yet that mind knows metamorphosis through choice, commitment to stylized forms of conduct, the passage of time, and through the potential options remaining open to him. In this way, the habits of folk psychology come to bear on the understanding of fictional worlds and their denizens. In pragmatic terms, we are no more disoriented by the disguisings of the Duke in this play than we are by friends at a Halloween party whose identities, for a time, escape our penetration. We have the computational competence for dealing with these operations, much as we can deal with stories-within-stories by seriatim narrators to a brain-designed limit. Disguise is but one layer of recursion. But while personhood in art is one crux in terms of the real, selfhood in the context of forgetfulness, volition, and the passage of time is another, proposing forms of insubstantiality and mutability within the ontology of the character. As with many of our second-nature interpretational skills, when brought to mind as computational practices, we may well wonder how we are able to perform them. The Duke, in that regard, is a collision of selves in ways that test our theories of other minds and the categories we attribute to them. Disguising merely to efficient and practical ends is one thing. Disguising as the outward manifestation of the transformed self is quite another. Identity is in the brain, and in the mind that that brain enables, the one genetically endowed, the other culturally fashioned. Will the real Duke Hercules stand up?

The Duke as Fawn can be even further problematized. All operations of the self in relation to consciousness are performatively embedded, interactive, habitual, "real" in that regard, yet initially contingent and ultimately flexible. Nor is there any ultimate escape from the theatre metaphor. As Owen Flanagan pointed out, "the brain is designed to fully cooperate in the constitution of one and only one self, although it is capable in extreme situations of cooperating in the constitution of multiple selves."[32] Perhaps that is the way out with Hercules/Fawn. Introspection may give us ideas about the moveable dimensions of the self. It may well be coupled to notions of continuity and identity, but the mind has a curious ability to preselect and simulate behaviours. It has been said that the human being is "a creature with a remarkably theatrical brain ... capable of perceiving what we now know as theatrical acts."[33] It means that minds are malleable, imaginative, and able to adopt play selves as part of their social strategies. It means, at the same time, that they are potentially devious and ruseful, and can seek advantages through self-misrepresentation. Whether this self is real or putative, ontological bedrock or social simulation, is an academic question. What matters is that the human brain may include among the gambits of planning its own transformation – or so it is led to think. Faustus proves that otherwise, and Fawn, too, must return to office. Group pressure and the discovery of one's deceit are powerful conditioners of our social continuity, willy-nilly. Trickery is not only the capacity to deceive but the capacity to compute the costs of being detected by one's collective victims. Folk psychology continues to shape the man who would shape his own soul and call him back to the terms of the psychological contract through which a continuity of selfhood is thrust upon him.

Yet dramatic worlds promote by metaphor a version of the self which is ever and always play-acting, and, by extension, that we are actors in our own lives according to schema, scripts, and roles from the very beginning, and hence that the self can be nothing other than a centralized illusion constituted by the actor in his or her chosen roles. That master metaphor is seductive; it is intrinsic to the theatre as a medium in which all characters are masks and agents of plot. Happy analogists have been ever ready to impose on the intentionality of the playwright this same association as a philosophical view, scooping up not only the making of character as artifice but the nature of the self as artifice. It is a case of reverse mimesis in which the order of art becomes the order of the constructed self. Thus, the conventions of theatre double as a theory of mind, making the art of acting the art of the self. When the character within the play assumes the competence of selfhood we have a first order challenge, but when the character assumes an alternate identity, a second order challenge arises because there is now artifice within artifice of a recursive or non-recursive kind – both difficult to manage.[34] The way out for the social constructivist is to go with the theatrical flow and constitute human nature in its image. This poses a direct challenge to folk psychology, because even artificial selves still manifest beliefs, desires, and emotions, but they now are said to belong to an aesthetic creation of a different order, necessitating aesthetic interpretation, taking us back to the evolutionary crux over parallel systems of aesthetic reception of reality. The question is whether the brain has an aesthetic psychology. Hence, the theatre seems to propose a difficult complication for folk psychology, for now there must be a primary level at which persons are represented as persons binding their actions to our best default explanations, while in parallel there is a secondary level at which all actions take place at indeterminate distances from propositional attitudes. Now we know something is wrong. Reversing this master metaphor entails the dismantling of the theatre as the maker of its own order of reality, and reinstalling the human brain as the maker of realities in its own designed image. The world of the theatre is decidedly provisional. If we lost that distinction, our lives would be lost in hallucinations or dreams. Shakespeare pretended as much on several occasions in making over reality in the image of the theatre.

What if we take some sorbet between courses and just chat a bit more about what transpires in the play? Fawn, as the self-made satirist, adopts the rather gentle strategy of the *eiron*, the calculating underdog whose innocence of manner and disarming ways leads braggarts and pedants to confession. His victims are induced to supply the information by which they are exposed. Strictly speaking, the *eiron* relies on tendentious questioning whereby his interlocutors expose their stupidities. The ultimate moment of truth is an elaborate courtly entertainment featuring a Ship of Fools of literary inspiration to which those who have been singled out for their folly willingly consign themselves. If the play has any particular defect, it is perhaps that the vices of these gulls are altogether too mono-dimensional and transparent, while the strategic confidence of Fawn is never seriously challenged. Even Duke Gonzago, the hosting Duke of Urbino, is made to join the fools for his pedantic mismanagement of his daughter's amorous escapades. Such interactions translate readily into themes concerning the categorical boasting, sexual predation and license, jealousy, insipidity, and derelict silence that characterize the respective fools.

Like Marcolphus, or the folk magus Faust, or Tyl Eulenspiegel, Fawn adopts a role of pure agency, giving the illusion that his identity is the sum of his trickster performances. He is talkative and has inventions for every occasion; he is affable, engaging, yet private, able to keep counsel, quick to seek his advantage, politic in building alliances with the court fool, and managerial in coordinating the final dramatic inset through which his gulls are one by one exposed and ridiculed. His mind is contained within his capacity to induce others to betray themselves through his action-scenarios leading to physical injury, public humiliation, or the loss of personal property in an economy of wit and ignorance. He, like Volpone, is defined by the logic of the confidence game. To take the full measure of the mental computations and cognitive efficiency of a person putatively reduced to this little measure would result in a complex and uniquely human profile. But the trickster is never so reduced. By dint of our mental machinery, we attribute plenary status even to the most marginal of creatures, as long as we can identify traces of those propositional states that are peculiar to our species. Marston relies on our complicity in seeing Fawn as an accomplished wit, alert, covertly motivated, suave, apt to meet every social style in kind, in essence the complete and perfect courtier. Trickster is the name we impose on him because of the modes he employs for social management, yet he is Fawn by name and trade, as deviously accommodating as Voltaire and Talleyrand – a work of urbane sophistication and admiration.

Marston's Hercules, as plenary personhood and trickster agent, both epitomizes and problematizes the application of our default psychological processing of the social world regarding the essence of selfhood in persons both literary and real. Hercules functions for us as a set of propositional networks and as a self in possession of individuating and unifying beliefs. He is at once the leader of a political state anxious to shed the burdens of office and a man disguised as an alternate self in adopting the logistics of the angry scourge tempered by strategic dissimulation. That we are able to make some progress in representing him to ourselves as an integrated being in multiple manifestations and to give some credit beyond mere theatrical conventions to his malleable selfhood is a marker of our capacities and impulses in attributing personhood liberally to those from whom we seek valid social information. There is nothing that Hercules does in *The Fawn* that is ultimately counterintuitive in practice. We have no difficulty in crediting his adopted self-manifestations – his mastery of the *eiron's* stance – in conjunction with his sustained interest in the dynastic future of the royal house through the management of his son's hesitant bride-quest. Above all, we attribute to him the cognitive and emotional competence to carry out his respective missions. In this very process, we demonstrate our phylogenetic penchant for knowing other minds in propositional terms. At the same time, there is a certain incorrigibility in our conviction that we can know our own mind and use it as a reliable basis for dealing with "the problem of other minds." These operations offer to ground our hermeneutic calibrations and the reception of art in the terms of embedded human cognition.

The psychological ontology of characters unexpectedly takes us back to the efforts of recent cognitivists to solve an even more ancient problem concerning the knowledge of other minds through the tilts and biases of the brain through which we fashion our impressions of alien intentionality. For the moment there is only a commonsense agreement about our practices, but little agreement about the truths and fallacies involved. Such a feature of human nature circumscribes by a wide margin the problem of literary reality, going all the way back to the matter of the Kantian-like categories by which we structure our impressions of a phenomenal world in terms we can call reality merely by convention. We consider one another as intention machines, the informing states of which can be divined only by a variety of causal computations, simulations, projections, and norms based on language and gesture. Homo sapiens are the only organisms with the competence to deal dialogically with their social environments in propositional terms and seek to read other minds in terms of their beliefs, desires, and moral values through binary grids both hedonic and ethical. This same analysis is extended to literary characters as a matter of course because they represent persons who are valueless to us unless we process them according to the usages of social reality. The principal value of the literary is therein grounded.

Folk psychology is an open category, accepting everything that we habitually do in categorizing other minds. By the law of parsimony, we never judge persons by all categories at once, but only by the categories requisite to explain the current action in propositional terms, including the stylized and schematic properties that constitute the trickster brain. Philosophers may complain about our limitations, but this generic way of seeing the social world defines and circumscribes our mental capacities and will always furnish the terms by which we not only evaluate persons but do so on the assumption that their brains are like ours in fundamental ways. The adaptive pressures that have selectively designed our mental architecture have constrained us to do so, to relatively efficient ends, in no other terms.

Hence, thinking about folk psychology is a way of thinking philosophically about the mental conditions involved in the representation of other minds, including those disclosed in the course of theatrical productions. Dramatic spectacle may enhance the process by raising to thematic awareness the trajectories of the self in action and by employing language to "mainline" the minds of others to the spectators, providing rare insights into propositional states often less accessible to us in daily life. Such approaches to character analysis may sharpen our awareness of Hercules as a composite figure appearing at a moment in the drama when, historically, certain "pattern" characters were enjoying experimental upgrading to more complex states of psychological agency and interiority in more fully realized contemporary social settings. Much of the genius of the Elizabethan theatre was in adapting the trickster of Roman and Italian comedy to the streets of London to address the flagrant values of urbane life, including the foibles Hercules purged from the Court of Urbino.

NOTES

INTRODUCTION

1 Jerome Barkow, "Beneath New Culture is Old Psychology: Gossip and Social Stratification," in *The Adapted Mind: Evolutionary Psychology and the Generation of Culture*, 627.

2 These arguments, and particularly those pertaining to the unreliability of explanatory "stories" based on imagining adaptations in ancestral environments, originated in the work of Richard Lewontin, such as his "Sociobiololgy as an Adaptationist Program," 5–14; and with Lewontin, Steven Rose, and Leon Kamin, *Not in Our Genes: Biology, Ideology, and Human Nature.*

3 For a one-stop resume of this speculative vortex, see Tim Crane, the "Mind-Body Problem" article in *The MIT Encyclopedia of the Cognitive Sciences*, 546–8. The challenge turns around matters of consciousness and material causation, but the bottom line in the entire debate never recedes: we can think, and material brains enable the production of these emergent states of mind.

4 Donald Symons, in a collection of essays openly committed to the vital necessity of returning to our readings of human conduct based on our evolutionary dispositions as a species, provides, at the same time, a clear discussion of the misuses of Darwinism, to which all modern evolutionary psychologists subscribe in their own terms. They underscore the systems of inclusive fitness and their adaptive "themes," transmitted through the genome as independent mechanisms which do not represent general goals but which make their blind contributions, functionally, to the creation of a community of behavioural strategies. Our brains are remarkably compartmentalized in the biases they create regarding food procurement, resource distribution, fighting, jealousy, and nourishing the young, and these are situationally activated throughout a lifetime, some of them designed to emerge only in the most specific of circumstances, such as feelings of nausea during the first term of pregnancy, or the sensation of falling in love. These features do not constitute a comprehensive determinism, for they do not

delete choices or prevent social organization through cultural and religious systems. Inversely, they can never be silenced and continue to shape our primary thinking, as basic to us as the promptings that make us involuntarily laugh or cry and thereby interpret our social surroundings. Genes do not determine specific behaviours. They merely "make possible novel behavioral means to the same old specific ends." "On the Use and Misuse of Darwinism in the Study of Human Behavior," 138, 139. That vital point is restated on behalf of Lena Cosmides and John Tooby – foundational contributors to the development of evolutionary psychology – by Matt Ridley in *The Agile Gene: How Nature Turns on Nurture*, 245–8: "They argued that the expressed behavior of a human being need not be directly related to genes, but the underlying psychological mechanisms could be." William Lycan, *Minds and Persons*, 107–22, situates this entire discussion through the now classic debate between the "compatibilists" and the "incompatibilists," the former being those who believe in the fundamental compatibility between limited free will, options, and agency, and a systemic human determinism. He claims that compatibilism will always win out, at least in all schools based on common sense. This debate, he points out, is as old as the ancient Stoics and was furthered by David Hume and Thomas Hobbes. The present discussion is a variant insofar as the entire crux is brought down to the computational options and systemic passions that inform the human brain, collaboratively or competitively. Mental operations are designed and determinative, yet apt for multitudinous applications. Yet there are moments in life in which those systems themselves become fatalistically rigid and at the very foundation of the tragic sense of life. See chapter 3.

5 The quotation by Williams heads up an article by John Tooby and Leda Cosmides, "Cognitive Adaptations for Social Exchange," 163; Jeremy Campbell, *The Improbable Machine*, 122.

6 The case is made most cogently or fails most spectacularly in Steven Pinker's *The Blank Slate: The Modern Denial of Human Nature*. The work sets out to demonstrate not only the conditions of what it is to be human on scientific grounds, but to reassure readers that our biological foundations in no way threaten our cultural diversity or the importance of our ethical values. As a polemic on the topic in hand, this work would be hard to surpass.

7 Daniel Dennett takes up this very point in *Freedom Evolves*. His long explanatory opening seeks to intercept the fallacy of lumping "responsible, cautious naturalists" – and he names several including himself – with "the few reckless overstaters," by foisting views upon the responsible which they have been "careful to disavow and to criticize," including the denial of consciousness and the freedom of the will, and the presumption of an in-

eluctable determinism following from the simple fact of genetic design (20). It is best to send readers to Dennett's text to consult his account of the origins of this resistance to Darwinian perspectives, their misplaced fears, and tenaciously "threadbare" arguments. Dennett associates them with Richard Lewontin, who sees himself as part of a "fire brigade" which must stop incorrect views about "IQ and race, now criminal genes, now the biological inferiority of women, now the genetic fixity of human nature," all of which must be doused "before the entire intellectual neighborhood goes up in flames." Lewontin, with Steven Rose and Leon Kamin, *Not in Our Genes*, 265. Dennett's best comeback is not only in showing the sound reasoning behind the emergence of our degrees of freedom as part of the evolutionary progress of our mental systems by which freedom can be imagined and practised, but also of showing how "*some* of our traditional ideas about free will are just plain wrong" and that those who hold them are a menace to the future of free will on our planet. He takes an entire book to reveal his point (21).

8 See Ellen Spolsky, *The Contracts of Fiction*, 90–102, 111–13, 154–5, 229–31.

9 Denis Dutton, in *The Art Instinct*, offers some profound insights into what we create and its relation to what our brains predispose us to like and to enjoy. We may create to shock and alienate, and to express our alienation, but these works find the power of their commentary on the human condition by agitating us with things designedly contrary to our hardwired predilections. Dutton explores the evolutionary platforms which contribute to our artistic production in making things colourful, comfortable, beautiful, representational, and natural, but also with the reception of art in experiential ways through the calculated violation of those same instinctual orientations. These same issues are explored by Jim Davies in *Riveted: The Science of Why Jokes Make Us Laugh, Movies Make Us Cry, and Religion Makes Us Feel One with the Universe*, little of which can be accounted for without the genetically grounded predilections of the brain to favour and respond to all the prepared stimuli of culture, designed in accordance with what it is to have a human brain. In this sense, culture can be meaningfully conceived only in the image of the mental instrument that makes, uses, and enjoys it.

10 Ridley, *The Agile Gene*, 247. The quotation by Cosmides and Tooby is from Ridley, 245–6, but comes originally from "The Psychological Foundations of Culture" in *The Adapted Mind*.

11 I will not interrupt here to pose, rhetorically, the same question in reverse. If evolutionary processes did not homogenize our natures over time, how could such universal traits from smiling all the way back to temporal sequencing, syntactical organization, the social uses of gossip – the list is

long and growing – be reinstantiated culturally the world over with each succeeding generation? How do parents first teach their children how to learn and remember? Memory-making can only be an innate faculty with a growth history of its own, genetically controlled. There are reasons why first memories only go back to, say, four years old. And there are reasons why we remember in the quantities and qualities that we do – all of it well beyond cultural instruction.

12 Elliot Sober, *Philosophy of Biology*.

13 Jonathan Kramnick, "Against Literary Darwinism," 316.

14 I call Jim Davies as my witness, whose book *Riveted* is concerned with the many ways in which our inherited natures predispose us to all manner of responses, preferences, biases, attractions, and addictions which shape our daily lives, and in ways so integrated to our experience as to almost escape notice. These tropisms of our natures define our cultures, and Davies is keen to show the common features that underlie all the many cultural diversifications of their expression. Yet no one could be more cautious: "I hope that this book has persuaded you that evolution has had *some* influence. Evolutionary explanations themselves are really compelling – people tend to give evolutionary explanations more credibility than they should." His concern is that readers themselves not become enthusiastic reductionists. "Culture affects preference." Where real judgment begins is in the apportioning of these influences, judiciously and cautiously (205).

15 Kramnick, "Against Literary Darwinism," 316.

16 Lewontin, "Sociobiology as an Adaptationist Program," 11.

17 Jonathan Cohen in "Vulcanization of the Human Brain," 13, responds to this challenge as follows: "the approach of drawing upon evolutionary explanations to understand neural and psychological mechanisms – sometimes referred to as evolutionary psychology – is highly controversial and has been criticized for offering nothing more than a series of 'just so' stories. However, even if scientists are never able to establish with certainty the specific evolutionary or developmental course of a neural or psychological mechanism, the evolutionary perspective can nevertheless be used to generate testable hypotheses that can lead to deeper insights into the nature of the mechanisms involved and the circumstances in which they are likely to be engaged." His entire study of the anomalies of rational and emotional decision-making is predicated on the sometimes awkward interface between these two independent brain systems, their asymmetry to be explained only in developmental terms, and in relation to the ancestral values which they were designed to deal with. Such hypothesizing provides critical insight into the challenges of modern moral, technological, and economic choices, choices which may far exceed in their demands our current

mental systems and dispositions. As the director of Princeton University's Centre for the Study of Brain, Mind and Behavior, his mandate is to use all the resources at his disposal, including MRI facilities, for correlating neural and behavioural patterns, in order to better understand the valences, impedances, and cues which activate these concurrent hedonic and rational decision-making systems, in conjunction with the dopamine-based reward systems activated deep in the brain stem.

18 An account of that massive migration out of Africa about 100,000 years ago down to about 15,000 to 35,000 years ago when the Bering Straits were at last crossed, and the many forward and retro movements between Europe and Asia is discussed and illustrated by Luigi Luca Cavalli-Sforza in *Genes, Peoples and Languages*, 92–6.

19 The example and its attendant historical application is derived from Cavalli-Sforza, *Genes, Peoples and Languages*, 43–4.

20 Bernard Williams, *Truth and Truthfulness*, 22.

21 Ibid., 24.

22 Kramnick, "Against Literary Darwinism," 317.

23 Brian Boyd, *On the Origin of Stories*, 69.

24 Davies, *Riveted*, 206.

25 Boyd, *On the Origin of Stories*, 73.

26 Kramnick, "Against Literary Darwinism," 318.

27 It would be supererogation here to replay the history of modularity, associated essentially with the work of Jerry Fodor, or to repeat the conditions of its short-lived success at the hands of Daniel Dennett and others, critiqued as it was in *Consciousness Explained* (260 and following). Important here is that modularity was but one provisional model among many for explaining consciousness according to its properties, and that its limitations were evident to many early observers and thus need not be taken as the chasm of error into which all literary Darwinians have fallen. Dennett was even reserved about his own "Multiple Drafts" model which, if it does not serve phenomenologically, serves in an explanatory capacity. Tilting at modularity is now tilting at windmills.

28 Patricia Smith Churchland, "How Do Neurons Know?" 57.

29 Kramnick, "Against Literary Darwinism," 322.

30 Quoted by Ridley, *The Agile Gene*, 72.

31 Donald, *A Mind So Rare*, 215.

32 Ibid., 206.

33 Ibid., 215.

34 Ibid., 207.

35 David S. Wilson, "Evolution and Social Constructivism," 20–37.

36 Dennett, *Freedom Evolves*.

37 Geary, "Folk Knowledge and Academic Learning," 503.

38 Ridley, *The Agile Gene*, 6.

39 Geary, "Folk Knowledge and Academic Learning," 509.

40 This is a relative point, for no one is denigrating the power and importance of language – our best and latest invention. But Antonio Damasio goes to lengths to demonstrate why the brain does not run on language, why thought is not language, or even linguistically framed. The brain runs on emergent sensations, properties, essences, representations, some of which are perceptual in nature, all of them, as brain properties, predating language by millions of years. *Self Comes to Mind*, 67–94; it is a point he made in *The Feeling of What Happens*, 107–8. Palmer underscores the point in *Fictional Minds*, 92.

41 Palmer, *Fictional Minds*, 147.

42 McGinn in *Mindsight* has written most cogently on the matter of the perceptual and imaginary modes.

43 Doležel, *Heterocosmica*, 63.

44 This is a very light application of what Dennett meant by such stances, which depends on the several levels at which we interpret intentionality, whether in terms of design and systems or in terms of beliefs and desires. For an efficient introduction to his thought, see Dennett's "Intentional Systems," 220–42.

45 This involves the now famous philosophical debate concerning the mind as a spiritual phenomenon ontologically different from the material body to which it is attached and the physicalist view of the mind as simply another emergent property of a material brain that happens to include a capacity for meta-reflection. As Edward Slingerland points out, it all goes back to the German division of the world into the "natural sciences" and the so-called sciences of the human spirit, *Geisteswissenschaften*, in "Who's Afraid of Reductionism? (377–8). His article tackles head on the problem of the "altar" of social constructivism and the maintenance of the mind-body divide through assaults on the putative "reductionism" of evolutionary psychology, which Slingerland dispels through a forceful and compelling defence.

46 Patricia Smith Churchland, "From Descartes to Neural Networks," 52.

47 The matter of the mind as a biological computer in relation to mechanical computers was taken up most engagingly by Jeremy Campbell in *The Improbable Machine*. It was there I read, *en passant*, how illogical the mind is in classical terms, but how efficient in estimative and approximate judgments – plenty good enough for what the creature must accomplish in its world, with plenty left over for play applications.

48 Robert Wilson, "Philosophy," 18.

49 Pinker discusses this evolutionary economy pertaining to altruists and cheaters in *How the Mind Works*, 202–4.

50 Barkow, "Beneath New Culture is Old Psychology," 628.

51 Zunshine, *Getting Inside Your Head*, 14.

52 Symons, "On the Use and Misuse of Darwinism in the Study of Human Behavior," 138.

53 Doležel, *Heterocosmica*, 73.

54 One of the most careful and thorough studies of this compelling matter is Richard D. Alexander's *The Biology of Moral Systems*, which will figure largely in future discussions.

55 Pinker is right in the main that "no sharp line divides thinking from feeling, nor does thinking inevitably precede feeling or vice versa." *How the Mind Works*, 373. But there are moments when, in that vice versa equivocality, we must come to moral and forensic judgments for reasons ranging from extenuation to tragic regret to the full weight of the law.

56 Robert Wright, *The Moral Animal*, 203. Chimpanzees form alliances, share, hug, groom, fight together, and react with violence to betrayal. Their repertory of social emotions should come as a reminder that we too are primates and have inherited an even more socially designed set of neural networks.

57 Damasio, *Looking for Spinoza*, 228.

58 This has been studied by Richard Alexander in his seminal work *The Biology of Moral Systems*.

59 This may be seen in the case of aggression, its social provocations, and the systemic responses that shape or take possession of thought. The system is biological with its varying levels of entry, just waiting for a critical mass of perceived provocation, especially with relation to insult or deprivation of legitimate resources either to the self or to kin or clan. Bernard Chapais, "Primates and the Origins of Aggression: Power and Politics among Humans," 214.

60 Suzanne Keen, "A Theory of Narrative Empathy," 207–36; Boyd, *On the Origin of Stories*, 64.

61 Among the protesters is Joseph Carroll in *Literary Darwinism*; for a more recent account, see Blakey Vermeule, in *Why Do We Care about Literary Characters?*

62 Herman, "Towards a Socionarratology," 233–4.

63 Dennett, "Intentional Systems," 227.

64 Adam Zeman, "Does Consciousness Spring from the Brain?" 290.

65 On recursive thought, see V.S. Ramachandran, *The Tell-Tale Brain*, 163. Lisa Zunshine, in *Why We Read Fiction* (27–30), relies heavily on cognitive theories of recursion and its evolutionary and adaptive origins to account

for the layered strategies of plotting, setting up this essential matter as the basis for "a genuine interaction between cognitive psychology and literary studies, with both fields having much to offer to each other" (27).

66 Boyd, *On the Origin of Stories*, 115.

67 Schank and Abelson, *Scripts, Plans, Goals, and Understanding*.

68 See also Pinker, *How the Mind Works*, 552.

69 Palmer, *Fictional Minds*, 176.

70 This topic is at the centre of a study of the "visual imagination" in V.S. Ramachandran and Sandra Blakeslee's *Phantoms of the Brain*, 85–104.

71 For further perspectives on the investment of our own cognitive categories in building up fictive worlds just short of the real, see David Lewis, "Truth in Fiction," 37–46.

72 John Searle, *The Rediscovery of Mind*, 135–6.

73 Jonathan Culler called this phenomenon "the unseemly rush from word to world" in ways far transcending the text. *Structural Poetics*, 130.

74 For a discussion of these terms, see Thomasson, "Speaking of Fictional Characters," 211.

75 Richard Gerrig and David Rapp, "Psychological Processes Underlying Literary Impact," 266–7.

76 Jerome Brunner, *Actual Minds, Possible Worlds*, 38.

77 Green, "Narrative Worlds and Self," 53–75. See also Keen, "A Theory of Narrative Empathy," 212.

CHAPTER ONE

1 The Faustian self in the context of Renaissance culture will be implicit in much that follows, for the contingencies in his mind were conditioned by the enticements of humanism, radical scientific theory, and the Christian world order, both theological and political. The Faustian experience in relation to the conditions of selfhood in their philosophical contexts is easily placed in the section entitled "Between Ancients and Moderns," in Jerrold Seigel's *The Idea of the Self: Thought and Experience in Western Europe since the Seventeenth Century*. Seigel distinguishes between the cosmic terms defining the self in medieval thought and the cosmological revolution of Copernicus and Newton, which "left these ideal harmonies in ruin" (54). He, in fact, pushes the "'discovery of the individual,'" typically "reserved for the fifteenth century back into the twelfth" (52). To the degree that the Faustian self-experience is measured out entirely in terms of *Zeitgeist*, ideological eras, and intellectual history, there is a well-established and persuasive argument to be made for the tragedy that arises from embracing, seriatim, the dominant but antithetical systems of early modern

thought, in the upward-downward directions of a moral universe. But in those arguments, Faustus is merely an allegorized agent in a drama about the history of ideas. The present study seeks to restore the felt qualities of personhood experienced by the protagonist.

2 This statement glosses over the distinctions that have been raised concerning personhood, self, and identity. "Selves are a subset of persons," in the words of Valerie Hardcastle; "You can be a person and not have a self, but you cannot be a self without being a person" (22). Selves are more aware, reflective, and cognizant of the identifying cluster of memories and life events that they advance in proof of who they are. Personhood pertains to the complex of properties from which self emerges: the capacity to desire, believe, reflect, plan and remember – the personhood that we attribute to others by dint of their having a brain like ours. You can respond to issues of identity as a self only when you can actually give a comprehensive and coherent account of a life in which you are the protagonist, to paraphrase Lynne Rudder Baker, *Persons and Bodies*, 14, 87. Identity may be that which is granted to us by others, or constructed through experiences of our own making and reckoning. Always there are the matters of false memory, confabulation, and so many other factors in building the autobiographical self. See also Bernard Baars, "The Director: Self as the Unifying Context of Consciousness," in *The Theater of Consciousness*, 142–53, and Rom Harré, *The Singular Self: An Introduction to the Psychology of Personhood*, 146.

3 Dawkins, "Selfish Genes and Selfish Memes," 142.

4 Rosaldo, "Toward an Anthropology of Self and Feeling," 140.

5 Goldman, *Simulating Minds*, 3.

6 Ekman and Friesen, "Constants across Cultures in the Face and Emotion." Discussed by Pinker in *How the Mind Works*, 215.

7 For further reflections on the ontology of personhood, see Hardcastle, *Constructing the Self*. The phrase itself I credit to my graduate student Patrick Juskevicius from his work on the ontology of fictional universes.

8 Humphrey, "The Uses of Consciousness," 78–9.

9 Harré, *The Singular Self*, 72.

10 See S. Cahill, "Toward a Sociology of the Person." It is still too early in the argument to deliver a record of the magnitude of research in recent years that seeks to describe our unique capacity to grant ontological substance to products of the imagination, and particularly where literary characters are involved. But the moment has come to shake up literary criticism with a reality check concerning what minds actually do with such characters by way of granting them the substance necessary to react with them in full social and psychological terms.

11 Brothers, *Friday's Footprint*, 4.

12 Unger, *Identity, Consciousness and Value*, 41.

13 Orgel, "What Is a Character?" 8. This is a radical form of what has come to be known as the "anti-realist" school among philosophers concerned with the ontology of imagined entities, such as literary characters, epitomized in the work of Kendall Walton. In essence, if we do not believe in any working fashion in the substance and reality of characters, yet come to care about them in the context of stories, then we must be involved in some elaborate and convention-bound form of pretend or make-believe to which we willingly submit ourselves. The extensive writing based on what is interpreted from words and texts in support of this view takes us down to the precise things that can be experienced in relation to patently make-believe sequences. Walton famously evokes the mud pies of children in which rocks represent raisins, and the degree to which interactive social worlds can be spun out of the convention of treating such pies as real. "Pictures and Make-Believe," 287. Walton then struggles to explain how we move on to statements of truth and falsehood in relation to the patently fictional of which we are uninterruptedly cognizant. All that must be part of the pretense as well. This mechanism must ultimately be elaborated to unwieldy proportions.

14 Orgel, "What Is a Character?" 9.

15 Again, the reader is spared a long detour into the extensive literature concerning the ontology of the imaginary, now from the "realist" position, but its existence must be signalled. There are cognitive and experiential design features of the brain that inform the way we "imagine" other persons in ontological terms, leaving only the kinds of reservations that we might attribute to the quality of the image that occupies the absence of vision where the optic nerve is situated at the back of the eye. The brain "fills it in" basically by not even looking for data from that sector. The point is, the way we grant ontological status to our representations amounts to a technical lie, but one so integral to the way we construct data that only the bitten scholastic would constantly challenge it. That holds equally true for characters. The arguments build themselves around the ways in which we treat these creations as social entities. We reason about them through inference, elaborate upon their lives without concern for explicit textual authority, assess their emotions and thoughts, and build worlds around them for the most adaptive of reasons. Our own brains are wired to do this because it is vital that we communicate with them on a reality basis. The experiences we gain in doing so are the more perfectly attuned to the meeting of their exact equivalents in real life. Brian Boyd advances similar claims in *On the Origin of Stories*. Melanie C. Green further supports this in "Nar-

rative Worlds and Self." Gordon Bower and Daniel Morrow in "Mental Models in Narrative Comprehension" go further in describing the way full representations are created through supplemental details based on mental schemata. Debora A. Prentice, Richard Gerrig, and Daniel Bailis discuss how we actively participate in the building of equivalents to ontological worlds in "What Readers Bring to the Processing of Fictional Texts." Peter van Inwagen calls the process "ascription" in "Creatures of Fiction." Amie L. Thomasson in "Speaking of Fictional Characters" emphasizes precisely the way in which we speak of such characters as real, thereby revealing the degree to which we have operatively acquiesced to their *de re* qualities.

16 Harré, *The Singular Self*, 115.

17 Vermeule, *Why Do We Care about Literary Characters?* 15.

18 That emotional reactions to stories are markers of the real in literature goes back to Aristotle, whose theory of tragedy turned around the felt communal response generated in spectators by the decline of the protagonist. See also Suzanne Keen, "A Theory of Narrative Empathy," who reiterates the position that our ability to feel for others as we might for ourselves is an embedded design of the brain that promotes the sensitivity needed to enhance social cohesion (209) – Aristotle in a cognitive key. Richard J. Gerrig and David N. Rapp carry the argument of the ontology of reader involvement as confirmation of the ontology of the represented to its greatest lengths in emphasizing not only the mental worlds we spin around stories to give them a "materialized" substance, but the degree to which we are "transported" by stories into a substitute world for which we exchange our own. "Psychological Processes Underlying Literary Impact," 267.

19 This paradox must be named because it is a leading idea in recent studies on selfhood. It is particularly foregrounded in the title of a book by Bruce Hood, *The Self Illusion*. His study sets up many of the *topoi* which the experienced self entails: a community of mental concerns which constitute the semblance of unity and oneness from which we operate and plan, the anomalies of determinism in the genetic configuration of that internal dialogue versus the self as the centre of the free will operator, and the degree to which the self is dialogically composed through social interaction, given that who we are is often composed of who we are expected to be. He deals with the working inaccessibility of the self, which is there to perform for the organism and not to polarize consciousness over its own operations, as well as with the self as a narrator and point of view. He covers all the topics concerning self as a precious and necessary illusion through which we, nevertheless, achieve a sense of interest, point of view, autobiographical singularity, drive, ultimate concern, and place in the world. Hence, illusion though it may be, like the characters also said to possess them, the

self remains the bedrock of identity, personhood, and life orientation. We may wish to avoid this endless paradox of the non-entity that nevertheless functions with full ontological clout and efficiency, but there seems to be no way out. See also Hardcastle, *Constructing the Self*, 4. This idea is as old as Friedrich Nietzsche, who stated in *The Will to Power* that "the subject is the fiction that many similar states in us are the effect of one substratum." He goes on to play with the idea of the multiplicity of subjects in one body (269). See also M.W. Katzko, "Unity vs. Multiplicity: A Conceptual Analysis of the Term 'Self' and Its Use in Personality Theory."

20 Harré, *The Singular Self*, 3, 4.

21 Michael Gazzaniga wrestles with the problem of binding whereby the brain manages to generate this unity of impression from those diverse operations originating throughout the brain that contribute their properties to the self experience, bringing him to posit an "interpreter" module associated with the left lobe furniture pertaining to language. There must be a final arbiter somewhere that rectifies, finalizes, and blends the unified self, if this does not enter into a misleading personification metaphor in its own right, because ultimately there is no playing space and no audience; there is merely an experienced presentiment of mind creating what it is like to be a self. *The Social Brain: Discovering the Networks of the Mind*.

22 Identity will always prove troublesome, if essential. It constitutes both who we take ourselves to be and who others take us to be. Identity is more bestowed than selfhood, in a sense. Thomas Henricks described identity and the self together as "a projection of personhood, a pattern of commitment and connection that carries individuals through events." We have intimations at once of the essence he is looking for, yet recognize at the same time the limitations of language for describing such matters. *Selves, Societies, and Emotions*, 47.

23 Nearly every scholar of the self points out this unifying feature, the singularity of perspective and experience which characterize the self as a point of view, yet invariably all scholars must tackle the question of the cultural and social forces which impinge upon self-formation. See Linda M. Breytspraak, *The Development of Self in Later Life*, 15.

24 Unger, *Identity, Consciousness and Value*, 44–50. If there are gaps in that time, we have no way of knowing them because consciousness cannot experience them. The nature of self and identity admits of no gaps, no time in which the organism is suspended. The default sensation of continuous time keeps us operational and smoothly functional, but it is connected to the incorrigibility that makes us "think that we have been right about our *past existence* relative to the great majority of possible scenarios" (46).

25 Or, in the words of Antonio Damasio, "consciousness begins when brains acquire the power, the simple power I must add, of telling a story," for that

telling is the beginning of a sense of the self. *The Feeling of What Happens*, 30. Of course the psychobiography does not really come to fullness much before adolescence, when the pressure to assert the self in a social world becomes imperative. Only then does a fully integrated version of the self appear. See T. Habermas and S. Bluck, "Getting a Life: The Emergence of the Life Story in Adolescence." On narrative as a vehicle of self-creation, see the essays in McAdams, Josselson, and Lieblich, *Identity and Story: Creating Self in Narrative*.

26 Jerome Brunner sees these as differences between the private self deemed free of cultural definition, and the public self shaped by cultural contexts. But such sealed categories are best seen in a far more fluid tertium quid, which he set out to define in *Actual Minds, Possible Worlds*, 68.

27 Damasio, *The Self Comes to Mind*, 287. On the vacillations of the protagonist, see Douglas Cole, *Christopher Marlowe and the Renaissance of Tragedy*, 121–47.

28 See D.P. McAdams, *Power, Intimacy and the Life Story*, 1–5.

29 McAdams, "The Psychology of Life Stories," explores "story" as a vehicle for registering the evolving conditions of selfhood, both as it is told and as the instrument for knowing one's own sense of being in time.

30 Further to narrative as a feature of the self in search of integration, see Harré, *The Singular Self*, 87. Here for the debate, as well, is the dissenting view of Brian Boyd, who does not embrace the near "truism" of speaking "of the self or of experience as fundamentally narrative," for he can imagine ways in which experience is construed without rendering it into a narrative form. *On the Origin of Stories*, 159. Perhaps. But the fact remains that the brain knows reality in spatial and temporal terms, which surround and causally organize the impression of events in temporal sequences, just as we know our own life experiences in temporal sequences, and these default constructions of reality generate thought states tantamount to narratives, because events are known seriatim in causal chains, and these narratives serve in turn as the *aides memoires* to self-construction.

31 On the illusion of self-worth sought from public opinion, see Devos, Huynh, and Banaji, "Implicit Self and Identity," 165. Regarding the supporting role of the emotions, the self is a teleonomic instrument, seeking beneficial ends with or without intentionality.

32 There are two studies worth signalling at this juncture, the debate between Sydney Shoemaker and Richard Swinburne in *Personal Identity*, in which Shoemaker takes a materialist and genetic approach to the brain's construction of the self, and Swinburne develops a modern theory of the soul. Todd Feinberg, in *Altered Egos: How the Brain Creates the Self*, begins with the clinical treatments of crippled and fragmented minds in which the unity of the self is lost as the basis for anatomizing the multiple operations

of mind which the self-operation draws together into an illusion of oneness, synchronicity, and unity of experience.

33 There are no shortages of studies now debating that very point: that our brains, made increasingly fit for our complex social and moral environments, by their emerging design, were predestined, repeatedly, to find religious solutions to matters of rewards and punishments, creation, the ultimate father, and the regulation of persons in groups. See, for example, Newberg, D'Aquili, and Rause, *Why God Won't Go Away: Brain Science and the Biology of Belief*.

34 This dual vision of the world has also been traced to the design of the brain and referred to in various ways, such as "the binary operator" of mind. Newberg et al., *Why God Won't Go Away*, 50–1.

35 Tomkins, "Script Theory," 170.

36 On the self as a series of conflicting positions see Hermans, "Voicing the Self," and McAdams, *Power, Intimacy, and the Life Story*, esp. 5–18.

37 Feinberg, *Altered Egos*, 30.

38 Clifford Geertz describes the narrative challenges in organizing all the cloudy randomness of materials and impressions that contribute to the unified self as an exercise that even God could not meaningfully organize. Such was his sense of the difficulties in accounting for the impression of the unified self that we typically enjoy as a component of the many-featured self. *After the Fact: Two Countries, Four Decades, One Anthropologist*, 2.

39 H. Gardner, *Creating Minds: An Anatomy of Creativity Seen through the Lives of Freud, Einstein, Picasso, Stravinsky, Eliot, Graham, and Ghandi*, 32.

40 Willis, "Doctor Faustus and the Early Modern Language of Addiction."

41 Davison, "'Houses of Voluntary Bondage': Theorizing the Nineteenth Century Gothic Pharmography."

42 Hermans, "Voicing the Self," 31–50.

43 Gregg, *Self-Representation*, xiv.

44 Hermans and Dimaggio, *The Dialogical Self in Psychotherapy*. Hermans is the founder of the "dialogic self" theory, which could be extended into a hermeneutic methodology for the reading of this play and of all characters, the formations of whose lives are presented in explanatory dramatic sequences. He sees the self as "an organized process of meaning construction," which implies that as long as there is life, reconstruction is also possible by changing the terms of the dialogue (psychotherapy). We are back to an old debate, but at this moment in the Faustian experience, the prospect of conversion through the weighing of inner voices resembles the hope of such therapy.

45 McAdams, *Power, Intimacy and the Life Story*, 5. See also R.F. Baumeister, *Identity: Cultural Change and the Struggle for Self*. Harré and Langenhore

take up the question of the positioning of the self in relation to a conversation, and the struggle toward a necessary integration of the self as a harmonious and single community in "Varieties of Positioning."

46 Baltes, "Life-Span Developmental Psychology." Peter T.F. Raggatt discusses the "inner-speech" of Mikhail Bakhtin in these terms, describing how the "I-positions" may change during the passage of time and will therefore activate new voices, each struggling to modify the dominant schema of the self. "Multiplicity and Conflict in the Dialogical Self: A Life-Narrative Approach," 18.

47 Baumeister, *Identity: Cultural Change and the Struggle for Self.* The debate between personal liberty and institutional conformity, between the constitutional self and the socially embedded self, merely adds perspectives to the familiar debate more typically constructed as the sinful egoist and obedience to divine revelation in the simplest of Everyman terms. This play invites a more generically human analysis, given the terms and frames of mind in which the protagonist expresses the dialogic struggle for the self. For further perspectives on the personal and cultural in the construction of the self, see Rosenwald and Ochberg, *Storied Lives: The Cultural Politics of Self-Understanding.* Repeatedly, the perspective will emerge that the self is a challenge in integration, a construction of the unified, harmonious, and "monophonic" self, critical to a sense of oneness and well-being, yet one that must negotiate among ideological stances. *Doctor Faustus* takes on that dimension as the debate between master narratives becomes more potent and insistent as the protagonist's life advances. Ultimately the conditions of thought challenge the illusion of integration, bringing Faustus to new choices concerning the orientation of the self, which he is powerless to resolve. On the integration of the self, see Peter Raggatt, "Multiplicity and Conflict in the Dialogical Self: A Life-Narrative Approach," 16–17.

48 Raggatt, "Multiplicity and Conflict in the Dialogical Self," 21.

49 Trivers, *Social Evolution*, 420.

50 How we formulate selective, partial, and subconsciously produced beliefs into self-defining absolutes is the subject of a recent study by Kathryn Schulz, *Being Wrong: Adventures in the Margin of Error*, and particularly her chapter on "Beliefs," in which she profiles the way we make models of the world without a full awareness of their repercussions, taking them as "mirrors in which the truth of the world passively appeared," thereby explaining Faustus as a kind of "naïve realist" because he could not imagine himself believing in things that are wrong – until it was too late (99–100).

51 Pinker, *How the Mind Works*, 424.

52 On this positioning of the self, with its disinclination to combine the incompatible elements of identity, see R.J. Lifton, *The Protean Self: Human Resilience in an Age of Fragmentation*, 50.

53 This is from Vermeule's discussion of the moral imperatives set out by Immanuel Kant in the treatment of other persons, that the plenary consideration of what it is to have personhood is ultimately a moral act in bestowing upon others their rightful moral worth. *Why Do We Care about Literary Characters?* 25.

54 Many dimensions of these schemata have been discussed, pertaining to the ways in which the mind establishes expectation profiles for many of the aspects of material and social reality, including the rituals of social conduct. Tricksters are in the business of understanding those schemata more clearly than their intended victims in order to play up to their expectations with hidden motives in mind. Goffman called these structures "frames" which become the interpretation boxes or models of what is happening, whether signs of aggression are real or merely a form of play. We use frames for reading the intentionality of others – intentions that can be faked and forged. The trickster becomes a specialist in abusing the interpretational value of such frames. See *Frame Analysis: An Essay on the Organization of Experience*, which is reinterpreted in Thomas S. Henricks's *Selves, Societies and Emotions: Understanding the Pathways of Experience*. It is everyone's desire to understand the nature of an activity in the same manner as all other participants in order to know how he or she should react and think. Such successful intelligence also pertains to safety and benefits for having accurately interpreted the environment or situation. Tricksters gain personal enjoyment and advantage in destroying this confidence by proving useless the employment of such frames (16).

55 Mahadev Apte, *Humor and Laughter*, 214.

56 Paul Radin, *The Trickster: A Study in American Indian Mythology*, with commentaries by Karl Kerényi and C.G. Jung, xxiv.

57 Jung, "On the Psychology of the Trickster Figure," in Radin, *The Trickster*, 201.

58 Dawkins, "Selfish Genes and Selfish Memes," 140–1.

59 Goffman, *Frame Analysis*, 83.

60 Whiten, "Machiavellian Intelligence Hypothesis," 495. Among the thinkers behind this idea are Nicholas Humphrey, "The Social Function of Intellect," and R.W. Byrne and A. Whiten, *Machiavellian Intelligence: Social Expertise and the Evolution of Intellect in Monkeys, Apes and Humans*.

61 Vermeule, *Why Do We Care about Literary Characters*, 53.

62 Michael Keefer offers an excellent discussion of these matters in *Christopher Marlowe's* Doctor Faustus: *A 1604-version edition*, xxiii–xxxiii.

63 There is a decided temptation at this point to read the play as a spiritual confrontation between two value systems, one that is backed by the adaptive biases behind the quest for power, possessions, respect, and the com-

petitive edge, and another which is described by Andrew Newberg in terms of the designed brain's incessant craving to invest in transcendental unity, "whether this ultimate reality actually exists, or is only a neurological perception generated by an unusual brain state." Newberg et al., *Why God Won't Go Away*, 165. This approach relates to a new literature proposing that the God phenomenon is an inevitable product of the human brain, and that the variations among world religions are little more than cultural diversifications upon common emergent values generated by the animistic, personifying, and mystical properties of thought. One might argue that Faustus, at this juncture, has produced the eleventh-hour God-factor according to the principles of brain design in much the same way that brains are designed to produce the self-factor. For samplings of this approach to the innate spirituality of the human mind, see Guthrie, *Faces in the Clouds: A New Theory of Religion*, Karen Armstrong, *A History of God*, and especially Pascal Boyer, *Religion Explained.* These are intriguing studies, suggesting that the God phenomenon is a feature of the way brains construct reality, and that it is hence a human universal easily converted to a psychological quest, and thus a factor in the life of the mind of the species, of whom Faustus is a representative. This would indeed place a new and important emblematic construction upon the unfolding of the Faustian experience and provide an explanatory mechanism for his final Angst. Vermeule broaches this idea in the discussion of universals, stressing the spirit of gossip by comparing it to the spirit of religion as two equally "thirsty" aspects of mind: "We humans are commonly said to have 'a God-shaped hole' in our souls. If you are a religious person, you can explain the hole by saying that God put it there in order to make it easier for us to receive Him. If you are a naturalist or an atheist, you believe the God-shaped hole is in our minds, not our souls. You then look for reasons that the concept of God might have evolved in our species." But she does not resolve the matter of the ontology of the God-hole in the brain. *Why Do We Care about Literary Characters?* 10.

64 Boyd, *On the Origin of Stories*, 64. This is the world order of the medieval Everyman, which is to say, the categorical condition of his being.

65 Unger, *Identity, Consciousness and Value*, 39.

66 Breytspraak, *The Development of Self in Later Life*, 104–5.

67 Soliloquies employed in this way are both a convention of theatre and a convention for representing the self through "descriptions of what is happening inside the subject's brain." Humphrey, *The Mind Made Flesh*, 72.

68 Petrarch was obsessed with this experience and what constituted the plenary transaction – one which he began on many occasions, yet abandoned as a totalizing transformation of mind which separated him from the beauties

of the humanist intellectual mindscape. You can read all about it in the *Secretum*, as well as in a chapter to follow in the present collection.

69 Perspectives from Henricks, *Selves, Societies, and Emotions* (29) are running in the background.

70 Unger, *Identity, Consciousness and Value*, 5. Unger speculates on the issue as it touches modern subjectivity and the extent to which the status of the soul continues to be a factor of identity (36).

71 Lüthi, *Once Upon a Time*, 106.

72 Hardcastle, *Constructing the Self*, 45.

73 Ibid., 24.

74 Schechtman, "Personal Identity and the Past," 21.

75 Frankfurt, *The Importance of What We Care About*, 170.

76 Giddens, *Modernity and Self-Identity: Self and Society in the Late Modern Age*, *passim*. There is a sense in which he becomes a wandering and unattached self now marking a new permeability of borders, breaking away from regional identity to become a kind of personal nation.

77 Berger, *The Heretical Imperative*, 26–7.

78 James, *The Principles of Psychology*, 202.

79 Raggatt, "Multiplicity and Conflict in the Dialogic Self" (19): "Many theorists of the self would concur that narratives of self are positioned in a matrix of social and moral relationships." This sentence alone might be employed as a point of departure for a complete analysis of the life trajectory of Faustus in the play.

80 LeDoux, "Emotional Colouration of Consciousness," 69. The girth of his bibliography on the topic is impressive.

81 Henricks, *Selves, Societies and Emotions*, 112.

82 Damasio, *The Feeling of What Happens*, 168–94.

CHAPTER TWO

1 Norris, *Fiction, Philosophy, and Literary Theory*, 171. He restates Shakespeare's "lack of a firm moral compass and failure (as Tolstoy likewise complained) to observe the most basic requirements of dramatic justice" (177).

2 Maguire, *Studying Shakespeare*, 32.

3 Greene, *Moral Tribes*, 59.

4 Symonds, "On the Use and Misuse of Darwinism in the Study of Human Behavior," 138, 139.

5 Several noteworthy efforts have been made to isolate, precisely, the common features that define this sub-genre of drama. William Lawrence was right, according to Ernest Schanzer, in confining "the problem in these plays to

the sphere of ethics," although he disagrees with Lawrence's interpretations – true to the debate. This, however, led him to suggest for a definition that the problem comedy is "a play in which we find a concern with a moral problem which is central to it, presented in such a manner that we are unsure of our moral bearing, so that uncertain and divided responses to it in the minds of the audience are possible or even probable." *The Problem Plays of Shakespeare*, 3, 6.

6 Salingar, *Shakespeare and the Traditions of Comedy*, 321.

7 Hillman, *William Shakespeare: The Problem Plays*, 8.

8 A note here to set out the scope of the conversation over the nature of the characters in this play could easily extend into a book. William Lawrence, in his study of the problem comedies published in 1931, gathered then a long list of contradictory views, many of them concerning Isabella, concluding only "that there is the widest diversity of opinion as to the heroine's true character." *Shakespeare's Problem Comedies*, 82. They are always the same complaints, her "rigid chastity," and her self-serving hypocrisy in convincing Mariana to take her place in bed, while others like Mrs Jameson celebrate her "grandeur, saintly grace, vestal dignity and purity" (82). All result from ethical readings of the play. Some critics may go in both directions, such as Sir Edmund Chambers, who finds the Duke "both repellant *and* of Divine Providence," as summarized by Schanzer, *The Problem Plays of Shakespeare*, 112. A favourite is Sir Arthur Quiller-Couch's assessment of Isabella, that there is "something rancid in her chastity; and, on top of this, not by any means such a saint as she looks. To put it nakedly, she is all for saving her own soul, and she saves it by turning, of a sudden, into a bare procuress." "*Measure for Measure*" (74), from his introduction to the play published by Cambridge in 1922. All these and so many others have scored with the best of convictions, pointing to one of the greatest problems associated with these plays. These views were easily located among the earlier critics, but anticipate in full the same range of convictions dressed in more recent critical garb.

9 On the complexity of our neural competence, Robert Trivers estimates that "as much as 60 percent of all our genes are active in the human brain, the most genetically diverse tissue in our body," and this because of the many kinds of neural productions required to secure our survival advantages. *The Folly of Fools*, 328.

10 William Flesch's analyses of the sophisticated and recursive levels to which we are able to compute reciprocity, including even the expected costs groups lay upon individuals in punishing the non-cooperators, as well as discriminating between first-, second-, and third-level defectors, is most revealing – all of which we perform to refined degrees and according to gut

values, without any conscious reflection (even our primate ancestors could do such social mathematics). See *[Comeuppance] Costly Signaling, Altruistic Punishment, and Other Biological Components of Fiction*, 49.

11 In 1971, Trivers published a seminal article entitled "The Evolution of Reciprocal Altruism." His insights have become axiomatic for nearly all researchers working in the field of evolutionary psychology. Moral awareness exacts a new dimension of computational thinking pertaining to moral debts, moral debtors, and cheaters. Richard Alexander hypothesizes that "most would agree that a moral life will inevitably call for *some* acts with net cost to the actor." *The Biology of Moral Systems*, 12. This is a critical part of this new level of social arithmetic. The benefits of cooperation can be achieved only by a species mentally programmed to share, exchange, delay gratification, trust, and earn trust. And if *reciprocal* altruism is as much as we can hope for, given the dictates of our *selfish* genes, net cost to the self will require some sophisticated tallying. This is a defining dimension of reciprocal economy, stretching far back in primate relations. Larger group structures become possible only when members can compute remote future benefits in relation to immediate costs, and in this, human nature clearly leads the way, yet also has its limits. The intriguing factor, nevertheless, pertains to the levels and qualities of calibration of which we *are* capable, and the degree to which these calculations have ratcheted up our social intelligence as we improve our cheating in danger of detection, and our cheating detection of others. Hence, the adaptive origin of altruism on a delayed reciprocal basis becomes a compelling explicator of the human condition. We begin to see that the costs entailed by group living become quantifiable in relation to options, and those options must include benefits, such as the reduction of aggression through gaming and ceremonies, cooperation in food production, protection from outsiders, sociability, furtherance of the commonweal, and commercial exchange. Herein resides the compunction to score in graduated binary values and to mentally list the degrees of cooperation and repayment. All such matters pertain to what a deliberating person can do and will probably do in some meaningful relationship to circumstances and in keeping with the emotions that those conditions have aroused. Gazzaniga, *The Ethical Brain: The Science of our Moral Dilemmas*, 167.

12 P.S. Churchland, *Braintrust: What Neuroscience Tells Us about Morality*. The hardwiring pertaining to caring instincts, according to Churchland, can also "carry more of the explanatory burden for common instances of cooperation than previously supposed" (96). Churchland provides the experimental evidence showing that hormones and neurotransmitters have a great deal to do with human sociability, bonding inclinations, care of offspring, group loyalty, and the functioning of human morality (59–60).

13 J. Greene, *Moral Tribes*, 185.

14 Churchland, *Braintrust*, 103–7. Churchland likewise casts a skeptical eye upon many of the applications humanists are now making through "inferring what behavioral traits were selected for in human evolution ... by a vivid imagination about the ancestral condition plus selected evidence about cross-cultural similarity, evidence that could be explained in many different ways" (114). She was anticipated, in this kind of cautious reservation, by Jesse Prinz in "Is Morality Innate?"

15 J. Greene, *Moral Tribes*, 136–7.

16 Paul Bloom, in *Just Babies: The Origins of Good and Evil*, provides a remarkable study of the ethical preferences of babies as young as six months who, through approbation/disapprobation signalling reveal the frames of an ethical intelligence based on perceptions of cooperation.

17 Simon Baron-Cohen considers the anomaly of empathy and why it is that we have such strong inclinations to invest feelings of concern for others. It would seem to be such "an imaginative leap in the dark, in the absence of much data," and yet he acknowledges that it remains "our most powerful way of understanding and predicting the social world." In that way, empathy becomes a part of our ethical being and a felt verification of our ethical approbation of others. "The Empathizing System," 476.

18 It is only fair to signal that such a Darwinian approach to thought production has not met with uniform endorsement. I am proceeding here on a provisional basis, grounded in the work of the many recent cognitive scientists and moral philosophers concerned with the biogenetic component of morals. Their work follows from the general premise that adaptive pressures selected the genes that built the neural platforms that permit the cognition that induces the ethical instincts that served our survival prospects under ancestral circumstances. Whether those instincts do so today is another question, but the kind of brain we have, if it has any adaptive features at all, remains the brain we have to work with now. Of course, there will be work to do to demonstrate the efficiencies, fallacies, and variable results that perplex our stories about reciprocity and justice. Moreover, an overly narrow interpretation of the adaptive, as though these structural biases of the brain turn us into automatons, must be constantly resisted, as well as simplistic analyses based on one-gene equals one-moral-value kinds of reductions. Ethical judgment is pegged to brain-based complex calculations, which allow us to reason out future benefits in relation to present sacrifices. We are not linked biogenetically to the golden rule or its opposite, although we can see in the eusocial structure of ant colonies just how far genetics can go in programming cooperators! Even self-interested altruism need not dictate ultimate constraints – a view challenged, for example, by the Good Samaritan principle, that if we help only those who

later may help us, then there is nothing adaptive about helping strangers, and particularly those who, by their hostile or ideologically driven religious or cultural preoccupations, will never reciprocate. That humans sometimes do this might very well suggest taking the entire adaptivist view back to the drawing board. Among those so inclined is Brian Zamulinski in his carefully reasoned *Evolutionary Intuitionism: A Theory of the Origin and Nature of Moral Facts*, whose perspectives are well worth consideration. For further critiques of adaptation and brain design, see R.C. Lewontin, S. Rose, and L.J. Kamin, who have called the retro-structuring of adaptive challenges and biogenetic solutions "just so" stories in *Not in Our Genes*. Alexander, in *The Biology of Moral Systems* (17–19), exposes their arguments and treats their book as a last hurrah for the blank slate approach to the human brain: as though the brain were learning neutral, and its design had no influence on the emergent properties of the thought which it produces. See also George C. Williams, *Adaptation and Natural Selection: A Critique of Some Current Evolutionary Thought*. This is an earlier study. Among the more recent protests is Jonathan Kramnick's "Against Literary Darwinism," which is discussed in the introduction. Even without that backstory, however, the debate over art, society, and moral invigilation remains all before us because, after all, this is what we must confess our reading interests to be largely about – moral invigilation – and that hence we should try to know what we can about how and why we do it.

19 Dan Sperber once protested: "what I find naïve is the belief that human mental abilities make culture possible and yet do not in any way determine its content and organization." He is dealing with a similar kind of argument, that if culture in general arises from the preoccupations of a designed brain, then those preoccupations must make themselves felt in cultural expressions and institutions. *On Anthropological Knowledge*, 73.

20 On intentional states, theory of mind, and refereeing the streams of data, see also Lisa Zunshine's chapter "Monitoring Fictional States of Mind" in *Why We Read Fiction*, 60.

21 Gazzaniga, *The Ethical Brain*, xv. Patricia Churchland in *Braintrust* uses the same term, "brain-based" (13), and not surprisingly Gazzaniga endorses her book on the front cover.

22 Cosmides and Tooby, *The Adapted Mind*, 115.

23 Pinker, *How the Mind Works*, 316.

24 Gazzaniga, in *The Ethical Brain* (14), investigates "intention," which "is an interesting ethical concept that we seem to understand intrinsically." We causally compute our own behaviour and that of others in this way, thereby turning intention into a major principle of ethics and theory of mind. We are hardwired to pay close attention to this mental activity and to rate it according to a schema of good and evil.

25 Alexander, *The Biology of Moral Systems*, 19.

26 Kramnick, "Against Literary Darwinism," 326. Kramnick will disagree with the Darwinian account of the brain's origins, but despite imagined difficulties and "missing links," the necessity remains that later brains were differentiated from prior brains by evolutionary processes, just as bodies were, and those evolutionary "choices" were installed according to the successes they represented in meeting environmental challenges, including the social, however those challenges were factored into ancestral life.

27 Flesch explores the complex principle of punishment aimed at non-reciprocators. It is the counterpart to cooperation, biogenetically embedded in the ethical computations of human communities – a frequent topic in literature. *[Comeuppance] Costly Signaling, Altruistic Punishment, and other Biological Components of Fiction*, 62 and following. Altruistic punishment usually falls to the lot of the Alpha male or would-be leader (Duke), who is in a position to absorb the associated costs.

28 Alexander discusses the emergence of human universals through "the evolutionary process that has given rise to all of our traits and tendencies or at least to the potentials for them." This is to propose that the foundations of our human psychology are due to the adaptive pressures that designed the brain. *The Biology of Moral Systems*, 14.

29 I bury a footnote here for daughter Sophie, my first intended reader (see the dedication), who will remember distinctly, to my embarrassment, a day in Narita airport. My queue through customs emerged at a snail's pace while she waited for me for nearly an hour on the other side. The rising anger caused by the line cutters and the mismanagement of the entire process, despite my long fight to maintain a rational calm, hit tirade force once I got through. This brings out the philosopher in me afterward: where did that come from, and how much of it was a feature of my true self? (Alas, all of it!)

30 Jonathan Cohen, in "Vulcanization of the Human Brain," examines the evolutionary composition of the brain as the basis for habitual thinking and conflict with older systems, what he calls "old brain mechanisms" involved in emotional judgments, based on platforms that organize earlier forms of cognitive efficiency as "rapid, stereotypical, and inflexible." Quoted in Stephen Hall, *Wisdom from Philosophy to Neuroscience*, 199. Hall takes up the issue, conceding that ethical evaluations include a form of neural conflict resolution, "whether two separate neural systems are fighting over the control switch of these decisions, or whether two different subsystems funnel competing information into a central evaluator" (200). That second system is, of course, the more recently developed analytical cortex of the reasoning brain.

31 On the brain's interpretation of light waves according to neural design and the construction of our impressions of reality, there is an engaging

perspective by V.S. Ramachandran in *The Tell-Tale Brain*, 51–4. His entire thesis has to do with the diversity of the idiosyncratic ways in which our brains structure our realities, and how those constructions correspond to what we can know about the phenomenal world beyond our senses. On a far broader scale than here, he plunges into the difficult questions pertaining to "the neurons that shaped civilization" (117–35). For that matter, he goes on to explain the neural origins of such universals as binary thinking in his chapter "The Artful Brain: Universal Laws" where he discusses contrast, isolation, perceptual problem solving, abhorrence of coincidence, orderliness, symmetry, and metaphor as foundational categories of evolutionarily defined mental orientations.

32 Andrew Newberg et al., in *Why God Won't Go Away* (50–1), define "the binary operator" as one of the brain's "most powerful tools for organizing reality" by enabling it to reduce "the most complicated relationships of space and time to simple pairs of opposites." This capacity to define things by what they are not is a vital feature of the brain's decision-making processes in matters ideological as well. Out of such operating biases, our cultural and religious dispositions emerge, inducing these authors to conclude that religious sensibilities are likewise architectural and biogenetic.

33 Brunner, *Actual Minds, Possible Worlds*, 95, citing Nelson Goodman's *Of Mind and Other Matters*. His entire study, as well as Goodman's, along with others in the cognitive revolution going as far back as the late 1950s, tackles the problem of mental action and mental systems, the certainty of the former necessitating the competence and functionality of the latter in producing them. This axiom of cognitive psychology asserts itself in the present approach to mind and the production of binaries.

34 Boyer, *Religion Explained*, 186. In this study, the author traces the ethical values which come back to us in forms projected upon supernatural beings, but which ultimately reflect a natural justice that arose in ancestral environments through eons of pressure to perform within, and conform to groups, but without totally destroying the deceptions we sometimes practise in order to get ahead, often at a cost to the group. That we watch one another in these terms is vital to our survival. Such watching has become behaviourally foundational to the adapted brain, the designs of which drive these evaluations. Duke Vincentio performs quite consistently in these terms throughout the play.

35 Regarding consciousness as a vehicle for creating and projecting the alternative courses available for managing conflict resolution and detecting cheaters and cooperators, see N.K. Humphrey, "The Social Function of Intellect," passim.

36 Brunner, *Actual Minds, Possible Worlds*, 8.

37 Ruse, *Darwinism Defended*, 275.

38 McGinn, *Shakespeare's Philosophy*, 178.

39 J.Q. Wilson, *The Moral Sense*, xii.

40 Damasio, *Self Comes to Mind*, 228.

41 Boyer, *Religion Explained*, 191.

42 P.S. Churchland in *Braintrust* conducts an engaging discussion of the search for an ethical universal such as, say, the Golden Rule, and the various ways in which it can be expressed positively as a universal imperative, and negatively as a guide to cooperation only with cooperators, taking the matter into the contrasting positions of Jeremy Bentham and John Stuart Mill. The Golden Rule can make us victims of an inflexible law, or hypocrites in wishing for others what we would not wish for ourselves. When there is freedom to impose conditions based on common sense, or the right to self-protection, in saying, for instance, that "both sides are decent, not twisted; that both sides have much the same set of moral values" (172), then the rule requires messy interpretation; we are introducing "moral content – content independent of the Golden Rule itself" (172).

43 A rather schematic approach to justice and mercy has been proposed for the play, placing Vincentio wholly on the side of mercy in his dealings with nearly everyone except Lucio who, for some, should also have been incorporated into a festive world. Muriel Bradbrook reviews that approach in "Authority, Truth, and Justice in *Measure for Measure*," largely to conclude that this work is "stiffened by its doctrinaire and impersonal consideration of ethical values," thereby leaving a moral substrate closer to depravity (398).

44 McGinn, *Shakespeare's Philosophy*, 184. His discussion of deception is on 181.

45 Greene, *Moral Tribes*, 34.

46 Donald Symons describes the community of behavioural strategies that make up our social intelligence, including "detecting cheaters in social exchanges." "On the Use and Misuse of Darwinism in the Study of Human Behavior," 138.

47 Lisa Zunshine raises the matter of narrative computational limitations concerning the levels of recursions that we can negotiate in *Why We Read Fiction*, 27–33. Writers dare not remove their stories from us through too many narrators lest we lose track of the voices and their respective inputs.

48 Geary, "Folk Knowledge and Academic Learning," 493. For a more elaborate development of these ideas, see D.C. Geary, *The Origins of Mind: Evolution of Brain, Cognition, and General Intelligence*.

49 Zunshine, *Why We Read Fiction*, 60, paraphrasing Leda Cosmides and John Tooby in "The Psychological Foundations of Culture."

50 Wright, *The Moral Animal*, 277.

51 One of the more graphic ways of illustrating the evolutionary pressure for cooperation, and why it nevertheless remains a crux in human behaviour, is the now celebrated Prisoners' Dilemma, or Prisoners' Game, in which, upon terms, two incommunicado inmates, each party to the crimes of the other, are invited to testify against the other, thereby increasing the other's prison term, while gaining freedom for himself. The choice comes with a counter-condition, however, that if both betray, then both will serve longer terms, while if both remain silent, they will each serve minimum terms. Sheer computational advantage may favour betrayal, but if both intuit the benefits of mutual cooperation, both may also benefit. This becomes a more obvious choice when the game is played over and over again, with each prisoner knowing the previous decisions of the other. Fundamentally the exercise settles into the pattern, not of the golden rule – do unto others as you hope they will do unto you – but of the reciprocator's rule: cooperate if the other does, betray if the other does (measure for measure). Computers, playing this game to Nth degrees, reconfirm, each time, the advantages of cooperation: help the other, who will help you in return according to biogenetically induced computational as well as emotional prompts. In fact, the most sophisticated programming calls for strategic acts of pardon for defection if the loss sustained may induce future cooperation according to the tit-for-tat rules of maximum mutuality. In variations on this game principle throughout ancestral history, the genes favouring cooperation gained the ascendancy, but tellingly not to the point of making us eusocial animals like bees or ants. We remain individuals with a capacity to betray if advantage seems assured, not to mention assailing out-group members – but that is another matter. Further to this topic, see Robert M. Axelrod, *The Evolution of Cooperation*, and Douglas R. Hofstadter, "The Prisoner's Dilemma, Computer Tournaments, and the Evolution of Cooperation." In more recent years, assessments of our moral natures through these ethical-dilemma games have become a staple of biogenetic approaches to morality because they reveal the platform prejudices of mental computation. Churchland in *Braintrust* takes up gambler's games (71–86), and Greene in *Moral Tribes* (29–60) nearly empties out all the permutations of the Prisoners' Dilemma. There are many others.

52 Alexander, *The Biology of Moral Systems*, 73.

53 Dawkins, *The Selfish Gene*, 141.

54 The point has been often made that the duke-in-disguise plot is entirely of Shakespeare's making, and that in turning the talion-minded magistrate of the sources into a friar in the shadows doing justice by "comforting anguished lovers and arranging for sham deaths," he has entirely altered a

play which, nevertheless, maintains justice as its focal concern. Leo Salingar, *Shakespeare and the Traditions of Comedy*, 319.

55 Salingar made the point half a century ago in *Shakespeare and the Traditions of Comedy* (312) that the term "problem comedy" may prove a misnomer because Shakespeare does solve the pressing moral issues, but not as readers might have expected, namely in legal terms, but rather through "theatrical ingenuity, instead of by systematic analysis or through discussion pointing to a general conclusion." That may not prove entirely true, but the theatrical procedures of the Duke are the operating principles for the achievements of the closure in matters social and implicitly ideological.

56 Greene, *Moral Tribes*, 41.

57 Wright, *The Moral Animal*, 195.

58 Barkow, "Beneath New Culture Is Old Psychology," 629. His only explanation for such compulsive practices reverts back to the "algorithms of the evolved mechanisms of our brains" (630).

59 Folk psychology is dealt with topically and en passant throughout these essays as the term adopted by cognitive philosophers to deal with the crux constituted of the default ways in which brains constitute perceptions of personhood and deal with other minds, their beliefs, desires, intentions, moods, and autonomy through untaught aptitudes that equip us to survive in social environments. It's features and faculties are much debated points, and there is widespread thinking that somehow we should be able to culturally bypass our default psychology, coupled with an even greater conviction that it is so embedded in human nature and so grounded in our emotional readings of the environment that it can never be escaped without the complete denial of human nature. Folk psychology is innate, "primitive or irreducible" in the words of Pinker, one of "the basic categories of the world," centred on the getting and returning of favours. *How the Mind Works*, 316, 403.

60 In effect, cultures will try to force duties and sacrifices upon individuals in the name of the group, and guilt may follow for dereliction, but the guilt arises from fear of detection as a non-cooperator. It is not an inner voice carrying our sense of justice or self-recrimination, nor is it engineered on behalf of the group. Guilt is self-interested and self-instructive in relation to assumed consequences.

61 *Civilization and Its Discontents* was first published as *Der Unbehagen in der Kultur* in 1930. Therein Freud looked at the pleasure principle associated with sexual gratification and the means whereby it is repressed by law and society as essential to civilized life, while at the same time an inevitable source of personal discontentment. This work, in its own right, could serve as a theoretical basis for an interpretation of *Measure for Measure*.

62 Greene, *Moral Tribes*, 46.

63 Alexander, *The Biology of Moral Systems*, 77. Or, in the words of Patricia Churchland, "shunning is one powerful form of punishment in highly social mammals." *Braintrust*, 81.

64 A remarkable parallel example occurs in the sixth "history" of Barnabe Riche's *Farewell to Military Profession*, "Of Gonsales and His Virtuous Wife Agatha." In brief, a husband decides to kill his virtuous and loyal wife in order to marry a courtesan. The pharmacist, who loves Agatha dearly, but who has always been shunned, is invited to prepare the poison. Instead, he administers a sleeping potion, which will later permit him to welcome her back to life inside her family monument (sound familiar?). But Agatha refuses him even then, choosing to honour her vows to a man who would have murdered her until such times as he is exposed and she is able to save his life by appearing alive at his murder trial. The high magistrate in this story spares his life for Agatha's sake, pronouncing what a monster he had been, concluding, "But I swear unto thee that if ever I may understand that thou dost use her henceforth otherwise than lovingly and kindly, I will make thee to thy grievous pain prove how severely I can punish such beastly and heinous facts, to the example of all others" (276).

65 Greene, *Moral Tribes*, 55.

66 Flesch, *Comeuppance*, 49.

67 McGinn, *Shakespeare's Philosophy*, 181.

68 Charles Darwin, *Notebooks 1836–1844*, 608.

69 Graeme Hunter, "Taming *The Tempest*," 129. There is, in fact, a remarkable set of analogies to be drawn between these two plays in terms of the forms of instruction they impose on their respective societies. Vincentio and Prospero act as teachers while temporarily removed from political power. They find themselves enabled by magic and trickery to this end before returning to their own political jurisdictions.

70 Wright, *The Moral Animal*, 208.

71 Merely as an aside, Angelo is a corrupt wooer, both in commodifying the girl as a sex object and in failing to keep his promise to save Claudio – a complete cad. But as a mental exercise, if he had agreed to save Claudio at Isabella's request and then asked her to marry him, counting on her gratitude, by how much would his situation differ from that of the Duke's, insofar as each had saved her brother?

72 Jeremy Campbell, *The Improbable Machine*, 196–9.

73 Dawkins, *The Selfish Gene*, 173.

74 Witness Edmund Spenser's Belphoebe in Bk. III of *The Faerie Queene*.

75 Wright, in *The Moral Animal* (61), reiterates the now widely circulated thesis that men and women have asymmetrical values because of the contrasting roles they have in begetting and caring for children. Evolution

invented the hedonics of sexual craving, romantic love, negotiating for security, and philandering. Once again, it comes down to promises and deceptions, to wariness and trickery. Even in these matters there has been a psychological arms race between conflicting desires and detection devices. Chastity is a critical state for women in negotiating loyalty for the sake of their offspring. Women can cheat. Men keep compulsive score. Suspicion may cause disastrous errors.

76 Dawkins, *The Selfish Gene*, 165.

77 Jacques Ferrand, *A Treatise on Lovesickness*, 336.

78 Dawkins, *The Selfish Gene*, 160.

79 The work of M.A. Conway and D.A. Bekerian in "Situational Knowledge and Emotions" is valuable here for establishing the combination of political and romance motifs in the play. They asked a group of respondents to list their prototypical situation of "grief, misery, sadness," and over half described the death of a loved one, and particularly one in a romantic relationship, followed by the decline into poverty, loss of honour, loss of professional prestige, or exclusion (exile). By extrapolation, the opposites would be romantic union and return to power and prestige, suggesting just how closely this play is aligned with the generic expressions of our fundamental emotions of success or failure in situational or experiential terms.

80 The allusion here is to what Ronald de Sousa called "paradigm scenarios," those through which we can explain universally how sorrow or joy is produced. Through these paradigms, we call to mind the clearest circumstances in which emotions are aroused incontestably, those which align themselves perfectly with unproblematic notions of good and bad, right and wrong, without need of a long backward summary of quirky memories and subtle interpretations upon which the present feelings might be built. This is merely another analytical way of problematizing the bed-trick, the religious disguising, the fraudulent confidentiality of a confessor. "The Rationality of Emotions," 142.

81 J.Q. Wilson, *The Moral Sense*, xii.

82 Wright, *The Moral Animal*, 5. A common-sense argument begins with the fact that human anatomy books show the architectural degrees to which our bodies are physiologically the same, eliminating much of the surprise that we might have that our brain systems are responsible for our common categories of thought production in precisely the same terms – genetic design (26).

CHAPTER THREE

1 Doležel, *Heterocosmica*, 81.

2 Damasio, *Self Comes to Mind*, 280.

3 From *The Death of Tragedy*, cited in Kerr, *Tragedy and Comedy*, 37.

4 By that reasoning, any retrofitting of Freudian, Jungian, Lacanian, or behaviourist theories onto character should also be disqualified, yet scholars have selectively entertained their ideas for their potential as universal principles.

5 On the universality of the emotions, see Hogan, *The Mind and Its Stories*, 15–16, but in fact throughout his entire study.

6 McEwan, "Literature, Science, and Human Nature," 19. McEwan explains further that "if there are human universals that transcend culture, then it follows that they do not change, or they do not change easily. And if something does change in us historically, then by definition it is not human nature that has changed but some characteristic special to a certain time and circumstance" (12).

7 Renaissance philosophers were constrained to explain by their own systems the origin of crimes of passion. Melancholy, fury, and demonic possession were the terms of choice for dealing with causally deranged minds because they were grounds for seeking medically confirmed acquittal. The defendant might be judged incapable of consciously debating motives, planning, and volitionally executing a criminal action. But judges were determined then as now not to absolve, by reason of insanity, anyone who, in a state of despondency or anger, strikes out in retaliation, for nothing could be more motivated. The author of *The Yorkshire Tragedy* mentions melancholy at the outset, but does not elaborate on the black fumes or the burnt biles capable of polluting the imagination and reason, or on a sudden mania leading to random acts of violence in a state of *dilucida intervalla*.

8 McEwan, "Literature, Science and Human Nature," 10.

9 Kerby, *Narrative and the Self*, 51.

10 The point has been made by many. Damasio takes narrative back to the formation of the "protoself" in *Self Comes to Mind*, 207.

11 Jon Elster, *Alchemies of the Mind*, 108.

12 Kerby, *Narrative and the Self*, 50.

13 Taylor, "Self-Interpreting Animals," 64.

14 This profile of emotional "meaning" can be found in variant forms in the writings of many cognitive philosophers and neuroscientists. The three authors who have most shaped the perspectives underlying the present study are Victor S. Johnston in *Why We Feel: The Science of Human Emotions*; Antonio Damasio in *The Feeling of What Happens: Body and Emotion in the Making of Consciousness*; and Aaron Ben-Ze'ev in *The Subtlety of Emotions*. All three are evolutionary functionalists, concerned with retro-engineering the evolution of the limbic system (an imprecise term that nevertheless remains useful in describing the midbrain and the production

of the twin sensations of feeling and emotion as a general category of brain activity). Johnston lays out a theory of the positive and negative affects as reflections of "the biological importance of sensory events" developing guiding memories of these events fortified by feelings. Thus feelings play an important role in "how, what, and when we learn and in determining how we reason about the world around us" (82). Damasio is likewise concerned with the generation of hedonic states in the basal forebrain, hypothalamus, and brain stem and the role of emotions in animals, but his primary concern is with consciousness, which is the necessary add-on that allows humans to also be aware of the emotions they are having – to "'feel' a feeling" (81). Damasio also writes in *Looking for Spinoza* (159), "I suspect that in the absence of social emotions and subsequent feelings, even on the unlikely assumption that other intellectual abilities could remain intact, the cultural instruments we know as ethical behaviors, religious beliefs, laws, justice, and political organization either would not have emerged, or would have been a very different sort of intelligent construction." He is not saying that emotions created these institutions, but that the emotional components of cooperation and empathy made their contributions and thus informed all legal and cultural institutions.

15 Taylor, "Self-Interpreting Animals," 75.

16 By illusion, I am not reducing the experience of the self to something without consequence. I am merely being cautious in my wording in relation to recent discussions of the self, as in Bruce Hood's *The Self Illusion: Why There Is No 'You' inside Your Head*, in which he examines the design propensities of the brain that construct the emergent properties of mind that constitute the self. Selfhood is merely a configuration of thought events that include our first person take on the world, a sense of our being in time, our autobiographical narrative, our values and beliefs, the self-esteem that keeps us motivated on behalf of our reproducing organism, and all else that makes up the self stance through which we negotiate environments and are brought to feel deeply about who we are.

17 Not all scholars concerned with the emotions are committed to explanations based on "evolutionary function," which prioritizes the evaluation of "information that may lead to increased attention towards essential matters of survival" – for example, Steven Pinker, Antonio Damasio, Victor Johnston, and several others to appear in subsequent footnotes. Tone Roald, *Cognition in Emotion: An Investigation through Experiences with Art*, 49. Their perspectives are of particular value in explaining the residual autonomy of the emotions as sub-cortical evaluators and the physiological distinction between midbrain and cortical emergent states, emotions, and thoughts, operating in parallel and imperfect convergences. Yet, all these

thinkers are more than aware of the intricate and steady interface between the two systems and the complex negotiations whereby thought elicits emotions and emotions evaluate thoughts. This is the troubling nexus at which the present study seeks to make critical distinctions concerning criminality, excited emotions, mental competence, qualities of judgment, ethical awareness, and self-blindness worthy of some compassion.

18 Elster, *Alchemies of the Mind*, 317, 332. Kathryn Schulz in *Being Wrong: Adventures in the Margin of Error* provides dozens of examples of this blame transfer to escape the tortured sensations of self-accusation and acknowledged responsibility. Among many examples is that of the Millerites who, in 1843, had agreed that the Second Coming of Christ would transpire in October, and then found themselves answering to thousands how the "Great Disappointment" could have happened, with Miller alone acknowledging anything like personal error, even though he remained certain of the imminence of the Rapture. Entire new denominations, including the Seventh Day Adventists, emerged from their complex rationalizations (201–19).

19 Oatley and Jenkins, *Understanding Emotions*, 313. Elster provides the example of Iago, who was caught between envy of one who had received an office he deemed an entitlement and anger against the man who had made the unjust award. In seeking redress, he justifies his assault on Othello's domestic life by entertaining the suspicion that Othello had seduced his own wife – a suspicion he was ready to entertain as a certainty, however weak the evidence (*Othello*, I.iii.387–91).

20 Dawkins, *The Selfish Gene*, 160. This is all, of course, predicated on the revolutionary and persuasively grounded thesis that, despite our many cherished ideals about human instinct and potential, we must also accommodate the axiomatic truth that "a body is really a machine blindly programmed by its selfish genes" (157).

21 Wilson and Daly, "The Man Who Mistook His Wife for a Chattel," 305.

22 Monica Calabritto, "A Case of Melancholic Humors and *Dilucida Intervalla*," 148.

23 On the lack of future perspective in the criminal mind, see R. Frank, *Passions within Reason*, 82.

24 See E.T. Higgins, "Self-Discrepancy: A Theory Relating Self and Affect." The dialogue within the self is entirely subsumed in this play, thereby enabling the protagonist to speak of one thing as he enacts another. Understood is some Freudian defence mechanism, some emotionally driven dissonance reduction procedure due to the intolerable levels of guilt, shame, and fear, for which there were no other imagined outlets. Yet even these conventional explanations seem to fall short of the mark concerning systemic design and a grounded theory of action.

25 Palmer, *Fictional Minds*, 117.

26 The point is contested that emotions "know" how to choose between options; rather, feelings provide "gut feelings" that abet in making choices by making one feel better or worse about the respective choices. Yet that too implies a "knowing" about what is better or worse. Damasio takes the evolutionary psychology approach in *Descartes' Error* that emotions know by design how to produce the feelings that best serve the interests of survival. The argument matters because all emotions may function by such "logic," thereby co-empting or overruling weak cognitive stances (169), an idea developed in *The Feeling of What Happens*, 42. See also Ronald de Sousa, *The Rationality of Emotion*, 194–5.

27 Damasio has written at some length to demonstrate that feelings are not extraordinary events attached to hyperbolical occasions only, although we think of the classic emotions such as anger, rage, and fear in these terms, but that all forms of thinking about the surrounding world are accompanied by hedonic judgments and intimations making themselves felt even if they are not registered as categories within consciousness. *The Feeling of What Happens*, 58. That perspective is seconded by Graham Richards: "we live, for most of the time, in a continuous and constantly shifting emotional atmosphere that infuses all our actions and experience." "Emotions into Words – or Words into Emotions," 51.

28 Carruthers, "Moderately Massive Modularity," 73.

29 Danto, *The Body/Body Problem*, 65.

30 Anscombe, *Intention*, 57.

31 Hardcastle, *Constructing the Self*, 106.

32 Cohen, in "Vulcanizing of the Human Brain" (esp. 13–15) tries to imagine the adaptive origins of our compound, multi-modal, asymmetrical hedonic and rational brain and the foundations of the interfacing between these two systems. For him, the add-on cortex over the instinctual and hedonic brain was a vital vulcanization of the base product in order to enhance our social performance by controlling our emotional surges. He imagines the environmental backstories pertaining to cooperation, dealing with cheaters, and related issues of reciprocity within survival groups as the pressures which incited selection for rational overriding mechanisms. But he also recognizes the "intelligence" of the emotions as indecision breakers at select moments vitally not incapacitated by the cortex. Insofar as the emotions still have powerful roles to play in decision-making situations, some of them quite paradoxical and hardly actuarial in terms of outcomes, as in the case of the present play, to better understand ourselves we can only pursue our investigation into the prompts, cues, and neural systems involved in discrete social configurations. These exchanges are so complex that accusations of simple determinism are clumsy to the point of irrelevance. Nevertheless, there is

a powerful sense, arising from Cohen's work at the Princeton Center for the Study of Brain, Mind and Behavior, that when, let us say, the anterior insula is seen to light up through imaging with greater intensity than the rational centres, prepare to deal with an individual worked up by injustice and ready to engage in punitive behaviour even at a cost to herself. That is a simple clinical observation, with or without the evolutionary backstory. Design may sometimes continue to be destiny.

33 Merikle, "Perception without Awareness," 795.

34 Raine et al., "Brain Abnormalities in Murderers Indicated by Positron Emission Tomography," and Raine et al., "Reduced Prefrontal and Increased Subcortical Brain Functioning Assessed Using Positron Emission Tomography in Predatory and Affective Murderers." See also Davidson et al., "Dysfunction in the Neural Circuitry of Emotion Regulation – A Possible Prelude to Violence."

35 Hardcastle, *Constructing the Self*, 153.

36 Pinker, *How the Mind Works*, 371. Our dilemma is whether any "event may occur … without reference to which the reason allegedly cannot be formulated," even though that formulation may not be based on a truth. Danto, *The Body/Body Problem*, 77.

37 The emergence of the self through the opportunities of reflexive awareness provided by the redundant sizing of the cortex is a retro-narrative taken up by many evolutionary philosophers, to which the emotions must be added as the hedonic sensors that guided priorities and emerging beliefs, and backed the organism's sense of the singularity of the self and its illusion of permanence, identity, individuality, ownership, and agency. Damasio discusses these matters in *The Feeling of What Happens*, 133–49. Gary Lynch and Richard Granger deal with many of these issues in *Big Brain: The Origins and Future of Human Intelligence*. One of the classic studies of the origin of identity through the increased capacities of the cerebral cortex is Karl Popper and John Eccles's work *The Self and Its Brain: An Argument for Interactionism*, and especially Popper's chapter, "Learning to Be a Self," 108–20. There he studies the extent to which the self owns its brain. The self in Daniel Dennett's classic work, *Consciousness Explained* (430), becomes "a center of narrative gravity" which, nevertheless, in its reflexive continuity generates a perfectly adequate sense of an existing, unique entity, another emergent state in the form of thought from our systemic, modular, integrated, architectural brain.

38 The classical emotions, those made manifest by wild stares, guffaws, sobbing, palpitations, tension, and adrenalin surges, became the signalling elements of plots built around the emotional interiors of characters socially betrayed. The principals we can name: envy, love, jealousy, pride, sorrow,

disgust, and revenge. Nineteenth-century thinkers refined the list of social emotions to include boredom or ennui, reverie, nostalgia, fatigue, anxiety, discouragement, cheerfulness, enjoyment, embarrassment, and guilt. From this list, we might choose for The Husband a variety of labels, such as jealousy, contempt, anger, self-loathing, revenge, despair descending into an even more abject state.

39 Nussbaum, *Upheavals of Thought*, 126.

40 Damasio, *Self Comes to Mind*, 278.

41 Ibid., 279.

42 Ibid., 283–4.

43 So many others have contributed richly to this drawn-out dialogue over the respective autonomies or control features of the multi-messaging brain. George Ainslie deals with rash decision-making through a psychological devaluation of the future in "Specious Reward: A Behavioral Theory of Impulsiveness and Impulse Control." Stephen Hall also discusses the at-one-time adaptive potential to enraged behaviour through discounted futures and the vestige of this "logic" in the emotional brain. Even if such moments are rare, underscoring a mere vestige of the ancient brain, their automaticity remains a potential. Hall built on the premise that two systems were in competition and that desperation programming remains latent. *Wisdom from Philosophy to Neuroscience*, 202. This is not unrelated to the "Ultimatum Game" in which a sum of money is granted to an individual if that person shares a portion of it with another person. If the shared sum is deemed too little, however, neither receives anything. It is not only a test of the amount below which another feels insulted, but the amount below which that person is willing to punish herself in losing what little she is offered in order to spite the other. That is a point at which emotional decision-making takes over from the close reasoned consideration of amounts and proportions. Alan Sanfey et al., in "The Neural Basis of Economic Decision-Making in the Ultimatum Game," reports on the magnetic resonance imaging profiles in brains confronted by unfair offers. In his assessment, the areas to watch are the dorsolateral prefrontal cortex, associated with goal maintenance and executive control, which is biased toward acceptance, and the bilateral anterior insula, which is increasingly biased toward rejection, proportionate to the degree of the insult. Two parts of the mind make their own calculations, but to different volitional ends.

44 Hardcastle, *Constructing the Self*, 156, 157.

45 De Sousa, comment from the *Times Higher Education Supplement* on the back cover.

46 Similar perspectives concerning the make-up of the human psyche are to be found in Damasio's *The Feeling of What Happens*, 40–1; in Palmer's

Fictional Minds, 19; and throughout Elster's *Alchemies of the Mind*, wherein it is argued that the reasoning faculties and emotional influences become increasingly impossible to separate.

47 Ben Ze'ev, *The Subtlety of Emotions*, 301.

48 Pinker, *How the Mind Works*, 390.

49 Cohen, "The Vulcanization of the Human Brain," 6, 7, 17.

50 Pinker, *How the Mind Works*, 364.

51 Ibid., 370.

52 The Husband of *The Yorkshire Tragedy* is by no means a unique portrait in this regard, for ancient and contemporary models of tragic berserkers were quite familiar to the writers and readers of the English Renaissance, most particularly through such plays as Seneca's *Hercules furens* and Cinthio's *Orbecche*. Of all the once furious and subsequently tranquilized protagonists to figure in Elizabethan fiction, Arsadachus, in Lodge's *A Margarite of America* (1596), stands out. He had been sent a probative box containing an element arising like smoke that would prove his fidelity to Margarita or drive him to frenzy. After inhaling, the prince became violent, first killing a loyal friend and companion in crime as an act of retribution. Then he murdered Diana, his mistress and mother of his children, in revenge for the death of his parents, then his own male child, and finally the innocent and devoted Margarita, still seeking to help him. It was done in a state of uncontrollable frenzy. Only after six hours of carnage did his fury subside, the entire episode to be "read" as a surging emotion of guilt, transfer, self-hating destruction, and eventual self-recovery after a few hours of sleep. Then, seeing the carnage all about him, he began to reflect on Hercules, the paltriness and incertitude of the human condition, the perversities of the mind, at last coming to a full recognition of his own monstrosity. His act of suicide as a form of repayment seemed hardly adequate, but served nevertheless as his escape from the dawning of an intolerable state of emotional torment. Arsadachus's prolonged soliloquy might have been transferred to The Husband of the present play with only minor factual adjustments required. *A Margarite of America*, 162–9.

53 Roald, *Cognition in Emotion*, 57.

54 Nussbaum, *Upheavals of Thought*, 398.

55 Ibid., 398.

56 Ibid., 410.

57 Unger, *Identity, Consciousness and Value*, 75.

58 Danto, *The Body/Body Problem*, 76.

59 In relation to the production of the emotions, there are the contributing organs of the brain with their design parameters. These process the neural messages upon which all limbic performance relies. The amygdala, the cingulate gyrus, the forward medial bundle, and the hypothalamus, to

name but four, are all systemic contributors, controlling the raising and quelling of hedonic states, the pleasure and pain centres, and the formation of memory, each organ a partial reader and appraiser of external changes in the environment apt to arouse feeling. They are determinative to the extent their operations shape cognitive and hedonic meaning and experience, yet "dialogic" to the degree the mind can override the effects of their operations.

60 Palmer, *Fictional Minds*, 118.

61 For a resumé of the options concerning beliefs, intentionality, and volition, see Dennett, "Intentional Stance," 412–13.

62 Lycan, *Minds and Persons*, 107–22. Lycan's discussion is the tip of an iceberg that would take us away from present concerns, yet his reasoning touches on the present matter. After considering all the options between the compatibilist and the incompatibilist positions, he concludes that the compatibilists will always win out, at least in the schools of common sense, insofar as they believe that limited free will, options, and agency remain compatible with determinism. This means defining free will in a limited way, narrowing the choices within theatres of action. The debate is as old as the ancient Stoics, and was furthered by David Hume and Thomas Hobbes. The present discussion is a variant insofar as the entire crux is brought down to the computational options and systemic passions that inform the human brain, collaboratively or competitively.

CHAPTER FOUR

1 Sir Philip Sidney, *An Apology for Poetry*, 104.

2 Ibid., 104.

3 Ibid., 106–7.

4 See Colin McGinn's discussion of this Humean idea and his rebuttal of the notion of "fading" in *Mindsight: Image, Dream, Meaning*, 11. Richard Thompson and Stephen Madigan concur that we may be led "to assume that memory is much like a tape or video recorder, holding a perfectly accurate record of what has been experienced. Nothing could be further from the truth." *Memory: The Key to Consciousness*, 6.

5 Bacon, *A Selection of His Works*, 317.

6 Samuel, *Memory*, 61.

7 For the concept of parsimony in relation to estimating the most likely of scenarios to have taken place in evolutionary time, see Dawkins, *The Ancestor's Tale*, 115–18.

8 Reuven Tsur, among many, has taken full cognizance of this phenomenon, speaking of selection, things well-defined, and low-differentiated objects forming the "ground." "By virtue of such an organization we achieve

cognitive stability," but we have bought this information economy "at the price of dumping some other information into the 'ground,' or 'shutting out' from the system altogether." *Toward a Theory of Cognitive Poetics*, 15–17.

9 Eric R. Kandel takes an entirely different approach in his studies of the biological conditions and constraints that determine storage, because fixing memories has to do with synaptic weighting and stability, which is in turn determined by the messenger RNA that regulates protein synthesis responsible for stabilizing synapses. Genetic design has determined the high level of memory formation through suppressor proteins, which are responsible for avoiding the clutter. Conversely, genes activated by CREB-1 proteins create new synaptic growth and maintain the terminals where memories persist. *In Search of Memory*, 275. What we remember comes down to the genetic designing of the systems.

10 I am intrigued by the idea of the *Panchatantra* or the *Katha Sarit Sagara* (The Ocean of the Streams of Stories) as collections of illustrative stories which, by their sheer bulk and complementarity, constitute something like a comprehensive handbook to life. Moreover, in compiling such anthologies from traditional lore, it is as though a story had already been told about every conceivable human challenge or situation, and that in mastering a cycle of stories, one cueing another and another by variations on life situations, fictive experience builds the complete person. It is fascinating to think of life itself as changes on situations, the one gravitating into another and another, until they constitute a plenary sociological portrait – an ocean of stories from many streams.

11 Rumelhart, "The Architecture of Mind," 225.

12 As my colleague Jim Davies reminded me, the debate about the manner in which the brain creates categories and employs them to identify incoming data has not been resolved. Prototypes are an average of all the instances that have been experienced, so that each new datum is evaluated according to its proximity to that prototype. The theory of exemplars holds that we store each instance we have seen and provide it with a category label so that new data are assigned to the categories belonging to the most similar exemplars. But that distinction need not delay us, insofar as both models necessitate an analogy-making capacity that enables new circumstances to be associated with the closest remembered instances and the fundamental experiences that established them.

13 Chaucer, *The Canterbury Tales*, 284, line 3443.

14 Rumelhart, "The Architect of Mind," 236.

15 Attention, in determining what is consciously experienced and what parts of external stimuli are examined, also has much to do with memory formation, the more so when it is emotionally excited. Attention, for the record,

seems to be regulated by dopamine and its receptors in the hippocampus. Kandel, *In Search of Memory*, 312.

16 On the crucial matter of percept and image, the mind modes of external sensation and internal imagining, and the ease and rapidity with which we leap from one to the other without confusing them (i.e., losing ourselves in hallucination), see McGinn, *Mindsight: Image, Dream, Meaning*. The entire book is concerned with this issue, but a good sense of it can be had from the first two chapters. The study deals with distinguishing the realities of the inner and outer lives, as it were, but McGinn also deals with the "realities" of fictive worlds and the full complement of understandings we can also bring to the image.

17 At this juncture, I must ride roughshod over the complex matter of a "theory that would offer some adequate account of what goes on when competent readers engage with a fictional text." Christopher Norris, *Fiction, Philosophy and Literary Theory*, 136. In his chapter "Will the Real Saul Kripke Please Stand Up? Fiction, Philosophy and Possible Worlds," Norris provides a succinct resume of the debate over the reality of fiction that has attracted the likes of Bertrand Russell, Gottlob Frege, John Searle, Jacques Derrida, Richard Rorty, Linda Hutcheon, Alexius Meinong, Saul Kripke, Hilary Putnam, David Lewis, and Colin McGinn. Its subtleties are rewarding, but an entirely new approach might be advanced simply by looking at the modes in which the brain processes data, those in which the data are emotionally felt as real but cognitively recognized as fiction, and the mind's capacity to process representations in the imaginative mode on an as-if-real basis simply because in doing so it ratchets up their experiential value as templates for dealing with future environmental contingencies, without coming close to confusing such representations with dreams or hallucinations (however Shakespeare or Calderon might seek to profile their fictive worlds). Alan Palmer, in *Fictional Minds*, 36–43, discusses characterization in light of the work of Uri Margolin (who wrote a series of articles from 1986 to 2003).

18 Thompson and Madigan, *Memory: The Key to Consciousness*, 31–3.

19 "Binding" is central not only to memory theory but to all constructions of the brain that rely on the gathering of diverse properties in order to produce a unified conscious event or emergent state. The challenge is to imagine a working model based on sound and "reasonable" neurobiological systems that would enable the brain to perform this complex work. For an overview, see John Hummel, "Binding Problem," and Tanya Reinhart, "Binding Theory," 85–8. Kandel also gives an account of the binding problem, dispersed synaptic storage of memory parts, and the reconstruction of complex memories. *In Search of Memory*, 298ff. See also Francis Crick, *The Astonishing Hypothesis*.

20 Teun A.van Dijk and Walter Kintsch, *Strategies of Discourse Comprehension*, chapter 4.

21 K.D. Forbus, D. Gentner, and K. Law, "MAC/FAC: A Model of Similarity-Based Retrieval," 141–2.

22 When attention addresses a good story, it is the order of words on the page and the compounding of events on a felt and lived basis that captivate and absorb consciousness according to the rate of the eye saccades across the page. Information is steadily being compounded, even as the eye moves forward to advance the story and the further unfolding of events. It is fascinating to think how story time, determined by the rate of those saccades, creates a temporal reality that must be blended with the rate of social events they deliver to our conscious experience. In the theatre, acting paces the social timing, but the story may be edited to speed up the delivery in relation to an implied historical time: weeks, months, even years in two hours of critical highlights. Critics in the sixteenth century were concerned that stage time have some correlation to story time in the interests of credibility, fearful that events over years squeezed down to two hours on the stage would lose all semblance of reality and coherence. Twenty-four hours of lived time reduced to those same two hours of performance, however, would pass inspection. But as we all know, those wily Elizabethans proved the continentals wrong about the ability of the (English) imagination to collapse chronological time to stage time. In reading, we allow that the saccades of the eyes over the words correlate to the timing of social events, creating the rhythm of the "now" moments as the story advances. These the reader "lives" as "real" time, just as theatregoers experience stage presentation time as "real" time. In both instances, a story "core" emerges, thanks to the brain's ability not only to keep track of past events but to hold them in a semi-conscious state while the conscious mind entertains only the on-coming events of the present.

23 Brian Boyd discusses, in relation to literature, the mechanisms of time-sequenced memories and the weightings given to them on the principle that the most recent events are more likely to be relevant and critical to the present than earlier memories, in *On the Origin of Stories*, 153.

24 A sense of the self is, of course, a feature of long-term memory building on the contexts of working memory and its disposing of temporal sequences. There has been considerable to excessive speculation about the critical moment in life at which the subject discovers its own selfhood as distinct from others – the all-determining epiphany of personhood. It is revealing, however, that neonate brains are not equipped to build time sequences and to formulate long-term memories because they are genetically programmed to apply all their learning resources to semantic data and motricity. The

ability to conceive of a self comes along at age four or five as a natural side effect to the reprogramming that permits the formation of long-term memories. See David Samuel, *Memory*, 73–4, and Anthony Paul Kerby, *Narrative and the Self*, 16.

25 Paul Churchland, *The Engine of Reason, the Seat of the Soul*, 100.

26 Thompson and Madigan, in *Memory*, 48–9, discuss the processes of working memory – transforming information, keeping track of changes, updating memory, and setting up comparisons – as operations tantamount to intelligence itself. They are frontal lobe functions and differ from person to person.

27 On the role of inference in memory reconstruction, see Ellis and Hunt, *Fundamentals of Human Memory and Cognition*, 163, 174.

28 On confabulation as a means of generating meaningfully complete representations of imaginatively reconstructed realities in normal subjects, see Baddeley, "The Psychology of Remembering and Forgetting," 49.

29 Brunner, in *Actual Minds, Possible Worlds*, 25–6, points out that authors strategize the gaps in their novelistic worlds in order to invite the fill-in contributions of readers. In this, a certain economy must be employed that relies on the reader's knowledge of the world. Inversely, to create even the illusion of a totality of reality through words would be painfully prolix and leave nothing to interpretation and imagination.

30 The insistent production of these confabulations for the sake of generating coherent, or non-threatening, or ego-comforting, or probability-based rationalized versions of events has huge implications for the evidential concerns of judicial systems, which are still based largely on eye-witness testimony and truth-swearing. The accuracy of such testimony, nevertheless, can be as low as 20 percent, even moments after witnessing a staged criminal event. Daniel Dennett added polemically scientific weight to this epistemic bomb in his chapter entitled "Dismantling the Witness Protection Program" in *Consciousness Explained*, 321–68. Kathryn Schulz dealt with the matter most engagingly in *Being Wrong: Adventures in the Margin of Error*, especially in her account of the 1985 rape of Penny Beerntsen (220–46). The staged criminal act experiment was first conducted in 1902 by Franz von Liszt, a professor of criminology at the University of Berlin, with results confirmed through the pedagogical repetition in psychology and criminology classes. *Being Wrong*, 223–4.

31 Michael Shermer's "Self-Justification Bias" is the "tendency to rationalize decisions after the fact to convince ourselves that what we did was the best thing we could have done." It comes down to cherry-picking data by amplifying positives and diminishing negatives. *The Believing Brain*, 263–4.

32 Dennett, *Consciousness Explained*, 319.

33 Campbell, in *The Improbable Machine*, 142, provides corroborating observations.

34 Dennett, *Consciousness Explained*, 113.

35 Ibid., 270.

36 Gazzaniga, *The Ethical Brain*, 122.

37 Flanagan, *The Science of the Mind*, 193.

38 Johnston, *Why We Feel*, 162.

39 Boyd, *On the Origin of Stories*, 157.

40 On schemas in general, see Ellis and Hunt, *Fundamentals of Human Memory and Cognition*, 180.

41 There is a complementary account of Sir Frederic Bartlett's contribution to the understanding of memory in Campbell's *The Improbable Machine*, 91–3, and another in Boyd's *On the Origin of Stories*, 156–8.

42 Tye, *Ten Problems of Consciousness*, 104

43 Ornstein and Erlich, *New World, New Mind*, 175. Brian Boyd likewise describes such memories as episodic clusters framed by action schemas and fleshed out by innocent confabulation. *On the Origin of Stories*, 134.

44 The matter of schemas as a priori patterns for knowing the world and storing experience has been much discussed. One such discussion is Campbell's in *The Improbable Machine* in his chapter entitled "Baker Street Reasoning" (86–96). We use patterns of worldly knowledge to process information. "Baker Street reasoning is logic without logic, which is what natural intelligence is all about." He is referring to the ways in which Sherlock Holmes proceeds through a close observation of detail, which he relates to his general knowledge of the world, that if the suspect had an anchor tattoo he has a better chance of being a sailor than a truck driver. "Knowledge structures of this kind go by the general name of 'schemas,' and they are so important they have been called the building blocks of thought, as fundamental to a theory of human reason as cell biology is to an understanding of the living system" (87). He goes on to relate them to the insights of Immanuel Kant concerning the schemas by which brains take the information delivered by the senses and turn it into knowledge.

45 Baars, *In the Theater of Consciousness*, 46, 58.

46 Dennett, *Consciousness Explained*, 264. Dennett sees memory as essentially latent, or only partially proactive, blundering about and trying on different solutions until a dominant model emerges. In this rather messy way, memory still contributes to a brain making up its mind on a better than average basis. Others will give memory a more defined and active role in bringing up experience, precepts, or guidelines toward working out problems (224).

47 Hofstadter, "Analogy as the Core of Cognition," 499, 538. See also MacIntyre, "Epistemological Crises, Dramatic Narrative, and the Philosophy of Science."

48 Dawkins discusses synaptic conductance in a massively redundant system through which patterned stimuli seek out matches, or near matches. The entire system is made possible through massive connectivity, or what Dawkins calls "the magic of large numbers." *The God Delusion*, 137.

49 We are back to schemas. Gentner and Markman, "Structure Mapping in Analogy and Similarity," 127.

50 Ibid., 131.

51 Ibid., 132.

52 Carr, "Narrative and the Real World: An Argument for Continuity," 7–25.

53 Mandler, "Organization and Memory," 82.

54 Norman, *Memory and Attention*, 83–5.

55 Campbell, *The Improbable Machine*, 90.

56 North, *The Moral Philosophy of Doni Popularly Known as The Fables of Bidpai*, 207. For a history of the text and its migration through seven languages before it came to North, including Hebrew, Latin, Spanish, and Italian, see the introduction, esp. 11–34.

57 North, *The Moral Philosophy of Doni*, 207–8.

58 Ibid., 208.

59 Jeremy Campbell, in *The Improbable Machine*, speaks of the memory as "organized in the brain in such a way that large amounts of relevant world knowledge are triggered almost instantly by a very small quantity of incoming data" (48). His principal point is that when data arrive, they trigger a plethora of schemata of knowledge about the world in order to interpret that data on a relational, analogical, associational basis. But this blends quickly with the way in which precepts pertaining to experience associated with stories also appear as data that trigger knowledge about the world through related stories and experiences on an analogical basis. This is the way the mind works, far more fundamentally than through computational logic, which is the underlying axiom of Campbell's study.

60 Cosmides and Tooby, "Consider the Source," 60–1.

61 North, *The Moral Philosophy of Doni*, 253–5.

62 Further to the matter of analogy and brain mapping as the foundation of planning, see Gentner and Markman, " Structure Mapping in Analogy and Similarity," 127–56, with an excellent bibliography. Analogy is studied as a "structure-mapping process" which is a critical means whereby humans are able to generate new predictions. They explain the principle stated above as a "relational focus" which developed simply because

"*one-to-one correspondence* limits any element in one representation to at most one matching element in the other representation" (131). They go on to imagine the kind of algorithmic system activity that constitutes the "structure-mapping engine," which in turn permits analogical inference and extended mapping in connection with categories and planning.

63 Zunshine, *Why We Read Fiction*, 41

64 Thompson and Madigan, *Memory*, 96.

65 Norman, *Memory and Attention*, 83–5.

66 Memory capacities, of course, vary, and lead us back around the readers' response circle to attention, cultural understanding, basic intelligence, and pedagogical motivation. Nevertheless, the human mode of memory making belongs to the species as a phylogenetic trait, working within specific frames of typical practice. We cannot say, from reader to reader, who retains and who does not retain the data of working memory pertaining to specific stories – or for that matter to names, or faces, or incidents in one's own biography. Yet whatever the level of competence, the faculty is universal.

67 Pinker, *How the Mind Works*, 125– 6.

68 Minsky, *The Society of Mind*, 159. Boyd in *On the Origin of Stories* (134) introduces the idea of a "neural convergence zone" where patterns meet other "higher-level information."

69 Simon, "Some Computer Models of Human Learning," 102.

70 North, *The Moral Philosophy of Doni*, 11.

71 Evans, *Spenser's Anatomy of Heroism*, 78.

72 Johnston covers the entire territory pertaining to the hedonic component in memory making in *Why We Feel*, 72 and following, within a larger discussion concerning how we learn.

73 Ibid., 75.

74 An excellent group of papers on this topic may be found in Ivic and Williams, *Forgetting in Early Modern English Literature and Culture*.

CHAPTER FIVE

1 Trevarthen, "Intersubjectivity," 415–18.

2 Bermúdez, "The Domain of Folk Psychology," 25

3 Ibid., 28.

4 Nussbaum, *Upheavals of Thought*, 119.

5 In *Upheavals of Thought*, Martha Nussbaum deals with this question in considerable depth, confirming the generic nature of human emotionality across cultures while profiling the many subtle ways our emotional manifestations are modified through cultural conditioning, language, and norms. "I have said that all known societies have some variety of the major

emotion-types: love, fear, grief, anger, jealousy, envy, compassion, and some others. But even at the level of the big generic categories we do not find a perfect one-to-one correspondence across cultures, since cultures organize in different ways the elements that individuate emotions from one another" (163–4). This local and personal relativity and subjectivity is a constant factor in the expression of emotions, but the centre holds that we are an emotional species and that the challenges of our common environment have designed our brains around a common genetic model. This phylogenetic instrument colours our worlds in uniform ways, thereby leaving a vocabulary of responses that pertains to the experiences of the race.

6 The entire issue of design by evolutionary principles is taken up at length in Daniel Dennett's *The Intentional Stance*. He comes to his peroration on the matter in chapter 8, "Evolution, Error, and Intentionality" in which he deals head-on with "Mother Nature" as a heuristic designer who, through mass perishing of her offspring, improves upon her genetic legacies. "While it can never be stressed enough that natural selection operates with no foresight and no purpose, we should not lose sight of the fact that the process of natural selection has proven itself to be exquisitely sensitive to rationales, making myriads of discriminating 'choices' while 'recognizing' and 'appreciating' many subtle relationships." Among them is the rise of the limbic system as a designed instrument for reading the environment in its uniquely felt terms. The nature of that design as an "intentional stance" of its own among the human mental faculties is at the foundation of what emotions mean and do, and ultimately what it means to cry (299). Steven Pinker joins in the campaign in *How the Mind Works*: "it is wrong to write off ... the emotions as evolutionary accidents – namely, their universal, complex, reliably developing, well-engineered, reproduction-promoting design" (525).

7 Patrick Hogan takes up these matters in *The Mind and Its Stories*. "The fact that some stories are highly esteemed in any given culture suggests that those stories are particularly effective at both tasks – representing the causes and effects of emotions as understood or imagined in that society and giving rise to related emotions in readers" (1). Not only are narratives effective in profiling emotions at work within characters, but effective in replicating those emotions through empathy. He assigns these effects to universals, pointing out that emotional representation cannot be divided by authors, periods, genres, schools, or movements because the emotions are common to the species, thereby supplying a firm basis for cross-cultural literary analysis (3). Indeed, "our most prominent stories are generated from the prototypical structures of our emotion concepts" (11). Mark Turner builds on similar principles in *Reading Minds*, 16.

8 This alludes in passing to a debate far too complex to enter into here concerning the degree to which adaptation selects for the individual as opposed to the group. In the case of bees, termites, and ants, selection has been entirely in favour of the group, as though the hives and nests themselves constitute one large biological entity, its parts mechanically disposed to advance the reproductive interests of the queen. The colony itself becomes a single body of harmonious and blindly cooperative parts, much as our brains are hives of cooperative parts in the generation of conscious unity and the singular self. Humans are adaptively selfish, at least for their own genes, but the adaptive value in the crying and laughing impulses may well be in their signalling to the community, making them in part eusocial insofar as mental states in individuals also pertain to the well-being of the group. We gossip, for example, not only to our own advantages but in relation to those with whom we share information on a reciprocal basis. Crying recognizes that the good of the self is heavily invested in the support of the group. Richard Alexander engagingly discussed this topic in *The Biology of Moral Systems*. The "ultrasociality" of the eusocial insects is approximated in humans through "reciprocity," which provides the foundation for our moral systems (63–9).

9 Tom Lutz recounts a number of literary and cinematic anecdotes concerning this manipulative but usually transparent employment of tears in *Crying*, 251–86. He concludes from his analysis of the 1987 film *Broadcast News* as a lesson in the false displays of emotions that "sincere tears are good and false tears are bad, and that people who display false emotions are not to be trusted emotionally" (262).

10 Alan Palmer in *Fictional Minds* describes mind states that are beacons to those looking on, despite the constant risk of misinterpretation, which is even truer of those who remain silent (113).

11 Oatley, "Emotions," 274.

12 Patrick Hogan describes the quality of concern he calls "categorical empathy" in relation to close-knit communities constituting the "collective self-definition of an in-group … that provides the bases for the social prototype of happiness as group domination." *The Mind and Its Stories*, 141. Lafew's response can also be evaluated in terms of Hogan's "situational empathy," which is based on shared experiences, especially suffering (150).

13 Pinker, *The Blank Slate*, 259. Damasio makes the same point in *Looking for Spinoza*, 163.

14 Empathic feeling toward those for whom we are disposed to care has been much debated; it originates in an innate and vitally adaptive human trait basic to the collective life. Trevarthan goes so far as to state that we are more aware of the feelings and intentions of others than of our own in "Intersubjectivity," 416. Palmer states that "empathy is the power of en-

tering into another's personality and imaginatively experiencing their experiences. It is an essential part of the reading process," and the theatrical experience as well. *Fictional Minds*, 138. Caring, for Patricia Churchland, is the primal condition of mind behind kin structures and moral systems. *Braintrust, passim.*

15 Pinker, *How the Mind Works*, 524.

16 Nussbaum, *Upheavals of Thought*, 319. Subsumed in her analysis of compassion and the self is her extended theory of the eudaimonism feature of emotions. Not only personal goals, but "mutual relations of civic or personal love and friendship, in which the object is loved and benefited for his or her own sake, can qualify as constituent parts of a person's *eudaimonia*" (32). Lafew's joy can be genuine joy for others for whom he cares.

17 It would be supererogation here to review the extensive literature both pro and con concerning mirror neurons and the regions of the brain – the voxels of area 44 in humans and R5 in monkeys – where these telltale neurons are situated. But for a constructively skeptical account of the entire phenomenon to date, see Patricia Churchland, *Braintrust*, 135–62.

18 Brian Boyd's study of the neural properties and brain design that pertain to the making of stories includes thoughts on the emotional bonding between readers and characters. "Social neuroscience has begun to discover how our minds can be affected by emotional contagion, by responding, even without registering consciously, to cues of specific actions or emotions in others." He builds on the research into intersubjectivity and the degree to which we watch one another and seek to read mental states from our earliest age. *On the Origin of Stories*, 192.

19 Pinker, *The Blank Slate*, 256.

20 Early in the last century Henri Bergson evoked this contrasting assessment of the social order as a kind of viewing from the exterior and an entry into the interior of things. He had to fall back on such words as "analysis" and "intuition," the former distanced and rational, the latter empathic and inclined to identification. We reason around or feel into, which, in effect, is a kind of controlled point of view. An account of this divided stance may be found in William Hirstein's *Brain Fiction*, 131, or in Leszek Kolakowski, *Bergson*, 24. Either choice entails a denial of conflicting evidence.

21 The argument concerning emotions and the building of communities entails a modification of the evolutionary process to include the interests of groups in the genetic design of the brain in a way that sustains group goals as an aspect of individual goals. In the words of Peter Carruthers, "selection began to operate in the group," the principles serving to "enhance group cohesion and collective action." "Moderately Massive Modularity," 74–5. That evolution can support group strategies through the programming of individuals is a principle easily demonstrated throughout the

animal kingdom, but precisely how human emotions came to work for the benefit of groups, and thus support social cohesion, will continue to engage debate.

22 Readers convinced by this argument that emotions define genres, and that romance tragic-comedy is the quintessential plot that deals with the fulfilment of human survival instincts through successful mating and the entry into the mutual relations necessary to the birthing and rearing of children, may turn to Patrick Colm Hogan's "third hypothesis on emotion and narrative" for corroboration. He is concerned with romantic union and its attendant stories – the romance imperative often combined with the acquisition of political power. These, together or separately, are the most common goals for the achievement of human happiness, and as such are shot through with the imperatives of human emotions. *The Mind and Its Stories*, 94–5. But he does not take these story types back to brain design, phylogenetic desires, or the genetic bias toward reproductive success. His purpose is comparative, to show that love tragic-comedy is the most common of plots among literatures around the world and "is quite consistent across cultures and historical periods" (102).

23 Robert Storey discusses this issue in *Mimesis and the Human Animal* (121), where he describes humans as "'wired' not only for narrative comprehension but also for the emotionally induced reception of narrative – and the cultural enfranchisement that it makes possible – at a 'sensitive' point in their lives." Narrative, through the emotions, becomes a means for enculturation and the enrichment of memory.

24 McGinn, *Mindsight*, 105.

25 Ibid., 109.

26 Ibid., 110.

27 Ibid.

28 The problem touches the question in hand only obliquely, and yet is fascinating. We know that we feel in relation to aesthetic pleasure – feelings more readily isolated in relation to music, dance, and related arts of pattern, colour, and kinesis. If there is no credible backstory for these pleasures, they must be borrowed from other systems. Story makes those discriminations more difficult to isolate: can we feel the satisfaction of genre conventions fulfilled in structural ways without feeling for the agents who enact that potential into a completed Gestalt through the commitments of their lives? Can there be a romance aesthetic independent of an emotional investment in the desires of the characters? Proven upon us, it creates an escape from feeling as characters feel, but made into a condition of art, it divorces us entirely from the ethical substrate of imaginary worlds, thereby bankrupting literature as a vehicle of felt learning.

29 Keith Oatley and Jennifer Jenkins take up this difficult question in their subchapter "Emotions in Drama, Ritual, and Psychotherapy." They discuss Aristotelian "katharsis" as the experiencing of a primary emotion through artistic prompting, and move on quickly to Scheff's dubious explanation that such experiences arise because we distance ourselves from the overwhelming emotions of damaging events in real life, building up "emotional arrears" that in turn distort our emotional lives. Art allows us to deal with these pent up emotions safely "at a best aesthetic distance." *Understanding Emotions*," 371. Their references are to T.J. Scheff's *Catharsis in Healing, Ritual, and Drama.* Characteristic of his theory is the view that "when we cry over the fate of Romeo and Juliet, we are reliving our own personal experiences of overwhelming loss, but under new and less severe conditions" (13). The theory has multiple failings, particularly concerning substitution of unrelated personal memories for the adequate stimuli to contagious emotion provided by the play over the fate of the characters and what we feel for *them*, and the matter of aesthetic distancing and the dilution of emotional intensity. But the discussion touches on the question under investigation concerning the emotions that are otherwise potentially separated aesthetically from the social emotions of the narrative circumstances. Not argued here, as it has been by others, is that even tears of joy are deceptive in that joy for others merely reminds an old man like Lafew that death is imminent, thus causing him to cry for himself alone. Such theories deny the power of fiction to help us "cultivate our socially adaptive capacity for entering mentally into the experience" of others. Carroll, "The Human Revolution and the Adaptive Function of Literature," 42.

30 Ben Ze'ev, *The Subtlety of Emotions,* 171.

31 Lawrence, *Shakespeare's Problem Comedies*, 48. He traces the plot design to folklore, which carries its own drives toward closure, drives which Shakespeare must have recognized as the fuses of his plot. The values of that folklore, in turn, tell us that Helena must be held noble and honourable in pursuing the union, that the "tricks" were entirely acceptable to Elizabethans, that Bertram's sudden change of heart was entirely in keeping with such storytelling conventions, and that there is nothing to indicate that their future life together "would be anything but happy." The problem lies only in the cynicism of modern perspectives.

32 Shakeapeare, *All's Well That Ends Well.* Hunter faces the anomalies squarely, pointing out our response to the personal in parallel with our awareness of structures, patterns, motifs, and the juxtaposition of "extreme romantic conventions with down-to-earth and critical realism" (xxxii–xxxiii). The as yet unrealized hope for a unified view of the play will arrive only with "the power of a new poetic vision" in which all of the characters,

heroic, immature, and corrupt, find their places within a comprehensive vision of the human (lvi).

CHAPTER SIX

1 Bacon, *A Selection of His Works*, 317.

2 The indefinable nature of the association whereby we simply "know" that two statements share in a common essence is not unrelated to the indefinable approach to the proverb as a simple form advanced by Archer Taylor in 1931, when he stated that "the definition of a proverb is too difficult to repay the undertaking; and should we fortunately combine in a single definition all the essential elements and give each the proper emphasis, we should not even then have a touchstone. An incommunicable quality tells us this sentence is proverbial and that one is not." *The Proverb*, 3. The problem is that individuals can never decide among themselves just which sentences are proverbial and which are not, so that ambiguities of presentation and meaning obstruct the chances of finalizing a definition. In any case, the present study is not concerned with defining the form, but with tracing the mental operations that bring recognized proverbs to experiential actualization. Concerning the intuition by which we recognize the shared fields binding parallel statements, moreover, there are estimates to be made based on the best retro-engineered guesses about the neurobiological production of meaning.

3 Thagard, *Mind*, 89.

4 Dennett, *Consciousness Explained*, 455. One such metaphor which persists in philosophical analysis concerns consciousness as a theatre – a metaphor which continues to prove its worth to explanatory ends, even though consciousness is nothing like a theatre. See Donald Beecher, "Mind, Theatres, and the Anatomy of Consciousness."

5 Pinker, *The Blank Slate*, 439.

6 For further information on these continental editions, see the introduction by Jan Ziolkowski to his edition of the Latin *Solomon and Marcolf*.

7 The manuscript filiations and publication history of this work are reviewed in the introduction to *The Dialogue of Solomon and Marcolphus*, ed. Donald Beecher with the assistance of Mary Wallis, 74–86. All citations of the proverbs are taken from this edition.

8 The English edition published by Leeu is, in fact, a selection from the larger cupboard of proverb pairs which have been associated with the tradition throughout its nearly 500-year history. The most extensive version is a handsome manuscript in the Würzburg University Library, produced in 1434 by the monks of Neumünster Monastery, which contains 142 pairs.

Several in the intervening years suffered corruption, making these earlier manuscripts invaluable for comparative purposes. The most extensive work on the history of the early manuscripts and the range of proverbs was carried out by Walter Benary in his edition of *Salomon et Marcolfus*, which is in turn reviewed in the introduction to the English edition employed throughout this essay: Beecher and Wallis, *The Dialogue of Solomon and Marcolphus*, 82–6.

9 Werner Lenk, "Zur Sprichwort-Antithetik im Salomon-Markolf-Dialog," 153.

10 This is a modern German transcription of his words by W. Benary, *Salomon et Marcolfus*, vii, taken from Paul Piper, *Die Schriften Notkers und seiner Schule*, 2: 522.

11 Schönbrunn-Kölb, "Markolf in den mittelalterlichen Salomondichtungen und deutscher Wortgeographie," 100.

12 Robert J. Menner, ed. *The Poetical Dialogues of Solomon and Saturn*. See also Mary Wallis, "Patterns of Wisdom in the Old English *Solomon and Saturn* II."

13 The question has been taken up by paremiologists, whether and to what extent proverbs help to bind groups together by embodying identifying values and mentalities. Alan Dundes, some forty years ago, noted the degree to which cultures have underlying assumptions and how folk ideas become important "building blocks" from which world views are constructed. "Folk Ideas as Units of Worldview," 96. Wolfgang Mieder, in a series of studies, develops multiple aspects of the relationship between proverbs as expressions of national character and cultural world view, as in his chapter entitled "'Good Old Yankee Wisdom' Proverbs and the Worldview of New England" in which he begins by reviewing the main contributions devoted to the question. "*Proverbs Speak Louder Than Words*," 145.

14 Wesling, *Joys and Sorrows of Imaginary Persons*, 124.

15 Danièle Letocha, "The Duty of Memory," 272–5.

16 Williams, "Proverbs and Ecclesiastes," 263.

17 *Altfranzösiche Sprichwörter*, No. 78.

18 Turner, *Reading Minds*, 70–2.

19 George Lakoff and Mark Johnson are among those who made the "embodied mind" (i.e., the mental orientations based on degrees of awareness of the nature of the body and the mind's role in directing and thinking in relation to it) the unique source for the many buried metaphors of action and direction connecting abstract ideas. They advance this perspective most cogently in *Philosophy in the Flesh* ("He carried out his ideas" or "Her salary rose last year"). Together with Mark Turner, they have gone far in establishing the degree to which image and action schemata, front to

back, left to right, up and down, hence bodily movement through space, as well as other things done with the body, such as grasping, sitting, standing, running, rising, swallowing, excreting, digesting, and related notions figure regularly in metaphorical ways in relation to abstract ideas. Because these movements and functions are so familiar, they are easily schematized and projected, thereby structuring much of our linguistic expression as "acts of a human brain in a human body in a human environment which that brain must make intelligible if it is to survive." *Reading Minds*, vii–viii. This way of thinking has been enormously valuable and influential in assessing one of the domains of our knowledge, in showing how sayings reflect action domains, and how schemata work to interconnect otherwise alien materials. But the body is not the only point of reference for our understanding of folk physics. Trees and rain fall as well as bodies. A much larger environment than that constituted by the body has defined the conditions upon which our knowledge of the world in based. Moreover, the schemata upon which we ground metaphor are not perfectly synonymous with the schemata which are embedded in the architecture of the brain, but arguably must be generated in relation to what that architecture "knows."

20 This brief assessment, of course, rides roughshod over the many scholarly analyses concerning how we define and catalogue the figurative employments of language, which is one of the acknowledged constants of proverbial sayings. Norrick, in *How Proverbs Mean* (101), acknowledges not only the interest in such studies in understanding "natural figurative meaning," but in extending "hypotheses about recurrent patterns of thought in the realm of cognitive psychology." That was prescient thinking, and the basis for the observation was his familiarity with the empirical psychological research of Richard P. Honeck and others in "Proverbs, Meaning, and Group Structure," and Andrew Ortony in "The Role of Similarity in Similes and Metaphors." There has been wide and longstanding appreciation that proverbs are often scenic and general, yet by their "natures" invite application to a target. That "the leopard cannot change his spots" is largely pointless unless a social context is invented at the same time that pertains to all who, given their natures, are unlikely ever to change. This can be reduced to a series of letters in a logic formula to clarify the equivalences and schemata whereby we somehow "see" the specific in the generic. This model works best for what Norrick calls "complete scenic, species-genus synecdochic proverbs" (107), by which he means proverbs setting a scene that by the synecdoche implicit in moving from species to genus (ironically Turner's "generic is specific") creates a new field of social application (as in the proverb example above). His analysis then moves on to include such variants as "metaphoric proverbs," "metonymic proverbs," "hyperbolic

proverbs," and "paradoxical proverbs." These are among the subsets of proverbial expression that the generic neurological connectivities render into meaning and allow us to discriminate by bringing their component properties to consciousness.

21 Hofstadter, "Analogy as the Core of Cognition," 499 and following.

22 This is the direction taken here to reduce the mystery, somewhat, that was iterated by Allen Paivio in "Psychological Processes in the Comprehension of Metaphor" (151) in which he concludes that no theory or reasoning or semantic or propositional structures can explain the relationship between proverbial expression and its reconceptualization as an applicable precept, nevertheless stating his faith that work in memory and cognition will provide clarification. That was the look of things in 1979 when the article was published in *Metaphor and Thought*.

23 The notion of unusual design as a way of inducing the sense of incongruity and arousal in readers which compels them to find meaning in ambiguity is by no means a new idea. It figures centrally in the work of D.E. Berlyne in *Conflict, Arousal, and Curiosity*. The same principle applies to riddles, with which proverbs have much in common, viewed as linguistic formulae intended to create initial "confusion" in setting oppositions and apparent contradictions within the "riddling description." See Georges and Dundes, "Toward a Structural Definition of the Riddle."

24 Andrew Brook, "Emmanuel Kant," 427, paraphrasing W. Sellars, *Science and Metaphysics*.

25 Francis Crick takes up the problem of binding or unity of emergent effect in *The Astonishing Hypothesis*.

26 Campbell, *The Improbable Machine*, 201.

27 Ibid., 149–51.

28 William F. Brewer, "Schemata," 729.

29 This is related to the lengthy debate over the matter of qualia, much discussed in cognitive science circles since the appearance of Thomas Nagel's "What Is It Like to Be a Bat?" and Frank Jackson's "Epiphenomenal Qualia" – the terms of their debate reshaped or "disqualified" by Daniel Dennett in *Consciousness Explained*, 369–411. Nagel set up in ingenious thought experiment about what it is like to be a bat, not merely in imagining ourselves as bats, but in knowing what a bat's own experiences are like. This was a launch pad for arguments against reductionism (which held that every brain phenomenon is viewed as nothing more than a function of its material operations). If there is something which a creature recognizes as the uniquely personal and experienced property of its mental activity, that would appear to be an entity distinct from neurological functions. But Dennett took umbrage to the special properties accorded to qualia,

simply because all consciously experienced thoughts and sensations are the emergent properties of neural networks having about them a "something it is like" to experience them. This is a chapter in the mind-body debate that need not delay us here because, in the end, most would agree that the experiential content of consciousness is brain manufactured and that those emergent states are, in a sense, what acceleration is to an automobile – a property for which the vehicle is designed but which is not the vehicle itself.

30 Johnston, *Why We Feel*, 14.

31 By way of corroboration, Andrew Newberg et al., in *Why God Won't Go Away*, seek the origins of our religious impulses in the habits and constitution of the brain by surveying many of these same schemata or default modes of knowing the world. The term they employ for these knowledge systems is "operators," which echoes the "little man in the machine" problems of earlier years by personifying the "makers" of these emergent properties of mind, but the categories of operators they supply are nevertheless revealing, among them the "holistic operator," the "reductionist operator," the abstractive operator," the "quantitative operator," the "causal operator," the "binary operator," and the "existential operator," each one of them describing a condition of knowledge implicit in the functional procedures of neural discrimination. The holistic operator is related to the problem of binding in the brain whereby all the divergent discriminations – orthography, syntax, images, concepts, and figurative projections – come together as a single impression. The reductionist analysis does precisely the opposite, enabling the brain to see the constituent parts of complete Gestalts. The "abstractive" enables the brain to arrive at "general concepts from the perception of individual facts" (49), while the "causal" interprets the sequences of reality as discrete causes and their effects, including the rather mystical property of causation itself as a bedrock principle of world order. The "binary" they describe as one of the brain's most powerful tools because it allows for efficient movement through the physical world by reducing complicated relationships of space and time into simple pairs in opposition, such as up and down or before and after, or left and right (50), not to mention good and bad, or praise and blame.

32 This feature is particularly characteristic of the Hebrew proverbs of the Old Testament, which is obliquely related to the paired proverbs in the Solomon and Marcolphus tradition. Thus the stock formula of, say, Proverbs 10:1, "A wise son makes a father glad; a foolish son is a sorrow to his mother," the first of a whole series of positive and negative observations, each attached to its own *topos*. See T. Anthony Perry in *Wisdom Literature and the Structure of Proverbs*, 73–4. Proverbs in contrasting registers, however, introduce considerably more variables which often pertain to matters

of class, hyperbole, parody, satire, and echo in the ostensible study of a single *topos*.

33 This is a slight rewording of Proverbs 14:1, "a wise woman builds her house; a foolish one tears down that which she has built with her hands."

34 Sam Glucksberg and Boaz Keysar have reported on the extent to which the brain seeks out metaphorical levels of meaning even when literal meanings were requested. The mind pursues these further levels of association and projection as though the search for metaphorical potential is an integral part of searching for literal meaning; it is not merely an add on dimension. The pursuit of meaning to this degree would seem to be a part of our paranoia concerning statements that may hold latent associative meanings vital to our personal well-being, principally because the schemata applied to other domains may produce usable data. The brain compares because all discriminations depend on eliminatory comparisons to the best fit. "Understanding Metaphorical Comparisons: Beyond Similarity."

35 Bernard-Henri Lévy, in *Hôtel Europe*, discusses the damage caused by identity politics, and proposes that analogical thinking is at the foundation of those divisive identity garrisons through consensus building around similitudes: "Il faut, à ce démon de l'analogie, répondre par l'évidence de la dissemblance, du désaccord, de la querelle" (We must answer the demon of analogy with the evidence of difference, disagreement, and quarrel), 143. Politically, his point is well taken, but that the analogy-making brain can shut itself down in the name of this enterprise is entirely doubtful; it is the default manner by which we process and make meaning of the novel in relation to the known, both intuitively and through experience.

36 This flexible associative capacity was undoubtedly "pressured" into the brain's architecture by the imprecision of incoming stimuli. Plasticity allows for the placement of diverse kinds of primary recognitions within categories. Things are known in terms of greater and lesser likenesses by analogy, and the expansion of the cerebral cortex permits such analyses to greater and greater levels of discrimination. Metaphor is an add-on capacity of analogy classification, whereby trees may become ballerinas dancing on the lawn by dint of an imagined resemblance in their movements, although ballerinas do not as readily become trees dancing on the stage. See Hofstadter, "Analogy as the Core of Cognition," (501), for further perspectives on the levels and depths of analogical processing.

37 Making strange was also studied by Reuven Tsur in *Toward a Theory of Cognitive Poetics* (4), where he refers to "defamiliarization" and "systematic disturbance" by slowing down normal cognitive processes and intensifying attention, much as riddles and parables are designed to do, for riddles

rely on similar orders of projection from literal to figurative or the inverse, just as parables require paradigm shifts from the generic to the specific.

38 Turner, *Reading Minds*, 43.

39 Hobbes studied this point earlier; it forms the foundational axiom behind Colin McGinn's sustained philosophical enquiry into the modes of percept and image, the relative experiencing of world sensations and mind-generated sensations in *Mindsight*.

40 There are far more scientific discussions of the matter, but instead I offer this tidbit off the internet: "At a purely chemical level, every experience humans find enjoyable – whether listening to music, embracing a lover, or savoring chocolate – amounts to little more than an explosion of dopamine in the nucleus accumbens as exhilarating and ephemeral as a firecracker." J. Madelaine Nash, Time Magazine senior science correspondent and 2004 Perlman Award winner for excellence in scientific writing.

41 See Ellen Dissanayake's *Homo Aestheticus*, "The Core of Art: Making Special," (39–63), where she discusses aesthetics as a condition of special thinking about objects designated for and inviting this mental modality – "the biological core of art," a kind of induced category of experience which our evolved mental architecture makes possible.

42 Hernandi and Steen, "The Topical Landscapes of Proverbia," 9, 7.

43 Mar and Oatley, *Perspectives on Psychological Science*.

44 Turner, *Death is the Mother of Beauty*, 145.

45 Schank, *Dynamic Memory*.

46 Lakoff and Johnson, *Metaphors We Live By*, 3.

47 Ibid., 193.

48 To be sure, our neural systems reconstruct, according to their operational constraints and habitual saliences, "versions" of external stimuli, but I prefer not to blend the reconstruction of "percepts" and "images" with the constructions of the "secondary" imagination by calling them all constructions into things which are not on the same footing. Colin McGinn in *Mindsight* (esp. 134–5) remarks just how profoundly and quintessentially the mind discriminates between perceptual and imaginative modes, and between the internal recollection of things which are remembered from the real world and things fabricated by the imagination. Metaphor is precisely the domain in which belief and imagination can temporarily meet in "as if" associations, thereby forming something like "metaphorical belief," which is a personal evocation not subject to empirical verification or denial. It is something of its own not to be confused with other modes for knowing and reconstructing the world.

49 Levin, "Standard Approaches to Metaphor and a Proposal for Literary Metaphor," 124.

50 Or so it seemed to Paivio, "Psychological Processes in the Comprehension of Metaphor," 152. Some progress has been made since then, but mostly in the manner pursued in the present article – the imputation of systems by the nature of their emergent properties, by using literature as the basis for a theory of mind.

51 Thagard, *Mind: Introduction to Cognitive Science*, 87.

52 Turner, *Reading Minds*, 6.

53 Jolles, *Formes simples*, 211.

54 For the anti-proverb and related modern deformations and adaptations of the proverb in advertising, mass media, cartoons, and the information world, see Mieder, "*Proverbs Speak Louder than Words*," *passim.*

CHAPTER SEVEN

1 Bergougnan, *Romans grecs: Les Éthiopiques ou Théagène et Chariclée*, xviii.

2 Bergougnan's reservations by no means represent a voice in the wilderness. Even Northrop Frye, whose critical insights have done more than any other to dignify and legitimate the genre through its links to myth, has registered its weaknesses and found himself challenged by the same lack of realism. "The romancer does not attempt to create 'real people' so much as stylized figures which expand into psychological archetypes." In this factor he seeks to find what virtue there is. "It is in the romance that we find Jung's libido, anima, and shadow reflected in the hero, heroine, and villain respectively. That is why the romance so often radiates a glow of subjective intensity that the novel lacks, and why a suggestion of allegory is constantly creeping in around its fringes." He sensed the patterns of truth, but could find their verification only in other literary artifacts and in the mythopeic intelligence. It was a significant stride. *The Anatomy of Criticism*, 304. Even in *The Secular Scripture*, he continues in a similar vein: "It looks, therefore, as though romance were simply replacing the world of ordinary experience by a dream world, in which the narrative movement keeps rising into wish fulfillment or sinking into anxiety and nightmare." These are the hyperbolical stances and emotions still floating free of explanatory principles (53).

3 Bergougnan, *Romans grecs*, xix.

4 Wolff, *The Greek Romances in Elizabethan Prose Fiction*; Hägg's several articles, but especially *The Novel in Antiquity*, with its flyleaf map of the ancient world; Morgan's "History, Romance and Realism"; Bowie's "Novels and the Real World" and "Apollonius of Tyana."

5 The *Aethiopica* has been called variously *Ethiopica*, *Theagenes and Chariclea*, and in the edition consulted for this study, *An Ethiopian Story* in

Collected Ancient Greek Novels. Page numbers and the spelling of names, such as Charikleia, are derived from this edition.

6 Quite apart from the preoccupations of the Renaissance morals police concerned with the corruption of ladies, in particular, by the reading of romance – see Michael McKeon, *The Origins of the English Novel: 1600–1740*, 520 – there has been the more global critique that romance is pure fantasy and hence not about anything that matters in the real world. René Pruvost, *Matteo Bandello and Elizabethan Fiction*, 325. In the words of Jean-Michel Ganteau, "Fantastic but Truthful: The Ethics of Romance," the genre is not realistic "because it eschews verisimilitude, prefers the exotic to the familiar and the far to the near ... In other words, romance is concerned with things foreign ... what escapes common experience." His approach to the rescuing of the genre was its ethical teaching, reversing what Virginia Woolf said of her *Orlando*, that it was "truthful, but fantastic," by calling romance "fantastic, but truthful" (225). The present essay defines "truth" in quite different terms.

7 Denis Dutton opens his chapter "Art and Natural Selection" with a parallel discussion of the remote but necessary relationship between the evolutionary constitution of the brain and the human capacities for culture, careful to point out that minds did not develop an "'innate concept of *carburetor* or *trombone*,'" (quoting Pinker, *How the Mind Works*, 20), but rather developed traits which have as their "goal *survival* or *reproduction*," and not something that merely improves our quality of life. It is an important precondition to thinking about mental traits and the emergent values that become a part of artistic expression and cultural creation. *The Art Instinct: Beauty, Pleasure and Human Evolution*, 86.

8 Antonio Damasio defines the "genomic unconscious" as those elements of instinct, automatic behaviours, drives, and motivations that originate in the genetic coding that creates the "distinctive features of our phenotype." He describes such dispositions in terms of "thematic scope," which is often pervasive, much of which prefigures our cultural and social lives, taking for examples the jealousy of Othello, the harsh punishment meted out to Anna Karenina, the incest taboo that haunts Oedipus, and the cognitive asymmetry distinguishing men from women in general. The genome, in this regard, accounts not only for differences but for much of "the sameness that hallmarks the repertoire of human behavior." *Self Comes to Mind*, 278, 79.

9 Sidney, *An Apology for Poetry*, 103.

10 Greene's *Menaphon*, edited by Brenda Cantar; all page references are to this edition.

11 Pinker, *How the Mind Works*, 483.

12 Irenäus Eibl-Eibesfeldt in *Human Ethology* (240) commented on the strange effect that beauty has, not only to draw contenders into the race for the fittest women but also to appall contenders out of fear of rejection. The more she is deemed beautiful, the more delicate the male strategizing, including the employment of gifts, compliments, and poetic celebrations, while always leaving a face-saving mode of exit.

13 This principle has given rise to an increased discussion of the interplay between natural selection and sexual selection in the making of the species. Natural selection is about the adaptations necessary for more efficient survival in hostile environments, but sexual selection is about the human features preferred through eons of bias in the selection of sex partners. Women, it is argued, have dominance in this field because, as nurturing mothers, they must choose mates who can protect and resource the production of offspring. Men have an eye for waist to hip ratios because they signal fertility, while women, typically, have chosen men who are taller, a bit older than they are, muscular, and physically symmetrical, not to mention kind, intelligent, in possession of resources, followed by other such traits as adaptability, generosity, industriousness, creativity, and a sense of humour. Here, diversity enters, but other, more concentrated choices concerning height and musculature have also entered the genome. Dutton offers a readable resumé of the place of sexual selection in the making of social and aesthetic clichés and social tendencies in "Art and Human Self-Domestication," *The Art Instinct*, 135–63.

14 There is wide agreement that mating negotiations are based on the asymmetry of the respective parents' investments in the offspring. The point will come up again. Because male fitness implies impregnating as many women as possible to pass on his seed, while female fitness implies seeking protection for herself and her children, given the high personal cost of bearing children, social bargaining, pledges, and promises inevitably ensue between prospective partners. Females are taught by their genes to negotiate not only for the best males they can get but for signs of protection and emotional support. Such negotiations are the principal "work" performed by the protagonists of romance. The scope and contours of this sequential economy resulting from the high consequentiality of sexual activity has been anatomized in recent years with increasing precision. See Anthony Walsh, *The Science of Love*, esp. 180ff. A full profiling also appears in Pinker's *The Blank Slate*, 252. He is reading Symons, *The Evolution of Human Sexuality*, who in turn is relying on the analytical trends developed and popularized by Tiger and Fox in *The Imperial Animal*, inter alia. Romance plots often enjoin upon eligible men the need to articulate their thoughts and feelings as an essential part of love negotiations.

15 Frye describes the many plot types arising from this father-daughter strife, noting that "the theme of the calumniated girl ordered out of the house with her child by a cruel father, generally into the snow, still drew tears from audiences of Victorian melodramas," but he had little notion of the psychological underpinnings. *Anatomy of Criticism*, 199.

16 Tiger and Fox, *The Imperial Animal*, 88–91.

17 This conflict between fathers and daughters has been assessed from many angles as a universal dilemma. Jerome Barkow points out that among the psychological characteristics underlying the social stratifications of hunter-gatherer peoples is the quest for prestige and social rank obtained by sacrificing the interests of daughters to the interests of exchange. Daughters were valuable to their fathers only for purposes of bartering for social advantage through arranged marriages. "Beneath New Culture Is Old Psychology," 633. Symonds clarifies, citing W.H. Davenport, that power in societies where other resources are scarce is often measured in terms of the custodial rights to women and the ability to offer them in marriage. Many societies function in terms of these "sexual futures" even as unborn females. This search for status on the part of older males often came into conflict with the powerful prompts of young females in securing their own reproductive futures. Symonds, "Sex in Cross-Cultural Perspective," 141.

18 See also R.L. Trivers, "Parental Investment and Sexual Selection," 136–79. He has a more mathematical approach to the divided interests of children and their parents in terms of genetic investment, with children being more oriented to their own offspring than to their parents, an asymmetry of investment that entails a kind of break-away from the past, and a frustrated parental generation that will leap forward to their real and emblematic future in their grandchildren. Thus, by the logic of gene interest, it is to be expected that children's priorities will always be tipped in favour of the future rather than the past. Children are programmed to seek auto-determination and to break from their own birth parents while negotiating with them for continued resources. *Menaphon* replicates this pattern.

19 James Wilson, *The Moral Sense*, cited in Pinker, *The Blank Slate*, 251.

20 Updike, "The Tried and the Trëowe," 201.

21 On the continuity of genetic design over evolutionary time, see Levine and Suzuki, *The Secret of Life*, 61.

22 Pinker, *How the Mind Works*, 315.

23 Pinker, *The Blank Slate*, 73.

24 A very long historical note might accompany any such list of the psycho-physiological disturbances caused by love. Briefly, it was a phenomenon pursued by writers from the ancient world down to the Renaissance, all of them recognizing the power of erotic desire or compulsive love to disorient both body and mind. Lists of the symptoms of love began with the poets –

Sappho provided one – but the despair of unrequited or over-fanaticized love soon became a medical *topos* because lovers afflicted with erotic melancholy or mania were at times in mortal danger, which only the beloved or a physician might counteract. Their treatises led to extensive speculations on causes, the power of love, the pathological influence of sight, the burnt biles, the corruption of the imagination, these things largely attributed to the pathological powers of the melancholy body. Their anatomization of love in terms of its manners, symptoms, and susceptibilities was brilliant and had a profound influence on authors seeking to give their portraits a patina of scientific authenticity. Love impulses were recognized universals, but for deeper causes, they relied on the analytical language pertaining to Venus and her Son with his bow and arrows whereby the vital passions were evoked. The present psycho-evolutionary explanation is something of a radical and comprehensive fine-tuning. Perhaps the most encyclopedic pre-modern treatment of the poetic and medical traditions of love is Jacques Ferrand's 1623 work, *A Treatise on Lovesickness*.

25 The evolutionary explanation of this excessive yet adaptive trait operates in the background of Frank Tallis's *Love Sick* (284): "Love has evolved over thousands of years, to ensure that we achieve our evolutionary destiny – the transmission of our genes into subsequent generations." He has an entire section on the "madness" of love which, throughout medical history, has suggested pathological origin, but which, in effect, is part of a larger design that favours sudden erotic bonding sufficient to ensure offspring. Mind states that enhanced such events were of course selected out and replicated in the brains of the children, who became increasingly vulnerable to this tendency. His chapter entitled "A Necessary Madness" cites W. Somerset Maugham, in *A Writer's Notebook*: "Love is only the dirty trick played on us to achieve continuation of the species."

26 That nature has arranged through the design of the midbrain for this rare-in-a-lifetime experience gives pause for reflection. Part of who we are as a species is our vulnerability to this sudden rush of obligatory commitment to a member of the opposite sex (traditionally), a surprise rush which, in its promptings, is life-defining. It is tantamount to an emotion-driven calculation or estimation concerning a mate deemed right for us by our mammalian brain, which is subsequently taken for a destiny that surpasses nearly all other considerations. Falling in love can be wonderful if reciprocated by an eligible candidate, but it is, in fact, a dictatorial move on the part of our genetic design to remove us by emotional force from the shopping phase to a committed phase according to a clock of its own design. We can only marvel at the criteria of its systemic agenda, and imagine the evolutionary just so tale by which this trait, like morning sickness for expectant mothers in their first term of pregnancy, got so precisely wired into the genome.

27 In these terms, James Chisholm in "Love's Contingencies" (49) recovers the survival logic of falling in love, with all the attendant anxieties, jealousies, and intense pleasures which accompany this adaptive trait. See also Helen Fisher, *The Anatomy of Love*.

28 What emotions are to intelligent human behaviour has been much investigated in recent years, nearly all studies seeking to align these hedonic emanations from the human midbrain with their adaptive advantages as readings of the environment and as swift reactions to a variety of contingencies. The question of how humans have incorporated this guidance system predicated on sensations of reward and discomfort into their subliminal and conscious lives in conjunction with cultural values and controlled volition has produced an extensive and speculative analysis resulting in a diversity of weightings. But there is nearly complete agreement that emotions play a significant role in choices of conduct and decision-making, and that they are far more active in prioritizing attention and cementing memories than was formerly imagined. Paramount is that "emotions evolved not as conscious feelings, linguistically differentiated or not, but as brain states and bodily responses. The brain states and bodily responses are the fundamental facts of an emotion, and the conscious feelings are the frills that have added icing to the emotional cake." LeDoux, *The Emotional Brain*, 302. In this way, the emotions are prioritized as a form of modular understanding of the world, and in their phylogenetic design they are at the foundation of universal attitudes and strategies. "There are universal emotional themes that reflect our evolutionary history, in addition to many culturally learned variations that reflect our individual experience. In other words, we become emotional about matters that were relevant to our ancestors as well as ones we have found to matter in our own lives." Paul Ekman, *Emotions Revealed*, 217. The properties of emotions are a large part of the shaping of human universals and a prime mechanism whereby they are enforced.

29 Symons, *The Evolution of Human Sexuality*, 253.

30 Anthony Walsh's anatomization of love in biogenetic terms provides a running commentary on all these tendencies. Sight replaced smell as the impetus for mating after females abandoned estrus. Something like rudimentary romantic love emerged when lovemaking was between identifiable individuals who cared for each other's pleasures and placed the memories of those pleasures within a time continuum. Thus the emotions came into play around bonding and desire involving a partner chosen by sight. These were preconditions to the formation of long-term bonds, and sexuality was a celebration withheld until the legitimacy of those feelings of desire were believed in by both partners and not feigned merely for sexual gratification.

The feigning, prating, and aestheticizing lover was one of the preoccupations of the romance heroine for precisely these reasons. *The Science of Love*, 185.

31 Tiger and Fox, *The Imperial Animal*, 24.

32 Ibid., 25.

33 Walsh in *The Science of Love* (210) offers a complete study of these characterological tendencies during courtship phases. Evolutionary biologists expect a fit between reproductive strategies that evolved eons ago and the human psyche operating in modern cultural contexts, although, as a hierarchical reductionist, he admits that the fit is sometimes a loose one. Statistically, men should fall for physical beauty, feel themselves in love faster than women, and should prefer younger women who promise health and fertility, while women will accept older men because their providing power is more proven. They will hold off longer and keep their chastity while sizing up prospects and assuring themselves of the man's loyalty and affection. These are innate strategies because those who followed them were reproductively adaptive. There are the variations we can all think of: men who marry for wealth, women who marry handsome physiques with poor wages, but the exceptions do not destroy the "rules"; romantic lovers are rarely far removed from the logic of the genes.

34 Frye, *The Secular Scripture*, 73.

35 Tiger and Fox, *The Imperial Animal*, 70.

36 Shirley Fisher, *Homesickness, Cognition, and Health*, 89.

37 Fred Davis, *Yearning for Yesterday*, 15.

38 Frye, *The Secular Scripture*, 176.

39 Dutton, *The Art Instinct*, 135.

40 Turner, *Reading Minds*, vii–viii. Art imitates life in these matters and assumes its forms, exfoliating from human biology.

41 Storey, *Mimesis and the Human Animal*, xvii.

42 Carroll, *Evolution and Literary Theory*, 131.

43 See Whissell, "Mate Selection in Popular Women's Fiction."

CHAPTER EIGHT

1 Cuddon, *The Penguin Dictionary of Literary Terms and Literary Theory*, 937.

2 Harmon and Holden, *A Handbook to Literature*, 504, 505.

3 Fowler, *A Dictionary of Modern Critical Terms* (234), states merely that it is "a lack of certainty … about what is going to happen, especially to characters with whom the reader has established a bond of sympathy."

4 Iser, "The Reading Process," 212.

5 *Suspense: Conceptualizations, Theoretical, Analyses, and Empirical Explorations* is central to the resumé presented here. The collection contains sixteen complementary essays by writers familiar with the work of the other contributors. Their work pertains to a common set of problems, which they approach both theoretically and experimentally. Much that is to be found on this topic outside this volume is written by many of these same contributors.

6 Noël Carroll's views are found in "Toward a Theory of Film Suspense" and "The Paradox of Suspense."

7 Zillman, "The Psychology of Suspense in Dramatic Exposition," 224–7.

8 Vorderer, "Toward a Psychological Theory of Suspense," 235.

9 Oatley, "A Taxonomy of the Emotions of Literary Response and a Theory of Identification in Fictional Narrative," 55, 57. See also P.E. Jose and W.F. Brewer, "Development of Story Liking."

10 For one of several arguments against identification theory, see William F. Brewer on "Character Sympathy" in "The Nature of Narrative Suspense and the Problem of Rereading," 116.

11 On this vital distinction, see Colin McGinn, *Mindsight*. As an opening salvo, "when we think about our mental images we should be struck by two things: (1) how *similar* they are to regular perceptions, and (2) how *different* they are from regular perceptions" (2). But one thing is certain, that on a day-to-day basis, an inability to make a rigorous if not flawless distinction between image and percept would constitute a radical flaw affecting our chances for survival.

12 Walton, *Mimesis as Make-Believe*, 253.

13 Ibid., 257.

14 Gerrig, "Suspense in the Absence of Uncertainty."

15 Gerrig, *Experiencing Narrative Worlds*, 201ff. See also Cupchik, and Laszlo, "The Landscape of Time in Literary Reception," and William F. Brewer, "The Nature of Narrative Suspense and the Problem of Rereading."

16 Ohler and Nieding, "Cognitive Modeling of Suspense-Inducing Structures in Narrative Films," 129.

17 Tooby and Cosmides, "The Psychological Foundations of Culture," 47–8.

18 Rabkin, *Narrative Suspense*, 85, 159.

19 Carroll, "Toward a Theory of Film Suspense," 77.

20 Gerrig, *Experiencing Narrative Worlds*, 80.

21 Trivers, "The Evolution of Reciprocal Altruism," 35–57. See also, Lauren Wispé, "The History of the Concept of Empathy."

22 Krebs, "The Challenge of Altruism in Biology and Psychology," 104.

23 Zillman, "Mechanisms of Emotional Involvement with Drama," 41. See also N.H. Frijda, *The Emotions*.

24 Patricia Churchland, *Braintrust*, 151–3.
25 Smith, *The Theory of Moral Sentiments*, 1.
26 Hoffmann, "Is Altruism Part of Human Nature?" 127.
27 Joseph Carroll, *Evolution and Literary Theory*, 160.
28 Pinker, *How the Mind Works*, 406.
29 P.E. Jose and W.F. Brewer, "Development of Story Liking."
30 Andrew Ortony, Gerald Clore, and Allan Collins, *The Cognitive Structure of Emotions*, 93.
31 Gerrig and Bernardo, "Readers as Problem-Solvers in the Experience of Suspense," 471.
32 Rabkin, *Narrative Suspense*, 13.
33 Ibid., 18.

CHAPTER NINE

1 See Aristotle, *On Poetry and Style*.
2 Watson, *The Lost Second Book of Aristotle's* Poetics, 156.
3 These writers were, of course, ignorant of the one-time existence of the second book of Aristotle's *Poetics* and its ancient and medieval commentators, to which mention will be made in subsequent pages.
4 The same challenge has been addressed in recent times and in numerous ways, as in an article by Jacques Guilhembet entitled "Esthétique de la comédie," in which the author states that everything said about the emotions and tragedy can be said about comedy given the symmetry of the genres (184), even though his working assumption is that what Aristotle meant by "catharsis" will always remain problematic. The writer then discusses all the ways in which comic events, such as the mocking banishment of troublemakers, is psychologically purgative, after working through Charles Mauron's theories about laughter as little epileptic seizures, *Psychocritique du genre comique*.
5 Della Porta, *The Sister*, 73.
6 Riccoboni, "From the Comic Art" (1585), 100.
7 Minturno, "From the Art of Poetry," 85.
8 Sidney, *An Apology for Poetry*, 135.
9 William C. Carroll, "Romantic Comedies," 183.
10 Jim Davies talks about "goal achievement" in relation to hope in his study *Riveted*, 60–1.
11 The medical overtones of this term have been much discussed, because catharsis suggests a purgative effect, a delivering of the body or mind of some property deleterious to health. Aristotle explains how art operates on the soul using this master metaphor: art eliminates fear through telling

stories that make the auditors fearful, perhaps by arousing compassion, or perhaps by making us no longer fear the event represented because it has already transpired. There are no easy solutions to this reading of the emotions as quantities of matter, insofar as emotions are evaluations of the status of the being in its environment, and not material pathogens in themselves. See Watson, *The Lost Second Book of Aristotle's Poetics*, 142.

12 These are difficult matters to put into words. One must imagine a neurological system designed for activation in relation to events in the environment which, by selection, activate particular response mechanisms in the emotional brain. Those mechanisms pertain to neurological designs that react only to specific ranges of activity in the environment. The backstory of laughter must demonstrate an alignment between the sensitized design and the range of social, linguistic, or situational stimuli that activate it, according to an embedded logic defined by adaptational benefits. Examples are mere guesswork of a retro-engineering kind, but observers are coming to consensus on many of these instances pertaining to tickling, jokes, information jags and release, surprising reversals, the social signalling of play, collusion in mockery and group control, the celebration of mastery and self-advancement over odds, all of which have laughter-producing potential. The challenge is to explain why the jags, transitions, and momentary disorientations represented by these forms shadow the environmental pressures that brought the laughter complex into being and secured it through genetic design.

13 Ramachandran and Blakeslee, *Phantoms in the Brain*, 200.

14 Pinker, *How the Mind Works*, 546.

15 Ramachandran and Blakeslee, *Phantoms in the Brain*, 201.

16 Robert Sheppard, "How We Think," 46.

17 The model of pathways from the cortex to the limbic system is perhaps complicated by the presence of bemusement centres in the frontal lobes linked to activity in the supplementary motor cortex. More research is required to coordinate the responses registered in this area with our responses to comedy. Ramachandran and Blakeslee report on experiments involving electronic stimulation to this region. Patients reported sudden changes in something like mental ambience, for the impression described was that everything had become ridiculous, including the examining surgeon, the room, even the instruments used in the experiment (*Phantoms in the Brain*, 201). It is instructive to see how such specifically local stimulation can produce a global ethos that colours all of experience. It is one thing to speak of laughter pathways, of kindling sequences established through qualities of repeated excitement. It is something else to say that the brain can be taken over by a sense of the ludicrous, an impression of the absurd in everything,

as though there were a specific centre for satire quite apart from the pathways of laughter. It may be that laughter must also disarm the mind in a way that allows a quality of infectiousness to take over. The centre exists, but no theory of laughter has yet taken into consideration the potential role of a "ridiculousness centre" at the same time. This may account for the incremental success of the raconteur of jokes, each joke seeming sillier than the former.

18 Pinker, *How the Mind Works*, 414–16.

19 Ibid., 546.

20 "The world puts limits on what these shortcuts can be [shortcuts meaning the psychological and genetically determined biases that organize the human experiencing of the world, including the laughter response], which constrains the human world of art, ideas, and religion. We like to focus on the differences between different cultures – their arts, their religions – but our underlying psychology requires them to occupy a rather restricted space of possibilities." Davies, *Riveted*, 203.

21 Dana Sutton, *The Catharsis of Comedy*, 57.

22 Wayne Shumaker, *Literature and the Irrational*, 192–3.

23 Albert Cook, "The Nature of Comedy and Tragedy," 487.

24 Ralph Piddington, *The Psychology of Laughter*, 168; see also Watson, *The Lost Second Book of Aristotle's* Poetics, 224.

25 Ramachandran, *The Tell-Tale Brain*, 39.

26 Robert Storey elaborates that laughter signalling fake aggression discerns friend from foe, building bonds through reassurance. *Mimesis and the Human Animal*, 159.

27 Boyer, *Religion Explained*, 129.

28 van Hooff, "A Comparative Approach to the Phylogeny of Laughter and Smiling," 215. He confirms the "mock-aggression or play" theory and the safety of laughter as well (235).

29 On laughter and mastery, see Howard R. Pollio, "Notes towards a Field Theory of Humor," and Thomas R. Shultz, "A Cognitive Developmental Analysis of Humor." Shultz states that "pleasure in mastery represents a primitive stage of humour" related to the mastery of data in the incongruity of jokes (30).

30 Storey, *Mimesis and the Human Animal*, 162.

31 Piddington, *The Psychology of Laughter*, 213. See also J.C. Gregory, *The Nature of Laughter*.

32 Morreall, "A New Theory of Laughter," and Holland, *Laughing: A Psychology of Humor*.

33 Closest to the Aristotelian font, but rather too detailed to place in resumé here, is the discussion by Walter Watson of "The Causes of the Laughable"

which, in following the commentaries on the second part of the *Poetics*, are subdivided into two categories of "wrongness," those pertaining to diction and those arising in incidents. *The Lost Books of Aristotle's* Poetics, 192–215. The nature of laughter-producing word play is seen in homonyms, synonyms, repetition, additions and subtractions, alterations, parody, the transfer of sounds and attributes, each one a dislocation from the norm based on the frames of usage and expectation that determine the "right" uses of language. Such wrongs are likewise extended into social usage based on conventional frames of decorum and expectation, pointing to the wrongheaded mentalities that lead to these laughable "errors," and ultimately to the attitudes of satire, censure, and self-superiority, or the indulgence of fools and marginals.

34 Vives, *The Passions of the Soul*, 59.

35 Ramachandran and Blakeslee, *Phantoms in the Brain*, 204. Further to the incongruity theory, see Morreall, "A New Theory of Laughter," 130. This theory originated in the work of Kant and Schopenhauer.

36 Bruce Katz attempted to think through this frame conflict approach to incongruity in terms of predictions activated, but replaced by unexpected contexts and outcomes that impose a moment of dissonance in the mind. Laughter comes at the moment of conversion from one explanatory paradigm to another, thereby joining disambiguation with pleasure. In slow motion, this profiles the suspense plot that threatens irresolution before reversing the frames and contexts. The critical matter is time and the degree to which conflictual frames can build toward the laughter response. "A Neural Resolution of the Incongruity-Resolution and Incongruity Theories of Humor."

37 Holland, *Laughing: A Psychology of Humor*, *passim*.

38 Koestler, *The Act of Creation*, 2. There are other jokes and forms of laughter that continue to turn around the management of power, arrogance, and dominance, all as forms of communication, social control through mockery, and the power of laughter to spread group consensus concerning power inequalities, even with conspiratorial overtones. Laughter may even represent a contagious signal for a spontaneous uprising because it conveys a oneness of mind in the interpretation of a political climate. Pinker, *How the Mind Works*, 551. Milder forms have to do with stories of defeat of the high-and-mighty that convey laughter. They can signal the subversive. This is entirely relevant to theatrical responses to plays, but does not apply to those concerned with reconciliation through which group cohesiveness escapes disintegration due to social dysfunctionality (particularly in leadership societies such as courts). Moreover, to complicate matters even more, there are times when laughter exposes the mean-spirited, or those who do

not join in common laughter. Thus laughter is a means for discipline by holding group members to a common consensus. De Sousa studies this aspect in *The Rationality of Emotion*. Comedy may even be the art form that dictates when it is wrong to find something funny, which may apply to tragicomedy, for to laugh ambiguously at the misfortunes of others, or at their sentimental good fortune, may cut the other way and expose the unethical side of laughter as a base response. This will come up again.

39 Beginning with the *Cligès* of Chrétien de Troyes, passing through Masuccio Salernitano's *Novellino*, and culminating in Lope de Vega's *Castelvins y Monteses*.

40 Marvin T. Herrick, *Tragicomedy*, 78.

41 Aristotle, *On Poetry and Style*, 28.

42 Castelvetro, *On the Art of Poetry*. Speroni discussed these matters in relation to the controversy over his tragedy *Canace*, Introduction.

43 Giambattista della Porta, *The Sister*, Introduction.

44 Northorp Frye, "The Argument of Comedy," 455. For Frye such celerity is the genius of New Comedy.

45 Faye Ran-Moseley, *The Tragicomic Passion*, 11.

46 Sidney, *An Apology for Poetry*, 136.

47 Lehman, "Comedy and Laughter," *passim*.

48 De Sousa, *The Rationality of Emotion*, 284.

49 Herrick, *Tragicomedy*, 224.

50 Rowley, Dekker, and Ford, *The Witch of Edmonton*.

51 Greene, *The Scottish History of James the Fourth*.

52 Frye, "The Argument of Comedy," 454. Frye isolates the comedy of the green world from the comedy of manners and ridicule because it represents "the triumph of life over the waste land"; it is an art form of deliverance (456).

53 Cited by Holland, *Laughing: A Psychology of Humor*, 29.

54 Ramachandran, *The Tell-Tale Brain*, 39.

55 Goodman, *The Structure of Literature*, 81.

56 Ibid., 5.

57 The theory of laughter associated with personal superiority is summed up by Morreall as an awareness of our advantage over the less knowledgeable or the physically infirm; a reaction of triumph, survival, or victory, which is nothing more than a controlled form of aggression of the kind described in the work of Konrad Lorenz. Morreall, ed., *The Philosophy of Laughter and Humor*, 129.

58 Goodman, *The Structure of Literature*, 92.

59 Shumaker, *Literature and the Irrational*, 26–7.

60 Morreall, "A New Theory of Laughter," 199.

61 Lehman, "Comedy and Laughter," 82.
62 De Sousa, *The Rationality of Emotion*, 276.
63 Shumaker, *Literature and the Irrational*, 267.
64 Nelson, *Comedy*, 2. He is, of course, thinking of romance comedy.
65 Storey, *Mimesis and the Human Animal*, 175.

CHAPTER TEN

1 The English translations from the letters are by Morris Bishop, *Petrarch and His World*. For the original texts, one should turn to *Petrarca: Le Familiari*. See also François Pétrarque, *L'ascension du Mont Ventoux*.

2 It is significant that Petrarch adopted this form of *sortes* for generating the impression that he had received a form of communication from the spirit world. It would have a future as late as the seventeenth century as the means of choice for demonstrating a state of election, particularly for those Protestants whose assurance had been cast into doubt by the logistics of the Calvinist doctrine of predestination. Dayton Haskin, *Milton's Burden of Interpretation*, 13–24. The practice of locating texts by chance, and by that means attributing oracular authority to them, or the force of truth in relation to questions posed, goes back to antiquity. Clearly it remained a popular form of truth seeking in the early Christian period because it was condemned by several Church councils in the fourth century. Haskin, *Milton's Burden of Interpretation*, 21. Augustine had been attracted to astrology in his earlier years, and also practised bibliomancy; he spoke of it in his *Confessions* (IV.3; p. 74) as a mere deception, something that appeared true only out of chance. "It was not surprising, then, that the mind of man, quite unconsciously, through some instinct not within its own control, should hit upon some thing that answered to the circumstances and the facts of a particular question." Augustine had to explain how the mysterious and chance appearance of a compelling biblical text leading to his conversion was not an act of divination. In Petrarch's case, the lines do not indicate deliverance, but only show the way. They, nevertheless, through the "truth" element invested in coincidence, figure in the conversion narrative as a spirit voice confirming the moment. It was, after all, in form and substance, a game of divination.

3 Giuseppe Billanovich may have been the first to argue for the later date and to accept as a consequence that the actual excursion was never made, in *Petrarca letterato*, I in *Lo scrittoio del Petrarca*, 193–8. He reiterates and expands his arguments in his "Petrarca e il Ventoso" (389–401), concluding that the piece could not have been written before Gherardo had entered the monastery in Montrieux (397), not to mention the literary allusions

employed that had not come to Petrarch's attention by 1336 (Petrarch discovered Cicero's letters only in 1345 in Verona, for example). Many later writers, including Davy Carozza and H. James Shey in their introduction to Petrarch's *Secretum*, 19, accept these arguments. Hans Baron, speaking for others, accepts the reasoning but does not concur that the original ascent was never made, allowing that the final version was modified to fit his concerns of the 1350s. *From Petrarch to Leonardo Bruni*, 18. For a summary of the dating, see Giles Constable, "Petrarch and Monasticism." 96n199.

4 Martinelli, *Petrarca e il Ventoso*, 149.

5 Arnaud Tripet, *Pétrarque ou la connaissance de soi*. For Courcelle, see note 10, and Billanovich, the end of note 13. See also Ernest Hatch Wilkins, *Life of Petrarch*.

6 Several of these positions will be described in subsequent paragraphs and footnotes.

7 Mazzotta, *The Worlds of Petrarch*, 84–5.

8 Baron, *From Petrarch to Leonardo Bruni*, 20.

9 Luciani, *Les "Confessions" de Saint Augustin dans les lettres de Pétrarque*, 66.

10 Courcelle, "Pétrarque entre Saint Augustin et les Augustins du XIVe siècle."

11 Constable, "Petrarch and Monasticism," 98.

12 Segrè, "Il *Secretum* del Petrarca e le *Confessioni* di Sant'-Agostino," and Enrico Carrara, "Francesco Petrarch, *My Secrets*," 248.

13 Carolyn Chiapelli suggests that Petrarch signals real progress, for what he confesses in the letter "is that although he may wander in uncertain ways, he knows that there is One Way to eternal peace." "The Motif of Confession in Petrarch's *Mont Ventoux*," 136. Robert Durling, in "The Ascent of Mt Ventoux and the Crisis of Allegory," is resoundingly skeptical in stating that the allegorical parts were self-cancelling (11) thereby "disarming ... the existential urgency" of the letter (13) leading to a breakdown in the symmetry between the Augustinian scene in the garden and Petrarch's "charade" on the mountain top, "undermining the very struggle for authenticity" on Petrarch's part, in turn making irony tantamount to the negative side of allegory (22–3). By contrast, again, the silence of the descent is a spiritual crescendo according to Jerrold Seigel's analysis of rhetoric and silence as a *topos* in the works of Augustine and Petrarch; the letter may be an illustration of the fact that "for Augustine, spiritual progress could be represented by a movement from speech to silence, from outer appearance to inner truth," a principle based on Augustine's "rhetoric of silence." "Ideas of Eloquence and Silence in Petrarch," 157. Those most opposed to the idea of conversion include Giuseppe Billanovich in *Petrarca letterato, lo scrittoio del Petrarca*, 1:195, and Georg Voight in *Pétrarque, Boccace et*

les débuts de l'humanisme en Italie, 128; their arguments originate in the desire to make Petrarch the first modern man and complete humanist to the exclusion of the Christian elements at the heart of humanism. See also Dieter Kramers, "L'ascension du Mont Ventoux," 122–31.

14 Tateo, "Interior Dialogue and Ideological Polemic in Petrarch's *Secretum*," 270. See also Charles Dejob, "Le *Secretum* de Pétrarque," 261–80.

15 Tateo, "Interior Dialogue and Ideological Polemic in Petrarch's *Secretum*," 271.

16 Fam. Let. XXIV.1, *Petrarch and His World*, 202. Also under Bishop, Morris.

17 The ascent motif is archetypal in nature, yet may have its specific origins in Saint Augustine's symbolism of the mountain in the spiritual life as an object of conquest through labour in *De beata vita* I.3. There is a noteworthy analogue in Hugo of St Victor's twelfth sermon, "De spiritualibus montibus et arboribus Israel," 177: 924–9.

18 Thouless, *An Introduction to the Psychology of Religion*, 105.

19 Petrarch's *De vita solitario* (1346) is another of his characteristically dual-purpose statements, for not only does it speak of his predilection for personal solitude as a place away from the business of cities where he could wander, study, or pass his time with a few select friends, but solitude was part of a religious ideal, a spiritual frame of mind which allowed him to direct his attention toward heaven. It is the silence in which the blessed life might be created, and a clue to the silence he imposed on himself on the return down the mountain, for in this work he speaks of the solitude that brings "a presentiment of future bliss." Constable, "Petrarch and Monasticism," 64.

20 Brown and Caetano, "Conversion, Cognition, and Neuropsychology," 147.

21 Ibid., 155.

22 James, *The Varieties of Religious Experience*, 335.

23 B.J. McCallister, "Cognitive Theory and Religious Experience," 338.

24 Theories of the self are only one component of the most recent cognitive studies on the nature of consciousness. Daniel Dennett deals with the question *en passant* throughout his justly celebrated *Consciousness Explained*, as does Bernard Baars, *In the Theater of Consciousness: The Workspace of the Mind*. Baars discusses the "director" feature of mind and the tendency to anthropomorphize this organizational capacity. It is a way of expressing the self as a complex configuration of potential information that is ready, upon prompting, to audit, phase, and modify, in conjunction with the emotions, the parallel modular preoccupations of consciousness. Dennett makes light of the idea that the brain must be a theatre to which it plays itself back as though in a performance, but the analogy persists as a way of understanding that consciousness plays itself for "someone," which may be another way of constructing the self.

25 A good deal more on the nature of the self is offered in the present collection in the chapter on *Doctor Faustus*.

26 Saint Augustine approaches the notion in his observation in Book X.16 of the *Confessions* that the mind can be present to itself by its own power (222).

27 This definition is shaped to accommodate the profile of Petrarchan conversion, to be sure, for a more plastic definition must include the mind's capacity to conduct operations it can attribute only to outside forces, or to imagine its completion in out-of-body experiences, states of rapture with or without the presence of a God, states of infinite socializing with a personal God, or the willed cessation of all conscious activity as the marker of nirvana. In keeping with the work of Karl R. Popper and John C. Eccles in *The Self and Its Brain*, there are mechanisms whereby consciousness is able to project multiple drafts of the future in relation to present circumstances that allow the mind to choose among options. Conversion is an elaborate form of provisional thinking that permits an experiential foretaste of an alternative course before it is chosen. This is closely tied to moral reasoning, which relies on the neural design capacities associated with the weighting of options in binary terms, and the making of choices in relation to duty, fear, benefit, and risk. Conversion, as with the making of moral decisions, is closely associated with self-worth, the emotions attached to shame, well-being, and a cluster of related affections. Bio- or neuroethics are taken up entirely with the systemic designs of the human brain that emerged in tandem with societies grounded in reciprocal altruism, systems which include the social emotions that guide ethical conduct. R. Sperry, in *Science and Moral Priority*, provides a point of departure, followed by many such studies from Richard Alexander's *The Biology of Moral Systems* (1987), to Patricia Churchland's *Braintrust: What Neuroscience Tells Us about Morality* (2011). These studies were explored at length in chapter 2 on the biogenesis of ethics in conjunction with Shakespeare's *Measure for Measure*. Conversion relies on these same interfaces between computational priorities, the emotions, and the finishing line of commitment and volition.

28 No specific mention has been made of the role of language in shaping the conversion experience because rhetoric prepared by an advocate is not a part of the Ventoux letter as it is in the *Secretum*. Nevertheless, self-talking and abundant use of metaphor are parts of the experience, and there have been certain leading images that have proven their effectiveness in preparing the mind for religious change. These have been studied by many recent scholars, including Ralph Metzner in "Ten Classical Metaphors of Self-Transformation," 47–62. Petrarch used emotional language as a way of exciting his own emotions as he wrote in order to bring himself to a full

cathartic experience, as he explained in his letter to Gasparo Squaro dei Broaspini, described below. Raymond Paloutzian in his *Invitation to the Psychology of Religion* (188) provides a resumé of the importance attached to language in the conversion processes, particularly in relation to the work of W. Proudfoot in *Religious Experience.*

29 Pruyser, *A Dynamic Psychology of Religion*, 146.

30 Damasio, *Descartes' Error, passim.* For a more detailed profile of the emotions and their interactions with the cerebral cortex in this collection, see chapter 3, above, on the intentionality of the emotions and the criminal mind.

31 Pinker, *How the Mind Works*, 370. Raymond Paloutzian studies the "Schachter" factor, which is the relativity of the interpretation of emotional sensations in accordance with the current preoccupations of the mind. Through "religious" suggestivity, all manner of feeling will be assigned religious meaning, and the more easily so the more the states of consciousness are "unusual." *Invitation to the Psychology of Religion*, 187.

32 Griffiths, *What Emotions Really Are*, 97.

33 The literature on this subject has expanded exponentially in recent years. An anthology much to be recommended is the *Handbook of Religious Conversion* containing seventeen articles on every aspect of the phenomenon. Excellent analyses of the affections and emotions experienced in the conversion process are to be found in studies by Raymond Paloutzian, *Invitation to the Psychology of Religion*, Paul Pruyser, *A Dynamic Psychology of Religion*, and particularly L.R. Rambo, *Understanding Religious Conversion* – a study in which the many components of the experience are grouped under seven headings corresponding to the generic phases of the overall narrative. See also Bernard Spilka, Ralph W. Hood Jr, and Richard L. Gorsuch, *The Psychology of Religion: An Empirical Approach*; Bruce T. Riley, *The Psychology of Religious Experience in Its Personal and Institutional Dimensions*; C.D. Batson, P. Schoenrade, and W.L. Ventis, *Religion and the Individual*; and G.E.W. Scobie, "Types of Christian Conversion," 265–71.

34 Fam. Let. IV.1, *Petrarch and his World.* (See also under Bishop, Morris).

35 Paloutzian describes the role of the emotions, in conjunction with the drive toward self-consistency, in preventing cognitive recidivism following the conversion experience. The emotions are first programmed to catapult the convert into a changed state, and then to hold the individual in that new state. *Invitation to the Psychology of Religion*, 187. There is a nearly ironic sense in which the conversion process resembles the falling in love sequence described in chapter 8 on the "truth" of romance, particularly concerning the ways in which the emotions fortify the sudden settling of the will upon a life mate.

36 Petrarch's *beata vita* can perhaps be no better defined than by Saint Augustine from whose writings it is derived. Debate arose in succeeding centuries among the Neoplatonists concerning the relation between knowledge and feeling in the beatified states they envisioned for themselves. Even this debate can be subjected to forms of modern analysis, as in Paul Pruyser's distinction between "activity affect" and what he calls "embeddedness affect." *A Dynamic Psychology of Religion*, 166. Joy in the former mode sustains and mobilizes the energies necessary to accomplish tasks in the world; it is the limbic colouring of the mind that leads out of acedia. In the latter mode, joy is a latent state waiting for activation through divine dispensation, prayer, rites, or spiritual exercises. The argument under examination is circular insofar as anticipated joy may be the active drive to prayer and rites. One sees this loop in the poetry of George Herbert, whose preoccupation with spiritual joy became the substance of his tortured muse. Clearly, for Petrarch, beatitude has emotional overtones if only as the absence of the unwanted passions of the pre-converted state, but the degree to which it assumed numinous properties is by no means certain.

37 Schleiermacher, *On Religion: Speeches to Its Cultured Despisers*, 36.

38 *Petrarch and His World*, 220.

39 William James invented his own term for this dimension of acedia, "anhedonia," or a state of passive joylessness, a "lack of taste and zest and spring." *On the Varieties of Religious Experience*, 125. He associated it with the "'misery-threshold'" that precedes conversion (117).

40 *Petrarch and His World*, 31.

41 *Petrarch, A Humanist among Princes*, 87.

42 Kristeller, "Augustine and the Early Renaissance," 347.

43 One might invoke, here, the conversion of Saint Thomas Aquinas, about which we would know more, for in 1273, at the age of forty-eight, after his work on the *Summa theologica* was complete, he fell under the blighting conviction that all his writing had been vanity, a thing of mere straw, compared to his recent vision, and that henceforth he would never write another word. It was a marked change, accompanied by spiritual phenomena of an unspecified kind, together with guilt concerning his great work in words, rhetoric, and logic, followed by a devotionally induced silence. The experience of conversion belongs to the Christian world order, of which Saint Augustine's account is but one eloquent version, and Petrarch's another. The importance is that they are all manifestations of a "well-defined course" in keeping with the "idea" of conversion. Thouless, *An Introduction to the Psychology of Religion*, 113.

44 Arendt, *The Human Condition*, 262ff.

45 Ibid., 263.

46 Ibid., 266.

47 *The Cambridge History of Renaissance Philosophy*, 345.
48 Ibid., 352.
49 Available in English as *On Pleasure*.
50 Kristeller, *Eight Philosophers of the Italian Renaissance*, 30–1.
51 Mazzotta, *The Worlds of Petrarch*, 5.
52 Fam. Let. VIII.7, *Petrarch and His World*. See also Bishop, Morris.

CHAPTER ELEVEN

1 Fawn, in his public and private selves, bears a patterned resemblance to Doctor Faustus with his suffering interior and trickster exterior, although the characters in station and purpose are otherwise quite different. He also resembles all those rulers and magistrates from Shakespeare's Henry V to the real King James I, whose failed attempt to circulate in the streets of London incognito may have given impetus to the designing of *Measure for Measure*. When mere deception of others through disguise progresses toward the overhauling of the self through that disguise, this theatrical instrument lends itself to questions of personhood, the self, and the social fashioning of the self into a converted or transformed self, whether temporarily or permanently. In that regard, personhood becomes linked to the psychology of conversion, and hence to the values examined in the preceding chapter pertaining to Petrarch.
2 Blakey Vermeule in *Why Do We Care about Literary Characters*, esp. 12–20, but throughout her study, deals with the cognitive and emotional engagements readers have with fictional characters and the profound levels of immediate and felt experience we derive from engaging with them through the ontological status we bestow on them. This, in turn, is based on the adaptive brain platforms of a quintessentially social species. It is why we gossip and why we read, thirsty as we are for social information that matters in real and immediate terms, even in fictive settings.
3 Revealingly, the protagonist of this play is also a duke of Ferrara who takes a travel leave, but unlike Hercules, and like Shakespeare's Duke Vincentio, returns to his own court in disguise to examine all the ills and enormities there, before making a recitation of all he has seen at the play's end.
4 In these matters I am relying on the critical introduction by Gerald A. Smith to *The Fawn*, and the introduction by J.W. Lever to *Measure for Measure*. As Smith states, "*The Fawn* was first played sometime between February 4, 1604, and March 12, 1606" (xi), the first date the earliest that the acting company was called "The Children of the Queen's Majesty's Revels" and the latter date that of its registration for publication. Evidence that it was written during the 1604 season or just prior is merely circumstantial, as Smith explains. The first confirmed date for the acting of *Measure for*

Measure is 26 December 1604, but "a number of allusions in the dialogue suggest that the play was composed and probably acted in the summer season of 1604" (Lever, xxxi) for reasons then explained in great detail, including the probability of James I's own incognito visit to the Exchange, or at least his attempt to do so (on 15 March 1604). W.W. Lawrence concurs regarding the unlikelihood of establishing influences among these plays, "especially since the dates of composition and production are in most cases so uncertain." *Shakespeare's Problem Comedies*, 188.

5 Lever mentions these and several other sources for "The Disguised Ruler" motif in his introduction to *Measure for Measure* (xliv–li), including the story of the Roman ruler Alexander Severus, prominent in Guevara's *Décadas de las vidas de los x. Cesares* (1539) and Sir Thomas Elyot's *The Image of Governaunce* (1541). He cites Marston's *The Malcontent* and *Fawn* in this regard, together with Middleton's *Phoenix*, stating how all three "presented fictitious Italian dukes who put off their conventional dignity with their robes of state and gave strident expression to the contemporary questioning of values" (xlvii). Of recent date is a full critical edition of Riche's *The Adventures of Brusanus, Prince of Hungaria*, with a discussion of the Shakespeare parallel, 85–91.

6 In order to move expeditiously through these preliminary ideas, I have taken the liberty of self-borrowing from three articles where readers will find full bibliographical information: D.A. Beecher, "The Courtier as Trickster in Jacobean Theatre," D.A. Beecher, "Intriguers and Tricksters: Manifestations of an Archetype in the Comedy of the Renaissance," and D.A. Beecher, "The Sense of an Ending: John Marston and the Art of Closure."

7 This statement is a reminder that even such fundamental psychological orientations as these require a designed brain with emergent properties qualified by the fixed dimensions of cognition and computation characteristic of our species. This little picnic invariably invites the rain of those who imagine the brain as a blank slate open to any kind of programming whatsoever through post-natal experience. But the categories by which we construct the personhood of others appear to be innate and remembered generationally by the genetic architectures which compose the brain precisely in order to generate these adaptive properties critical to our understanding of social environments. The nature/nurture debate has been elucidated in the introduction and need not be reiterated here. But the question of design and the shaping of human nature is always critical, for if our natures are so circumscribed, then the ontology of personhood as a mental category will be much influenced by the values and impedances of that design.

8 The discussion to follow has been informed by several specialized investigations into the nature of folk psychology and the problem of other minds,

but as a preliminary to those more detailed studies, there are the many shorter articles of great concision and well-selected bibliographies in *The MIT Encyclopedia of the Cognitive Sciences*. Among those consulted for the present work are: Stephen Stich, "Eliminative Materialism" (265–7), Alvin Goldman, "Epistemology and Cognition" (280–2), Lynne Rudder Baker, "Folk Psychology" (319–21), Daniel Dennett, "Intentional Stance" (412–13), Andrew Whiten, "Machiavellian Intelligence" (495–6), Barbara von Eckardt, "Mental Representation" (527–9), Robert Stalnaker, "Propositional Attitudes" (678–9), Christopher Cherniak, "Rational Agency" (698–9), Robert van Gulick, "Self-Knowledge" (735–6), Robert M. Gordon, "Simulation vs. Theory-Theory" (765–6), Daniel Gilbert, "Social Cognition" (777–8), and Alison Gopnik, "Theory of Mind" (838–41).

9 This latter issue I deal with extensively in the essay in this collection on Doctor Faustus.

10 Michael Gazzaniga, in *The Ethical Brain* (xv), outlines a "brain-based" approach to philosophy and psychology as the critical pre-condition to understanding the human sense of the world.

11 The argument to follow posits that from a cultural perspective we cannot even entertain the proposals of the eliminative materialists and remain in business. Their point is that default psychology based on beliefs and desires is probably not the way the brain really works, for which, see Paul Churchland, "Eliminative Materialism and the Propositional Attitudes," wherein he suggests that we would be better to abandon beliefs and propositional attitudes altogether as the basis for our working psychology and turn to neurobiological explanations. See Stich's "Eliminative Materialism" (265–7). But given the fact that our minds work, at least ostensibly, in terms of beliefs and propositions concerning our own and the minds of others, much of the analysis to follow is based on this default epistemology, referred to as "folk psychology," simply because most observers agree that it will never be replaced; it is the product of emergent properties of our brains determining the categories of experience through which we measure reality.

12 The essence of Jerry Fodor's thought on these topics can be read in two of his articles appearing in *Mind and Cognition: A Reader*: "Why There Still Has to Be a Language of Thought" (282–99), and "Banish DisContent" (420–38). Paul Churchland can be read in that same volume, coauthored with his wife Patricia Smith Churchland, "Stalking the Wild Epistemic Engine" (300–11). For a commentary on Fodor's thoughts, see Hollibert E. Phillips, *Vicissitudes of the I*, 114–16. Paul Churchland is at his most accessible in *The Engine of Reason, the Seat of the Soul*. See in particular "The Neural Representation of the Social World" (123–50) and "The Puzzle of Consciousness" (187–226). For a commentary on Churchland, see Phillips, *Vicissitudes of the I*, 118–24.

13 In this direction lies the huge debate over materialist reductionism and the menace of a new dualism that brings back distinctions between mind content as having its equivalent in the functions of the brain, while producing thoughts and sensations of a different order that cannot in themselves be reduced to neurobiological happenings. See Daniel C. Dennett, *The Intentional Stance*, especially "Folk Psychology as a Source of Theory" (43–57). "There are different reasons for being interested in the details of folk psychology. One is that it exists as a phenomenon, like a religion or a language or a dress code, to be studied with the techniques and attitudes of anthropology. It may be a myth, but it is a myth we live by, so it is an 'important' phenomenon in nature" (47).

14 Phillips, in *Vicissitudes of the I*, takes the issue around the hermeneutic circle in arguing that what happens in mind must be happening in the brain, by means still to be determined, that propositional states are common to the race of humankind, and that they cannot be dispensed with because the cost is too incalculably high. All subjects of enquiry from history and ethics to literature and art are based on propositional states and attributions, no matter how imprecise (135).

15 Baker, "Folk Psychology," 319. This is one of the most debated aspects of the entire folk theory, whether in attempting to know other minds we proceed fundamentally by theorizing about other minds, or whether we simply assume that other minds are like our own, and that hence we can know them by introspection – in short, by asking ourselves what we would be doing or thinking in their place. I have looked at numerous articles on the topic, including Stanley Tambiah's "Relations of Analogy and Identity" and Jerome Brunner's "Frames of Thinking." There is a more extensive investigation in Stephen P. Stich's *Deconstructing the Mind*. Major sections of the book are devoted to the topic, such as "Connectionism, Eliminativism, and the Future of Folk Psychology" and "How Do Minds Understand Minds? Mental Simulation versus Tacit Theory"; these contain terms that will reappear in the body of this article. See also the article by David Martel Johnson, "Taking the Past Seriously: How History Shows that Eliminativists' Account of Folk Psychology is Partly Right and Partly Wrong."

16 Andrew Brook and Robert J. Stainton, *Knowledge and Mind*, 206.

17 This is an argument advanced by Jerry Fodor in "Banish DisContent" where he states that propositional-attitude psychology works so well as to be almost invisible, and that to dispense with it would require a massive paradigm shift. Folk psychology is "implicit" in nature and has proven its evolutionary advantages in providing "prediction and control" (421).

18 Pinker, *How the Mind Works*. This is a long book guided by the principle that humans are what they are by a long process of selection and adaptation, and that the equipment that we have today for computation,

perception, the appreciation of beauty, social management, and much more is based on the specialized uses of more basic operations to create interim states and processors. Thus "natural selection is the only evolutionary force that acts like an engineer, 'designing' organs that accomplish improbable but adaptive outcomes" including the ways in which, as a species, we read other minds (36). He refers to Richard Dawkins and George Williams. The importance for our purposes is that folk psychology, too, is selective and adaptive, prioritizing our attentions to those matters and actions of greatest relevance to our own survival. For Peter Carruthers' contribution, see "Moderately Massive Modularity," 67–89.

19 Carruthers, "Moderately Massive Modularity," 75. As Ian Hacking points out in "Normal People," we possess a kind of theory of others based on social norms, for without such norms there would be a far less efficient basis for predicting behaviour. Norms are of course a philosophical minefield, but on the same basis that folk psychology asserts itself by the logic of what we must cognitively perform to the ends of social survival, "normalizing attitudes" emerge as the basis for making social attributions, predictions, and moral evaluations. So much of society is based on regularizing practices, and while, in the postmodern world we may have convinced ourselves that deviancy and subversion are the forces of progress and liberation, nevertheless, homo sapiens as a social animal will continue to conform in order to ensure inclusion. Normalcy is a mode of thought in its own right, a mental habit human minds resort to as a theorized base for social orientation. Concern about being abnormal is a driving human preoccupation (61).

20 Carruthers, "Moderately Massive Modularity," 73.

21 For a discussion of tacit theory and its place in the cognitive sciences, see Stich, *Deconstructing the Mind*. This is an important issue because in the common exploratory strategies of mind in recent decades, there has been a constant appeal to various residual forms of knowledge as the basis for knowing, speaking, or calibrating: "typically, a knowledge structure will be a set of principles or rules that constitute a recipe or program enabling people to carry out the activity in question by exploiting more basic capacities in a systemic way" (121). A critical approach to these assumptions began with Fodor's article "The Appeal to Tacit Knowledge in Psychological Explanation." The question is whether we do what we do by using recipes or programs, and if so, what are those programs like, and where do they come from? The issue here is whether we employ programs for decoding other minds, and the extent to which the criteria of these programs structure the representation of personhood.

22 The concept of the "normal" would seem sub-scientific, but just as we make categories of "things" in the physical world, we, in all probability, use sim-

ilar categories in our orientation to the social world. Hacking, in "Normal People" (59–71), puts forward a strong case for such operative categories as an essential precondition to all manner of judgment and inference, moral and otherwise. Clearly, these categories are plastic and malleable, yet operative in the same sense that folk psychology is operative. Nevertheless, they differ in kind from knowledge bases and languages of mind.

23 Dennett, "True Believers," 155.

24 Churchland's discussion of the self is attached to his theories of consciousness in *The Engine of Reason, the Seat of the Soul*, especially 187–226.

25 Baars, *In the Theater of Consciousness*, 142.

26 Dennett, *Brainstorms*, 151ff. There are many more lists of this kind. Stephen Clark in "Non-Personal Minds" (185) speaks of persons as those who recognize other persons, distinguish them from their own introspective reflections, remember their own pasts, imagine versions of the future, communicate with others propositionally and deliberatively, hold themselves and others to account for what they do, and formulate theories about the behaviour of others and of the impersonal world by a variety of cognitive processes. They are also capable of being amused by incongruities or the disruption of social expectations, forming friendships, reflecting on goals and their motives, and deceiving others for motive or sport. All such lists provide perspectives into the categories employed in circumscribing personhood.

27 Owen Flanagan's account of "The Narrative Structure of Self-Representation" is tellingly cogent. See *Consciousness Reconsidered*, 197ff.

28 Popper and Eccles, *The Self and Its Brain.* Of special interest is the section "Learning to Be a Self" (108–59). Telltale phrases relating to the present discussion include "the changing self which yet remains itself appears to be based on the changing individual organism which yet retains its individual identity" (114), or "a unified centre must inhabit some of the possible ways of behavior and only allow one single way at a time to proceed" (128). All of these tactics are tied to Popper's theory of provisional planning in the imagination, followed by analysis, even emotionalized trial versions, before a final course of action is chosen. Homo sapiens, the problem-solving animal, proceeds in this way: "our active attention is focused at every moment on just the relevant aspects of the situation, selected and abstracted by our perceiving apparatus, into which a selection programme is incorporated; a programme which is adjusted to our available repertoire of behavioural responses" (128). This is central to the trickster agent, indeed his supreme accomplishment. That same provisional planning in anticipation of action, and the establishment of propositional fields, is the mark of personhood, which comes into being by phases in young infants (115). Owen Flanagan in

The Science of the Mind also takes up the capacity of the mind to project in advance of acting in dangerous real time as one of our most valuable adaptive qualities (319). To a certain degree this is related simply to abilities to learn, for the self process is a "memory system capable of continuously projecting past matches of hedonic value or disvalue onto novel states of affairs and then on the basis of new outcomes updating the models and hypotheses it contains" (325).

29 But before we become too sanguine about our ability to intuit the *modus operandi* of consciousness, it is well to recall that motivated self-interest is a strong incentive to finesse the self-knowledge quest. We are notoriously wilful in the rationalizations of our self convictions, willing to assign to ourselves a spiritual ontology, inclined to overestimate the seamless continuity and rationality of the running narrative, inclined to engage in selective amnesia regarding painful memories, or inversely to dwell morbidly on topics leading to low self-esteem and depression. The self, at its best, is temperamental, biased, idiosyncratic in its attention priorities, and adversely malleable in its standards. It is surrounded by instincts, genetically predetermined in its design and capacities, significantly determined by cultural conditioning, limited by the properties of language, and confined to the limited range of its own experiences.

30 I have dealt with the misguided attribution of sex change anxiety, on medical grounds in any case, in "Concerning Sex Changes: The Cultural Significance of a Renaissance Medical Polemic."

31 Jung, "On the Psychology of the Trickster Figure," goes on to say that "the trickster is a collective shadow figure, an epitome of all the inferior traits of character in individuals. And since the individual shadow is never absent as a component of personality, the collective figure can construct itself out of it continually" (209).

32 Flanagan, *The Science of the Mind*, 353.

33 Gordon Armstrong, *Theatre and Consciousness*, 6.

34 David Blostein enters the debate regarding role-playing and selfhood by suggesting that in the case of Malvole, the ousted duke of *The Malcontent*, we cannot know, whereas in the case of Fawn, Marston makes "the moral reference point clear." John Marston, *Parasitaster or the Fawn*, 14–15. But the terms of comparison between role-playing, the self, and moral reference points are not entirely clear.

BIBLIOGRAPHY

Abrams, M.H., with Geoffrey Harpham. *A Glossary of Literary Terms*. Boston: Thompson, 2005

Ainslie, George. "Specious Reward: A Behavioral Theory of Impulsiveness and Impulse Control." *Psychological Bulletin* 82 (1975): 463–96

Alexander, Richard D. *The Biology of Moral Systems*. Hawthorne, NY: Aldine de Gruyter, 1987

Alte Französiche Sprichwörter. Edited by Adolf Tobler and Julius Zacher. *Zeitschrift für deutsches Altertum* 11 (1859), No. 78

Alter, Robert, and Frank Kermode. *The Literary Guide to the Bible*. Cambridge: Harvard Belknap Press, 1987

Anscombe, G.E.M. *Intention*. Oxford: Blackwell Publishing, 1956

Apte, Mahadev. *Humor and Laughter: An Anthropological Approach*. Ithaca: Cornell University Press, 1985

Arendt, Hannah. *The Human Condition: A Study of the Central Dilemmas Facing Modern Man*. New York: Doubleday Anchor, 1959

Aristotle. *On Poetry and Style*. Translated by G.M.A. Grube. Indianapolis: The Bobbs-Merrill Co., 1958

Armstrong, Gordon Scott. *Theatre and Consciousness: The Nature of Bio-Evolutionary Complexity in the Arts*. New York: Peter Lang, 2003

Armstrong, Karen. *A History of God*. New York: Ballantine Books, 1993

Augustine, Saint. *Confessions*. Translated by William Watts. 2 vols. Cambridge: Harvard University Press (1912) 196l

– *Confessions*. Translated by R.S. Pine-Coffin. London: Penguin, 1961

– *De beata vita*. Translated by Ruth Allison Brown. Washington, DC: Catholic University of America Press, 1944

Axelrod, Robert M. *The Evolution of Cooperation*. New York: Basic Books, 1984

Baars, Bernard. *In the Theater of Consciousness: The Workspace of the Mind*. New York: Oxford University Press, 1997

Bacon, Francis. *A Selection of His Works*. Edited by Sidney Warhaft. Toronto: Macmillan of Canada, 1965

Baddeley, Alan. "The Psychology of Remembering and Forgetting." In *Memory, History, Culture, and the Mind*, edited by Thomas Butler, 33–60. Oxford: Blackwell's, 1989

Baker, Lynne Rudder. *Explaining Attitudes: A Practical Approach to the Mind*. Cambridge: Cambridge University Press, 1995

– "Folk Psychology." In Wilson and Keil, *The MIT Encyclopedia of the Cognitive Sciences*, 319–21

– *Persons and Bodies: A Constitution View*. Cambridge: Cambridge University Press, 2000

Baltes, P.B. "Life-Span Developmental Psychology: Some Converging Observations on History and Theory." In *Life-Span Development and Behavior*, edited by P.B. Baltes and O.G. Brim Jr, 2: 255–79. New York: Academic Press, 1979

Barkow, Jerome. "Beneath New Culture Is Old Psychology: Gossip and Social Stratification." In Barkow, Cosmides, and Tooby, *The Adapted Mind*, 627–37

– Leda Cosmides, and John Tooby, eds. *The Adapted Mind: Evolutionary Psychology and the Generation of Culture*. New York: Oxford University Press, 1992

Baron, Hans. *From Petrarch to Leonardo Bruni: Studies in Humanistic and Political Literature*. Chicago: University of Chicago Press, 1968

Baron-Cohen, Simon. "The Empathizing System: A Revision of the 1994 Model of the Mindreading System." In Ellis and Bjorklund, *Origins of the Social Mind*, 468–92

Batson, C.D., P. Schoenrade, and W.L. Ventis. *Religion and the Individual*. New York: Oxford University Press, 1993

Baumeister, R.F. *Identity: Cultural Change and the Struggle for Self*. New York: Oxford University Press, 1986

Beecher, Donald. "Concerning Sex Changes: The Cultural Significance of a Renaissance Medical Polemic." *Sixteenth Century Journal* 36, no. 4 (2005): 991–1016

– "The Courtier as Trickster in Jacobean Theatre." *University of Ottawa Quarterly* 53 (1983): 199–209

– "Intriguers and Tricksters: Manifestations of an Archetype in the Comedy of the Renaissance." *Revue de littérature comparée* 61, no. 1 (1987): 5–31

– "Mind, Theatres, and the Anatomy of Consciousness." *Philosophy and Literature* 30, no. 1 (2005): 1–16

– "The Sense of an Ending: John Marston and the Art of Closure." In *Le spectacle dans le spectacle dans le théâtre anglais de la renaissance*, edited by Jean-Marie Maguin and François Laroque, 2: 369–79. *Collection Astraea* 4, no. 1 (2 vols.). Montpellier: Université Paul Valéry, 1992

Benary, Walter, ed. *Salomon et Marcolfus*. Heidelberg: Carl Winters Universitätsbuchhandlung, 1914

Ben-Ze'ev, Aaron. *The Subtlety of Emotions*. Cambridge: MIT Press, 2000

Berger, Peter L. *The Heretical Imperative: Contemporary Possibilities of Religious Affirmation*. Garden City, NY: Doubleday/Anchor, 1979

Bergougnan, Élie, trans. *Romans grecs: Les Éthiopiques ou Théagène et Chariclée*. Paris: Garnier [1932]

Berlyne, D.E. *Aesthetics and Psychobiology*. New York: Appleton Century Crofts, 1971

– *Conflict, Arousal, and Curiosity*. New York: McGraw-Hill, 1960

Bermúdez, José Luis. "The Domain of Folk Psychology." In O'Hear, *Minds and Persons*, 25–48

Billanovich, Giuseppe. "Petrarca e il Ventoso." *Italia medioevale e umanistica* 9. Padua: Antenore (1966): 389–401

– ed. *Petrarca letterato*. Vol. 1 of *Lo scrittoio del Petrarca*. Rome: Edizioni di storia e letteratura, 1947

Bishop, Morris, trans. *Petrarch and His World*. Bloomington: Indiana University Press, 1963

Bloom, Paul. *Just Babies: The Origins of Good and Evil*. New York: Crown Publishers, 2013

Boston, Richard. *An Anatomy of Laughter*. London: Collins, 1974

Bower, Gordon, and Daniel Morrow. "Mental Models in Narrative Comprehension." *Science* 247, no. 4938 (1990): 44–8

Bowie, E.L. "Apollonius of Tyana: Tradition and Reality." *Aufsteig und Niedergang der römischen Welt* 2, no. 16, edited by H. Temporini and W. Haase, 1652–99. Berlin: Walter de Gruyter, 1978

– "Novels and the Real World." In *Erotica Antiqua*, edited by B.P. Reardon, 91–6. Bangor, Wales, 1977

Boyd, Brian. *On the Origin of Stories: Evolution, Cognition, and Fiction*. Cambridge: Harvard University Press, 2009

Boyer, Pascal. *Religion Explained: The Evolutionary Origins of Religious Thought*. New York: Basic Books, 2001. Also as *Religion Explained: The Human Instincts that Fashioned Gods, Spirits, and Ancestors*. London: William Heinemann, 2001

Boyle, Marjorie O'Rourke. "Gracious Laughter: Marcilio Ficino's Anthropology." *Renaissance Quarterly* 52, no. 3 (Autumn 1999): 712–39

Bradbrook, Muriel. "Authority, Truth, and Justice in *Measure for Measure*." *Review of English Studies* 17 (October 1941): 385–99

Brewer, William F. "The Nature of Narrative Suspense and the Problem of Rereading." In Vorderer, Wulff, and Friedrichsen, *Suspense*, 107–27

– "Schemata." In Wilson and Keil, *The MIT Encyclopedia of the Cognitive Sciences*, 729

Breytspraak, Linda. M. *The Development of Self in Later Life*. Boston: Little, Brown and Company, 1984

Brook, Andrew. "Emmanuel Kant," In Wilson and Keil, *The MIT Encyclopedia of the Cognitive Sciences*, 427–8

– and Robert J. Stainton. *Knowledge and Mind: A Philosophical Introduction*. Cambridge: MIT Press, 2000

Brooke, Nicholas. *Horrid Laughter in Jacobean Tragedy*. London: Open Books, 1979

Brothers, Leslie. *Friday's Footprint: How Society Shapes the Human Mind*. New York: Oxford University Press, 1997

Brown, W.S., and C. Cateneo. "Conversion, Cognition, and Neuropsychology." In Malony and Southard, *Handbook of Religious Conversion*, 147–58

Brunner, Jerome. *Actual Minds, Possible Worlds*. Cambridge: Harvard University Press, 1986

– "Frames of Thinking: Ways of Making Meaning." In Olson and Torrance, *Modes of Thought*, 93–105

Byrne, R.W., and A. Whiten. *Machiavellian Intelligence: Social Expertise and the Evolution of Intellect in Monkeys, Apes and Humans*. Oxford: Oxford University Press, 1988

Cahill, S. "Toward a Sociology of the Person." In *Erving Goffman*, edited by G. Fine and G. Smith, 3: 17–42. Thousand Oaks, CA: Sage, 2000

Calabritto, Monica. "A Case of Melancholic Humors and *Dulcidia Intervalla*." *Intellectual History Review* 18, no. 1 (2008): 139–54

The Cambridge History of Renaissance Philosophy. Edited by Charles B. Schmitt and Quentin Skinner. Cambridge: Cambridge University Press, 1988

Campbell, Jeremy. *The Improbable Machine: What the Upheavals in Artificial Intelligence Research Reveal about How the Mind Really Works*. New York: Simon and Schuster, 1989

Carozza, Davy, and H. James Shey, eds. *Petrarch's* Secretum. New York: Peter Lang, 1989

Carr, David. "Narrative and the Real World: An Argument for Continuity." In Hinchman and Hinchman, *Memory, Identity, Community*, 7–25

Carrara, Enrico. "Francesco Petrarch, *My Secrets*" (Originally published in *Francesco Petrarca "Il mio segreto*." Florence: Sansoni, 1943). In Carozza and Shey, *Petrarch's* Secretum, 245–9

Carroll, Joseph. *Evolution and Literary Theory*. Columbia: University of Missouri Press, 1995

– "The Human Revolution and the Adaptive Function of Literature." *Philosophy and Literature* 30, no. 1 (2006): 33–49

– *Literary Darwinism: Evolution, Human Nature, and Literature*. New York: Routledge, 2004
Carroll, Noël. "The Paradox of Suspense." In Vorderer, Wulff, and Friedrichsen, *Suspense*, 71–92
– "Toward a Theory of Film Suspense." *Persistence of Vision* 1 (1994): 65–89
Carroll, William C. "Romantic Comedies." In *Shakespeare: An Oxford Guide*, edited by Stanley Wells and Lena Cowen Orlin, 175–85. New York: Oxford University Press, 2003
Carruthers, Peter. "Moderately Massive Modularity." In O'Hear, *Minds and Persons*, 67–89
Cassirer, Ernst. *The Individual and the Cosmos in Renaissance Philosophy*. Translated by Mario Domandi. New York: Harper and Row, 1963
– *Language and Myth*. Translated by Susanne K. Langer. New York: Harper and Bros, 1946
Castelvetro, Lodovico. *On the Art of Poetry*. Edited by Andrew Bongiorno. Binghamton, NY: Medieval and Renaissance Texts and Studies, 1984
Cavalli-Sforza, Luigi Luca. *Genes, Peoples and Languages*. London: Allen Lane-Penguin, 2000
Chapais, Bernard. "Primates and the Origins of Aggression: Power and Politics among Humans." In *Understanding Behavior: What Primate Studies Tell Us about Human Behavior*, edited by James Loy and Calvin Peters, 190–228. Oxford: Oxford University Press, 1991
Chaucer, Geoffrey. *The Canterbury Tales*. Edited by V.A. Kolve and Glending Olson. New York: W.W. Norton, 2005
Cherniak, Christopher. "Rational Agency." In Wilson and Keir, *The MIT Encyclopedia of the Cognitive Sciences*, 698–9
Chiapelli, Carolyn. "The Motif of Confession in Petrarch's *Mont Ventoux*." *Modern Language Notes* 16 (1978): 131–6
Chisholm, James. "Love's Contingencies: The Developmental Socioecology of Romantic Passion." In *Romantic Passion: A Universal Experience?* edited by William Jankowiak, 42–56. NewYork: Columbia University Press, 1995
Churchland, Patricia Smith. *Braintrust: What Neuroscience Tells Us about Morality*. Princeton: Princeton University Press, 2011
– "From Descartes to Neural Networks." In Irvine and Russell, *In the Agora*, 51–3
– "How Do Neurons Know?" In Irvine and Russell, *In the Agora*, 54–9
Churchland, Paul. "Eliminative Materialism and the Propositional Attitudes." *Journal of Philosophy* 78 (1981): 67–90
– *The Engine of Reason, the Seat of the Soul*. Cambridge: MIT Press, 1996
– and Patricia Smith Churchland. "Stalking the Wild Epistemic Engine." In Lycan, *Mind and Cognition*, 300–11

Clark, Stephen. "Non-Personal Minds." In O'Hear, *Minds and Persons*, 185–209

Cohen, Jonathan. "Vulcanization of the Human Brain: A Neural Perspective on Interactions between Cognition and Emotion." *Journal of Economic Perspectives* 19, no. 4 (2005): 3–24

Cole, Douglas. *Christopher Marlowe and the Renaissance of Tragedy*. Westport, CT: Praeger, 1995

Constable, Giles. "Petrarch and Monasticism." In Bernardo, *Francesco Petrarca*, 53–99

Conway, M.A., and D.A. Bekerian. "Situational Knowledge and Emotions." *Cognition and Emotion* 1, no. 2 (1987): 145–91

Cook, Albert. "The Nature of Comedy and Tragedy." Chapter 2 of *Dark Voyage and the Golden Mean*, 1949. Reprinted in Lauter, *Theories of Comedy*, 475–96

Cooper, Lane. *An Aristotelian Theory of Comedy*. New York: Harcourt Brace, 1922

Cosmides, Leda, and John Tooby. "Consider the Source: The Evolution of Adaptations for Decoupling and Metarepresentations." In *Metarepresentations: A Multidisciplinary Persepctive*, edited by Dan Sperber, 53–116. New York: Oxford University Press, 2000

– "The Psychological Foundations of Culture." In Barkow, Cosmides, and Tooby, *The Adapted Mind*, 19–136

Courcelle, Pierre. "Pétrarque entre Saint Augustin et les Augustins du XIVe siècle." *Studi Petrarcheschi* (1961): 58–71

Crane, Tim. "Mind-Body Problem." In Wilson and Keil, *The MIT Encyclopedia of the Cognitive Sciences*, 546–8

Crick, Francis. *The Astonishing Hypothesis*. New York: Charles Scribner's Sons, 1994

Cuddon, J.A., ed. *The Penguin Dictionary of Literary Terms and Literary Theory*. London: Penguin, 1991

Culler, Jonathan. *Structural Poetics: Structuralism, Linguistics and the Study of Literature*. Ithaca: Cornell University Press, 1975

Cupchik, G.C., and J. Laszlo. "The Landscape of Time in Literary Reception: Character Experience and Narrative Action." *Cognition and Emotion* 8 (1994): 297–312

Damasio, Antonio. *Descartes' Error: Emotion, Reason, and the Human Brain*. New York: Grosset, Putnam, 1996

– *The Feeling of What Happens: Body and Emotion in the Making of Consciousness*. San Diego: Harcourt, 1999

– *Looking for Spinoza: Joy, Sorrow, and the Feeling Brain*. Orlando: Harcourt, 2000

– *Self Comes to Mind: Constructing the Conscious Brain*. New York: Vintage Books, 2012

Danto, Arthur. *The Body/Body Problem*. Berkeley: University of California Press, 1999

Darwin, Charles. *Notebooks 1836–1844*. Edited by Paul Barrett et al. Ithaca: Cornell University Press, 1987

Davenport, William. "Sex in Cross-Cultural Perspective." In *Human Sexuality in Four Perspectives*, edited by Frank Ambrose Beach and Milton Diamond, 113–63. Baltimore: Johns Hopkins University Press, 1977

Davidson, J.R., K.M. Putnam, and C.L. Larson. "Dysfunction in the Neural Circuitry of Emotion Regulation – A Possible Prelude to Violence." *Science* 289 (2000): 591–4

Davies, Jim. *Riveted: The Science of Why Jokes Make Us Laugh, Movies Make Us Cry, and Religion Makes Us Feel One with the Universe*. New York: Palgrave Macmillan, 2014

Davis, Fred. *Yearning for Yesterday: A Sociology of Nostalgia*. New York: The Free Press/Macmillan Publishing, 1979

Davison, Carol Margaret. "'Houses of Voluntary Bondage': Theorizing the Nineteenth Century Gothic Pharmography." *Gothic Studies* 12, no. 1 (May 2010): 68–85

Dawkins, Richard. *The Ancestor's Tale: A Pilgrimage to the Dawn of Evolution*. Boston: Houghton Mifflin and Co., 2004

– *The God Delusion*. Boston: Houghton Mifflin and Co., 2006

– "Selfish Genes and Selfish Memes." In *The Mind's I: Fantasies and Reflections on Self & Soul*, edited by Douglas Hofstadter and Daniel Dennett, 124–44. New York: Basic Books, 1981

– *The Selfish Gene*. Oxford: Oxford University Press, 1976

Dejob, Charles. "Le *Secretum* de Pétrarque." *Bulletin italien* 3 (1903): 261–80

Della Porta, Giambattista. *The Sister*. Translated by Donald Beecher and Bruno Ferraro. Ottawa: Dovehouse Editions, 2000

Dennett, Daniel. *Brainstorms: Philosophical Essays on Mind and Psychology*. Cambridge: MIT Press, 1978

– *Consciousness Explained*. Boston: Little, Brown and Co., 1991

– *Freedom Evolves*. New York: Viking Penguin, 2003

– *The Intentional Stance*. Cambridge: Bradford Books, MIT Press, [1987] 1995

– "Intentional Stance." In Wilson and Keil, *The MIT Encyclopedia of the Cognitive Sciences*, 412–13

– "Intentional Systems." In *Mind Design: Philosophy, Psychology, Artificial Intelligence*, edited by John Haugeland, 220–42. Cambridge: MIT Press, 1981

– "True Believers: The Intentional Strategy and Why It Works." In Lycan, *Mind and Cognition*, 150–68

de Rocher, Gregory. *Rabelais' Laughers and Joubert's "Traité du Ris."* Tuscaloosa: University of Alabama Press, 1979

Descartes, René. *Les Passions de l'âme*. Paris: J. Vrin, 1970

de Sousa, Ronald. *The Rationality of Emotion*. Cambridge: MIT Press, 1990
– "The Rationality of Emotions." In *Explaining Emotions*, edited by Amelie O. Rorty, 127–52. Berkeley: University of California Press, 1980
Devos, Thierry, Que-Lam Huynh, and Mahzarin Banaji. "Implicit Self and Identity." In *Handbook of Self Identity*, edited by Mark Leary and June Price Tangney, 153–75. New York: The Guilford Press, 2012
The Dialogue of Solomon and Marcolphus. Edited by Donald Beecher. Ottawa: Dovehouse Editions, 1995
Dissanayake, Ellen. "The Core of Art: Making Special." In Dissanayake, *Homo Aestheticus: Where Art Comes from and Why*, 39–63. New York: The Free Press/Macmillan, 1991
Doležel, Lubomir. *Heterocosmica: Fiction and Possible Worlds*. Baltimore: Johns Hopkins University Press, 1998
Donald, Merlin. *A Mind So Rare: The Evolution of Human Consciousness*. New York: W.W. Norton, 2001
Dundes, Alan. "Folk Ideas as Units of Worldview." In *Toward New Perspectives in Folklore*, edited by Américo Paredes and Richard Bauman, 93–103. Austin: University of Texas Press, 1972
Durling, Robert M. "The Ascent of Mt Ventoux and the Crisis of Allegory." *Italian Quarterly* 69 (1974): 7–28
Dutton, Denis. *The Art Instinct: Beauty, Pleasure, and Human Evolution*. Oxford: Oxford University Press, 2009
Dutton, Richard. *Modern Tragicomedy and the British Tradition*. Norman: Oklahoma University Press, 1986
Eibl-Eibesfeldt, Irenäus. *Human Ethology*. New York: Aldine de Gruyter, 1989
Ekman, Paul. *Emotions Revealed: Recognizing Faces and Feelings to Improve Communication and Emotional Life*. New York: Times Books, 2003
– and Wallace Friesen. "Constants across Cultures in the Face and Emotion." *Journal of Personality and Social Psychology* 17 (1971): 124–9
– and Richard Davidson, eds. *The Nature of Emotion: Fundamental Questions*. New York: Oxford University Press, 1994
Ellis, Bruce, and David Bjorklund, eds. *Origins of the Social Mind: Evolutionary Psychology and Child Development*. New York: The Guilford Press, 2005
Ellis, Henry, and R. Reed Hunt. *Fundamentals of Human Memory and Cognition*. Dubuque, IA: W.C. Brown, [1972] 1989
Elster, Jon. *Alchemies of the Mind: Rationality and the Emotions*. Cambridge: Cambridge University Press, 1999
Feinberg, Todd. *Altered Egos: How the Brain Creates the Self*. Oxford: Oxford University Press, 2001
Feldman, Sandor. "Crying at the Happy Ending." *Journal of the American Psychoanalytic Association* 4 (1956): 477–85

Ferrand, Jacques. *A Treatise on Lovesickness.* Edited by Donald Beecher and Massimo Ciavolella. Syracuse, NY: Syracuse University Press, 1990

Festinger, L. *A Theory of Cognitive Dissonance.* Stanford, CA: Stanford University Press, 1957

Fisher, Helen. *The Anatomy of Love.* New York: Norton, 1992

Fisher, Shirley. *Homesickness, Cognition, and Health.* London: Lawrence Erlbaum, 1989

Flanagan, Owen. *Consciousness Reconsidered.* Cambridge: MIT Press, 1992

– *The Science of the Mind.* Cambridge: MIT Press, 1984

Flesch, William. *[Comeuppance] Costly Signaling, Altruistic Punishment, and Other Biological Components of Fiction.* Cambridge: Harvard University Press, 2007

Fodor, Jerry. "The Appeal to Tacit Knowledge in Psychological Explanation." *Journal of Philosophy* 65 (1968): 627–40

– "Banish DisContent." In Lycan, *Mind and Cognition*, 420–38

– *In Critical Condition.* Cambridge: MIT Press, 1998

– "Why There Still Has to Be a Language of Thought." In Lycan, *Mind and Cognition*, 282–99

Forbus, K.D., D. Gentner, and K. Law. "MAC/FAC: A model of similarity-based retrieval." *Cognitive Science* 19 (1995): 141–2

Fowler, Roger, ed. *A Dictionary of Modern Critical Terms.* London: Routledge, 1973

Francesco Petrarca, Citizen of the World. Proceedings of the World Petrarch Congress, 1974. Edited by Aldo S. Bernardo. Albany: State University of New York Press, 1980

Frank, R. *Passions within Reason.* New York: Norton, 1988

Frankfurt, Harry G. *The Importance of What We Care About.* Cambridge: Cambridge University Press, 1988

Freud, Sigmund. *Civilization and Its Discontents.* London: Penguin, [1930] 2002

Frijda, N.H. *The Emotions.* Cambridge: Cambridge University Press, 1986

Frye, Northrop. *The Anatomy of Criticism.* New York: Atheneum, 1967. First published by Princeton University Press, 1957

– "The Argument of Comedy." In *English Institute Essays*, edited by D.A. Robertson Jr, 58–73. New York: Columbia University Press, 1949

– "The Argument of Comedy." In Lauter, *Theories of Comedy*, 450–60

– *The Secular Scripture.* Cambridge: Harvard University Press, 1976

Ganteau, Jean-Michael. "Fantastic but Truthful: The Ethics of Romance." *Cambridge Quarterly* 32, no. 3 (2003): 225–38

Gardner, H. *Creating Minds: An Anatomy of Creativity Seen through the Lives of Freud, Einstein, Picasso, Stravinsky, Eliot, Graham, and Ghandi.* New York: Basic Books, 1993

Gazzaniga, Michael. *The Ethical Brain: The Science of Our Moral Dilemmas*. New York: Harper, 2005

– *The Social Brain: Discovering the Networks of the Mind*. New York: Basic Books, 1985

Geary, David C. "Folk Knowledge and Academic Learning." In Ellis and Bjorklund, *Origins of the Social Mind*, 493–519

– *The Origins of Mind: Evolution of Brain, Cognition, and General Intelligence*. Washington, DC: American Psychological Association, 2005

Geertz, Clifford. *After the Fact: Two Countries, Four Decades, One Anthropologist*. Cambridge: Harvard University Press, 1995

Gentner, Dedre, and Arthur B. Markman. "Structure Mapping in Analogy and Similarity." In Thagard, *Mind Readings*, 127–41

Georges, Robert A., and Alan Dundes. "Toward a Structural Definition of the Riddle." *Journal of American Folklore* 76 (1973): 111–18

Gerrig, Richard J. *Experiencing Narrative Worlds: On the Psychological Activities of Reading*. New Haven: Yale University Press, 1993

– "Suspense in the Absence of Uncertainty." *Journal of Memory and Language* 28 (1989): 633–48

– and David Rapp. "Psychological Processes Underlying Literary Impact." *Poetics Today* 25, no. 2 (Summer 2004): 265–81

– and Allan B.I. Bernardo. "Readers as Problem-Solvers in the Experience of Suspense." *Poetics* 22 (1994): 459–72

Giddens, Anthony. *Modernity and Self-Identity: Self and Society in the Late Modern Age*. Stanford, CA: Stanford University Press, 1991

Gilbert, Daniel. "Social Cognition." In Wilson and Keil, *The MIT Encyclopedia of the Cognitive Sciences*, 777–8

Girard, René. "Perilous Balance: A Comic Hypothesis." *Modern Language Notes* 87 (1972): 811–26

Glucksberg, Sam, and Boaz Keysar. "Understanding Metaphorical Comparisons: Beyond Similarity." *Psychological Review* 97 (1990): 3–18

Goffman, Erving. *Encounters: Two Studies in the Sociology of Interaction*. Indianapolis: Bobbs-Merrill, 1961

– *Frame Analysis: An Essay on the Organization of Experience*. Cambridge: Harvard University Press, 1974

Goldman, Alvin. "Epistemology and Cognition." In Wilson and Keil, *The MIT Encyclopedia of the Cognitive Sciences*, 280–2

– *Simulating Minds: The Philosophy, Psychology, and Neuroscience of Mindreading*. Oxford: Oxford University Press, 2006

Goodman, Nelson. *Of Mind and Other Matters*. Cambridge: Harvard University Press, 1984

Goodman, Paul. *The Structure of Literature*. Chicago: University of Chicago Press, 1954

Gopnik, Alison. "Theory of Mind." In Wilson and Keil, *The MIT Encyclopedia of the Cognitive Sciences*, 838–41

Gordon, Robert M. "Simulation vs. Theory-Theory." In Wilson and Keil, *The MIT Encyclopedia of the Cognitive Sciences*, 765–6

Gottschall, Jonathan, and David Sloan Wilson. *The Literary Animal: Evolution and the Nature of Narrative*. Evanston: Northwestern University Press, 2005

Green, Melanie C. "Narrative Worlds and Self." In *On Building, Defending, and Regulating the Self: A Psychological Perspective*, edited by A. Tesser, J.V. Wood, and D.A. Stepel, 53–75. New York: Psychology Press, 2005

Greene, Joshua. *Moral Tribes: Emotion, Reason, and the Gap between Them and Us*. New York: Penguin Press, 2013

Greene, Robert. *Menaphon*. Edited by Brenda Cantar. Ottawa: Dovehouse Editions, 1996

– *The Scottish History of James the Fourth*. Edited by J.A. Lavin. London: Ernest Benn Ltd, 1967

Gregg, G.S. *Self-Representation: Life Narrative Studies in Identity and Ideology*. New York: Greenwood Press, 1991

Gregory, J.C. *The Nature of Laughter*. New York: Harcourt, 1924

Griffiths, Paul E. *What Emotions Really Are: The Problem of Psychological Categories*. Chicago: University of Chicago Press, 1997

Grodal, Torben. *Moving Pictures: A New Theory of Film Genres, Feelings, and Cognition*. Oxford: Clarendon Press, 1997

Gruner, Charles. *Understanding Laughter: The Working of Wit and Humor*. Chicago: Nelson-Hall, 1978

Guilhembet, Jacques. "Esthétique de la comédie." *Littérature classiques* 27 (1996): 184–94

Guthrie, Stewart Elliott. *Faces in the Clouds: A New Theory of Religion*. New York: Oxford University Press, 1993

Habermas, T., and S. Bluck. "Getting a Life: The Emergence of the Life Story in Adolescence." *Psychological Bulletin* 119 (2000): 748–69

Hacking, Ian. "Normal People." In Olson and Torrance, *Modes of Thought*, 59–71

Hägg, Tomas. *The Novel in Antiquity*. Oxford: Basil Blackwell, 1983

Hall, Stephen. *Wisdom from Philosophy to Neuroscience*. New York: Alfred Knopf, 2010

Hardcastle, Valerie Gray. *Constructing the Self*. Amsterdam: John Benjamins, 2008

Harmon, William, and C. Hugh Holden. *A Handbook to Literature*. Upper Saddle River, NJ: Prentice-Hall, 1996

Harré, Rom. *The Singular Self: An Introduction to the Psychology of Personhood*. London: Sage Publications, 1998

– and L. van Langenhore. "Varieties of Positioning." *Journal for the Theory of Social Behaviour* 21, no. 4 (1991): 393–407

Haskin, Dayton. *Milton's Burden of Interpretation*. Philadelphia: University of Philadelphia Press, 1994

Heliodorus. *An Ethiopian Story*. Translated by J.R. Morgan. In Reardon, *Collected Ancient Greek Novels*, 349–588

Henricks, Thomas. *Selves, Societies, and Emotions: Understanding the Pathways of Experience*. Boulder: Paradigm Publishers, 2012

Herman, David. "Towards a Socionarratology: New Ways of Analyzing Natural Language Narratives." In *Narratologies: New Perspectives on Narrative Analysis*, edited by David Herman, 218–46. Columbus: Ohio State University Press, 1999

Hermans, Hubert J.M. "Voicing the Self: From Information Processing to Dialogical Interchange." *Psychological Bulletin* 119, no. 1 (1996): 31–50

– and G. Dimaggio. *The Dialogical Self in Psychotherapy*. New York: Brunner-Routledge, 2004

Hernandi, Paul, and Francis Steen. "The Tropical Landscapes of Proverbia: A Cross-Disciplinary Travelogue." *Style* 33 (Spring 1999): 1–20

Herrick, Marvin T. *Tragicomedy*. Urbana: University of Illinois Press, 1962

Higgins, E.T. "Self-Discrepancy: A Theory Relating Self and Affect." *Psychological Review* 94 (1987): 319–40

Hillman, Richard. *William Shakespeare: The Problem Plays*. New York: Twayne Publishers, 1993

Hinchman, Lewis, and Sandra Hinchman, eds. *Memory, Identity, Community: The Idea of Narrative in the Human Sciences*. Albany: State University of New York Press, 1997

Hirstein, William. *Brain Fiction: Self-Deception and the Riddle of Confabulation*. Cambridge: MIT Press, 2005

Hoffmann, Martin. "Is Altruism Part of Human Nature?" *Journal of Personality and Social Psychology* 40 (1981): 121–37

Hofstadter, Douglas. R. "Analogy as the Core of Cognition." In *The Analogical Mind: Perspectives from the Cognitive Sciences*, edited by Dedre Gentner et al., 499–538. Cambridge: MIT Press, 2001

– "The Prisoner's Dilemma, Computer Tournaments, and the Evolution of Cooperation." In Hofstadter, *Metamagical Themas: Questing for the Essence of Mind and Pattern*, 715–38. New York: Basic Books, 1985

Hogan, Patrick Colm. *The Mind and Its Stories: Narrative Universals and Human Emotion*. Cambridge: Cambridge University Press, 2003

Holland, Norman. *Laughing: A Psychology of Humor*. Ithaca: Cornell University Press, 1982

Honeck, Richard P. "Proverbs, Meaning, and Group Structure." In *Cognition and Figurative Language*, edited by R.P. Honeck and R.R. Hoffman, 127–61. Hillsdale, NJ: Erlbaum, 1980

Hood, Bruce. *The Self Illusion: Why There Is No 'You' inside Your Head.* London: Constable & Robinson Ltd, 2012

Horney, Karen. *Neurosis and Human Growth.* New York: Norton, 1951

Hrdy, Sarah. *Mother Nature.* New York: Ballantine Books, 2000

Hugo of St Victor. Twelfth sermon, "De spiritualibus montibus et arboribus Israel." *Patrologiae cursus completes. Series Latina.* Edited by Jacques-Paul Migne. Paris, 1844–64

Hummel, John. "Binding Problem." In Wilson and Keil, *The MIT Encyclopedia of the Cognitive Sciences*, 85–6

Humphrey, Nicholas K. *The Mind Made Flesh: Essays from the Frontiers of Psychology and Evolution.* Oxford: Oxford University Press, 2002

– "The Social Function of Intellect." In *Growing Points in Ethology*, edited by P.P.G. Bateson and R.A. Hinde, 303–21. Cambridge: Cambridge University Press, 1976

– "The Uses of Consciousness." In Humphrey, *The Mind Made Flesh*, 65–85

Hunter, Graeme. "Taming *The Tempest*." In Irvine and Russell, *In the Agora*, 127–38

Irvine, Andrew, and John Russell, eds. *In the Agora: The Public Face of Canadian Philosophy.* Toronto: University of Toronto Press, 2006

Iser, Wolfgang. "The Reading Process: A Phenomenological Approach." In *Modern Criticism and Theory*, edited by David Lodge, 212–28. London: Longman, 1988

Ivic, Christopher, and Grant Williams, eds. *Forgetting in Early Modern English Literature and Culture: Lethe's Legacies.* London: Routledge, 2004

Jackson, Frank. "Epiphenomenal Qualia." *Philosophical Quarterly* 32 (1982): 127–36

James, William. *The Principles of Psychology.* Chicago: The Encyclopedia Britannica, 1952

– *The Varieties of Religious Experience.* New York: Mentor, [1902] 1964

Jeanson, Francis. *Signification humaine du rire.* Paris: Seuil, 1950

Johnson, David Martel. "Taking the Past Seriously: How History Shows that Eliminativists' Account of Folk Psychology is Partly Right and Partly Wrong." In *The Future of the Cognitive Revolution*, edited by D.M. Johnson and Christina E. Erneling, 366–75. New York: Oxford University Press, 1999

Johnston, Victor S. *Why We Feel: The Science of Human Emotions.* Cambridge, MA: Perseus Books, 1999

Jolles, André. *Formes simples.* Paris: Seuil, 1972. Originally published in Tübingen by Max Niemeyer Verlag, 1930

Jose, P.E., and W.F. Brewer. "Development of Story Liking: Character Identification, Suspense, and Outcome Resolution." *Developmental Psychology* 20 (1984): 911–24

Jung, Karl. "On the Psychology of the Trickster Figure." In Radin, *The Trickster*, 195–21

Kandel, Eric R. *In Search of Memory: The Emergence of a New Science of Mind*. New York: W.W. Norton and Co., 2006

Katz, Bruce. "A Neural Resolution of the Incongruity-Resolution and Incongruity Theories of Humor." *Connection Science* 5, no. 1 (1993): 59–75

Katzko, M.W. "Unity vs. Multiplicity: A Conceptual Analysis of the Term 'Self' and Its Use in Personality Theory." *Journal of Personality* 71, no. 1 (2003): 83–114

Keefer, Michael. *Christopher Marlowe's* Doctor Faustus: *A 1604-Version Edition*. Peterborough, ON: Broadview Press, 1991

Keen, Suzanne. "A Theory of Narrative Empathy." *Narrative* 14, no. 3 (2006): 207–36

Kerby, Anthony Paul. *Narrative and the Self*. Bloomington: Indiana University Press, 1991

Kerr, Walter. *Tragedy and Comedy*. London: Bodley Head, 1967

Koestler, Arthur. *The Act of Creation*. New York: Macmillan, 1964

Kolakowski, Leszek. *Bergson*. Oxford: Oxford University Press, 1985

Kramers, Dieter. "L'ascension du Mont Ventoux." *Actes du Congrès International Francesco Petrarca, Père des Renaissances, Serviteur de l'Amour et de la Paix*. Avignon: Aubanel, 1974, 122–31

Kramnick, Jonathan. "Against Literary Darwinism." *Critical Inquiry* 37 (Winter 2011): 315–47

Krebs, Dennis. "The Challenge of Altruism in Biology and Psychology." In *Sociobiology and Psychology: Ideas, Issues and Applications*, edited by Charles Crawford, Martin Smith, and Dennis Krebs, 81–118. Hillsdale, NJ: Erlbaum, 1987

Kristeller, Paul O. "Augustine and the Early Renaissance." *Review of Religion* 8 (1944): 339–58

– *Eight Philosophers of the Italian Renaissance*. Stanford: Stanford University Press, 1964

Lakoff, George, and Mark Johnson. *Metaphors We Live By*. Chicago: Chicago University Press, 1980

– *Philosophy in the Flesh: The Embodied Mind and Its Challenge to Western Thought*. New York: Basic Books, 1999

Lauter, Paul, ed. *Theories of Comedy*. Garden City, NY: Doubleday and Co., 1964

Lawrence, William. *Shakespeare's Problem Comedies*. London: Penguin Books, [1931] 1969

Lazarus, Richard. "The Stable and the Unstable in Emotion." In Ekman and Davidson, *The Nature of Emotion*, 79–85

LeDoux, Joseph. *The Emotional Brain: The Mysterious Underpinnings of Emotional Life*. New York: Simon & Schuster, 1996

– "Emotional Colouration of Consciousness: How Feelings Come About." In *Frontiers of Consciousness: Chichele Lectures*, edited by Lawrence Weiskrantz and Martin Davies, 69–130. Oxford: Oxford University Press, 2008

Lehman, Benjamin Harrison. *Comedy and Laughter. University of California Publications; English Studies* 10: 81–101. Berkeley: University of California Press, 1954

Lenk, Werner. "Zur Sprichwort-Antithetik im Salomon-Markolf-Dialog." *Forschungen und Fortschritte* 39, no. 5 (1965): 151–5

Letocha, Danièle. "The Duty of Memory: The Contradictions in Conrad Celtis's *Oratio* (1492)." In *Ars Reminiscendi: Mind and Memory in Renaissance Culture*, edited by Donald Beecher and Grant Williams, 267–88. Toronto: CRRS Publications, 2009

Letters from Petrarch. Translated by Morris Bishop. Bloomington: Indiana University Press, 1966

Levenson, Robert W. "The Search for Autonomic Specificity." In Ekman and Davidson, *The Nature of Emotion*, 252–7

Lever, J.W., ed. *Measure for Measure*. Regents Renaissance Drama Series. London: Edward Arnold, 1964

Levin, Richard. *The Multiple Plot in English Renaissance Drama*. Chicago: University of Chicago Press, 1971

Levin, Samuel. "Standard Approaches to Metaphor and a Proposal for Literary Metaphor." In Ortony, *Metaphor and Thought*, 124–35

Levine, Joseph, and David Suzuki. *The Secret of Life: Redesigning the Living World*. Toronto: Stoddard, 1993

Lévy, Bernard-Henri. *Hôtel Europe*. Paris: Bernard Grasset, 2014

Lewis, David. "Truth in Fiction." *American Philosophical Quarterly* 15, no. 1 (1978): 37–46

Lewontin, Richard. "Sociobiology as an Adaptationist Program." *Behavioral Science* 24 (January 1979): 5–14

– Steven Rose, and Leon Kamin, eds. *Not in Our Genes: Biology, Ideology, and Human Nature*. New York: Pantheon, 1984

Lifton, R.J. *The Protean Self: Human Resilience in an Age of Fragmentation*. New York: Basic Books, 1993

Lipps, Theodore. *Psychological Studies*. Translated by Herbert C. Sanborn. Baltimore: Williams & Williams, 1926

Livingston, Paisley. *Literary Knowledge: Humanistic Inquiry and the Philosophy of Science*. Ithaca: Cornell University Press, 1988

Lodge, Thomas. *A Margarite of America*. Edited by Donald Beecher and Henry D. Janzen. Toronto: CRRS Publications, 2005

Lohrey, Andrew. *The Meaning of Consciousness*. Ann Arbor: University of Michigan Press, 1997

Luciani, Evelyne. *Les "Confessions" de Saint Augustin dans les lettres de Pétrarque*. Paris: Études Augustiniennes, 1982

Lüthi, Max. *Once upon a Time: On the Nature of Fairy Tales*. Translated by Lee Chadeayne and Paul Gottwald. Bloomington: Indiana University Press, 1976

Lutz, Thomas. *Crying: A Natural and Cultural History of Tears*. New York: Norton, 1999

Lycan, William G. "Free Will and the Burden of Proof." In O'Hear, *Minds and Persons*, 107–22

– ed. *Mind and Cognition: A Reader*. Oxford: Blackwell, 1990

Lynch, Gary, and Richard Granger. *Big Brain: The Origins and Future of Human Intelligence*. New York: Palgrave MacMillan, 2008

Machline, Vera Cecilia, "The Contribution of Laurent Joubert's *Traité du Ris* to Sixteenth-Century Physiology of Laughter." In *Reading the Book of Nature: The Other Side of the Scientific Revolution*, edited by Allen Debus and Michael Walton, 251–64. Kirksville, MO: Truman State University Press, 1998

MacIntyre, Alasdair. "Epistemological Crises, Dramatic Narrative, and the Philosophy of Science." In *Paradigms and Revolutions*, edited by Gary Cutting, 54–74. Notre Dame: University of Notre Dame Press, 1980

Maguire, Laurie. *Studying Shakespeare: A Guide to the Plays*. Malden, MA: Blackwell Publishing, 2004

Malony, H.N., and S. Southard, eds. *Handbook of Religious Conversion*. Birmingham, AL: Religious Education Press, 1992

Mandler, G. "Organization and Memory." In *The Psychology of Learning and Motivation: Advances in Research and Theory*, edited by K.W. Spence and J.T. Spence. London: Academic Press, 1967

Mar, Raymond A., and Keith Oatley. "The Function of Fiction Is the Abstraction and Simulation of Social Experience." *Perspectives on Psychological Science* 3, no. 3 (2008): 173–92

Marlowe, Christopher. *Doctor Faustus*. Edited by John D. Jump. London: Methuen, 1968

Marston, John. *The Fawn*. Edited by Gerald A. Smith. London: Edward Arnold, Ltd, 1965

– *The Malcontent*. Edited by Martin Wine. London: Edward Arnold, Ltd, 1965

– *Parasitaster or The Fawn*. Edited by David Blostein. Manchester: Manchester University Press, 1978

Martinelli, Bortolo. *Petrarca e il Ventoso*. Introduction by Enzo Noè Girardi. Bergamo: Minerva Italica, 1977

Mauron, Charles. *Psychocritique du genre comique*. Paris: José Corti, 1964

Mazzotta, Giuseppe. *The Worlds of Petrarch*. Durham, NC: Duke University Press, 1993

McAdams, D.P. *Power, Intimacy and the Life Story: Personological Inquiries into Identity*. New York: Guilford Press, 1985

– "The Psychology of Life Stories." *Review of General Psychology* 5, no. 2 (2001): 100–22

– Ruthellen Josselson, and Amia Lieblich, eds. *Identity and Story: Creating Self in Narrative*. Washington, DC: American Psychological Association, 2006

McCallister, B.J. "Cognitive Theory and Religious Experience." In *Handbook of Religious Experience*, edited by R.W. Hood Jr, 312–52. Birmingham, AL: Religious Education Press, 1995

McEwan, Ian. "Literature, Science, and Human Nature." In Gottschall and Wilson, *The Literary Animal*, 5–19

McGhee, Paul. *Humor: Its Origin and Development*. San Francisco: W.H. Freeman, 1979

McGinn, Colin. *Mindsight: Image, Dream, Meaning*. Cambridge: Harvard University Press, 2004

– *Shakespeare's Philosophy: Discovering the Meaning behind the Plays*. New York: HarperCollins, 2006

McKeon, Michael. *The Origins of the English Novel: 1600–1740*. London and Baltimore: The Johns Hopkins University Press, 1987

Menard, Danièl. *La Renaissance et le rire*. Paris: Presses universitaires de France, 1995

Menner, Robert J., ed. *The Poetical Dialogues of Solomon and Saturn*. London: Oxford University Press, 1941

Merikle, Philip. "Perception without Awareness: Critical Issues." *American Psychologist* 47 (1992): 792–5

Metzner, Ralph. "Ten Classical Metaphors of Self-Transformation." *Journal of Transpersonal Psychology* 12 (1980): 47–62

Mieder, Wolfgang. "'Good Old Yankee Wisdom' Proverbs and the Worldview of New England." In Mieder, *"Proverbs Speak Louder Than Words": Folk Wisdom in Art, Culture, Folklore, History, Literature, and Mass Media*, 143–67. New York: Peter Lang, 2008

Minsky, Marvin. *The Society of Mind*. New York: Simon and Schuster, 1986

Minturno, Antonio Sebastiano. "From the Art of Poetry." In Lauter, *Theories of Comedy*, 74–86

Moldoveanu, Mihnea, and Nitin Nohria. *Master Passions: Emotion, Narrative and the Development of Culture*. Cambridge: MIT Press, 2002

Morgan, J.R. "History, Romance and Realism." *Classical Antiquity* 1 (1982): 221–65

Morreall, John. "A New Theory of Laughter." *Philosophical Studies* 42 (1982): 243–54

– ed. *The Philosophy of Laughter and Humor*. Albany: State University of New York Press, 1987

Morris, Desmond. *The Naked Ape: A Zoologist's Study of the Human Animal*. London: Jonathan Cape, 1967

Nagel, Thomas. "What Is It Like to Be a Bat?" *Philosophical Review* 83 (1974): 435–51

Nelson, T.G.A. *Comedy: An Introduction to Comedy in Literature, Drama, and Cinema*. Oxford: Oxford University Press, 1990

Neve, J.L. *A History of Christian Thought*. 2 vols. Philadelphia: Muhlenberg Press, 1946

Newberg, Andrew, Eugene D'Aquili, and Vince Rause. *Why God Won't Go Away: Brain Science and the Biology of Belief*. New York: Ballantine Books, 2001

Nietzsche, Friedrich. *The Will to Power*. Translated by Walter Kaufmann. New York: Vintage, 1968

Norman, Donald A. *Memory and Attention: An Introduction to Human Information Processing*. New York: Wiley, 1976

Norrick, Neil R. *How Proverbs Mean: Semantic Studies in English Proverbs*. Berlin: Mouton Publishers, 1985

Norris, Christopher. *Fiction, Philosophy, and Literary Theory: Will the Real Saul Kripke Please Stand Up?* London: Continuum International Publishing, 2007

North, Sir Thomas. *The Moral Philosophy of Doni, Popularly Known as The Fables of Bidpai*. Edited by Donald Beecher, John Butler, and Carmine Di Biase. Barnabe Riche Society Publications, 14. Ottawa: Dovehouse Editions, 2003

Nussbaum, Martha. *Upheavals of Thought: Intelligence of Emotions*. Cambridge: Cambridge University Press, 2001

Oatley, Keith. "Emotions." In Wilson and Keil, *The MIT Encyclopedia of the Cognitive Sciences*, 273–5

– "A Taxonomy of the Emotions of Literary Response and a Theory of Identification in Fictional Narrative." *Poetics* 23 (1994): 53–74

– and Jennifer Jenkins. *Understanding Emotions*. Cambridge: Blackwell Publishing, 1996

O'Hear, Anthony, ed. *Minds and Persons*. Cambridge: Cambridge University Press, 2003

Ohler, Peter, and Gerhild Nieding. "Cognitive Modeling of Suspense-Inducing Structures in Narrative Films." In Vorderer, Wulff, and Friedrichsen, *Suspense*, 129–47

Olson, David, and Nancy Torrance, eds. *Modes of Thought: Explorations in Culture and Cognition*. Cambridge: Cambridge University Press, 1966

Orgel, Stephen. "What Is a Character?" In Orgel, *The Authentic Shakespeare and Other Problems of the Early Modern Stage*, 101–8. New York: Routledge, 2002

Ornstein, Robert E., and Paul R. Erlich. *New World, New Mind: Moving toward Conscious Evolution*. New York: Doubleday, 1989

Ortony, Andrew. "The Role of Similarity in Similes and Metaphors." In Ortony, *Metaphor and Thought*, 186–201

– ed. *Metaphor and Thought*. Cambridge: Cambridge University Press, 1979

– Gerald Clore, and Allan Collins, eds. *The Cognitive Structure of Emotions*. Cambridge: Cambridge University Press, 1988

Overbury, Sir Thomas, and others. *Characters*. Edited by Donald Beecher. Barnabe Riche Society Publications, 15. Ottawa: Dovehouse Editions, 2003

Paivio, Allen. "Psychological Processes in the Comprehension of Metaphor." In Ortony, *Metaphor and Thought*, 150–71

Palmer, Alan. *Fictional Minds*. Lincoln: University of Nebraska Press, 2004

Paloutzian, Raymond F. *Invitation to the Psychology of Religion*. 2nd ed. Boston: Allyn and Bacon, 1996

Panksepp, Jaak. "At the Interface of the Affective, Behavioral and Cognitive Neurosciences: Decoding the Emotional Feelings of the Brain." *Brain and Cognition* 52 (2003): 7

Perry, T. Anthony. *Wisdom Literature and the Structure of Proverbs*. University Park: Pennsylvania State University Press, 1993

Petrarca letterato, Vol. 1. See Billanovich

Petrarch [Francesco], A Humanist among Princes. [An Anthology of Petrarch's Letters]. Edited by David Thompson. New York: Harper and Row, 1971

Petrarch and His World. [A Selection of the Letters]. Translated by Morris Bishop. Bloomington: Indiana University Press, 1963

Petrarch's Secretum. Translated by David Carozza and H. James Shea. New York: Peter Lang, 1989

Pétrarque. *L'ascension du Mont Ventoux*. Translated by Denis Montebello. Introduced by Pierre Dubrunquez. Rezé: Séquences, 1990

– *L'ascension du Mont Ventoux*. Translated by Paul Bachmann. Preface by Eve Duperray. [Fontaine-de-Vaucluse]: Musée Pétrarque, 1996

– *Mon secret, ou du mépris du monde*. Translated by Pompée Mabile. Angers: P. Lachèse et Dilbeau, 1886

Phillips, Hollibert E. *Vicissitudes of the I: An Introduction to the Philosophy of Mind*. Englewood Cliffs, NJ: Prentice Hall, 1995

Piddington, Ralph. *The Psychology of Laughter: A Study in Social Adaptation*. New York: Gamut Press, 1963

Pinker, Steven. *The Blank Slate: The Modern Denial of Human Nature.* London: Allen Lane (Penguin), 2002

– *How the Mind Works.* New York: W.W. Norton, 1997

Piper, Paul. *Die Schriften Notkers und seiner Schule.* Vol. 2. Freiburg im/B & Tübingen: J.C.B. Mohr, 1982–83

Plato. *Symposium.* Translated by Robin Waterfield. Oxford: Oxford University Press, 1994

Pollio, Howard R. "Notes towards a Field Theory of Humor." In *Handbook of Humor Research,* vol. 1, *Basic Issues*, edited by P.E. McGhee and J.H. Goldstein, 213–30. New York: Springer Verlag, 1983

Popper, Karl R., and John C. Eccles. *The Self and Its Brain: An Argument for Interactionism.* New York: Springer Verlag/London: Routledge, 1977

Prentice, Debora A., Richard Gerrig, and Daniel Bailis. "What Readers Bring to the Processing of Fictional Texts." *Psychonomic Bulletin & Review* 4, no. 3 (1977): 416–20

Prinz, Jesse. "Is Morality Innate?" In *The Evolution of Morality.* Vol. 1 of *Moral Psychology*, edited by Walter Sinnott-Armstrong, 367–406. Oxford: Oxford University Press, 2008

Proudfoot, W. *Religious Experience.* Berkeley: University of California Press, 1985

Pruvost, René. *Matteo Bandello and Elizabethan Fiction.* Paris: Champion, 1937

Pruyser, Paul. *A Dynamic Psychology of Religion.* New York: Harper and Row, 1955

Quiller-Couch. Sir Arthur. "*Measure for Measure.*" In *Discussions of Shakespeare's Problem Comedies*, edited by Robert Ornstein, 73–7. Boston: D.C. Heath, 1961

Rabkin, Eric S. *Narrative Suspense.* Ann Arbor: University of Michigan Press, 1973

Radin, Paul. *The Trickster: A Study in American Indian Mythology.* Commentaries by Karl Kerényi and C.G. Jung. New York: Schocken Books, (1956) 1972

Raggatt, Peter T.F. "Multiplicity and Conflict in the Dialogical Self: A Life-Narrative Approach." In McAdams, Josselson, and Lieblich, *Identity and Story: Creating Self in Narrative*, 15–35

Raine, A., M.S. Buchsbaum, and L. LaCasse. "Brain Abnormalities in Murderers Indicated by Positron Emission Tomography." *Biological Psychiatry* 42, no. 6 (1977): 495–508

– et al. "Reduced Prefrontal and Increased Subcortical Brain Functioning Assessed Using Positron Emission Tomography in Predatory and Affective Murderers." *Behavioral Sciences and the Law* 16 (1998): 319–32

Ramachandran, V.S., *The Tell-Tale Brain: A Neuroscientist's Quest for What Makes Us Human*. New York: Norton, 2012

– *The Tell-Tale Brain: Unlocking the Mystery of Human Nature*. London: Windmill Books/Random House, 2012

– and Sandra Blakeslee. *Phantoms in the Brain: Probing Mysteries of the Mind*. New York: William Morrow, 1998

Rambo, L.R. *Understanding Religious Conversion*. New Haven: Yale University Press, 1993

Ran-Moseley, Faye. *The Tragicomic Passion: A History and Analysis of Tragicomedy and Tragicomic Characterization in Drama, Film, and Literature*. New York: Peter Lang, 1994

Reardon, B.P., ed. *Collected Ancient Greet Novels*. Berkeley: University of California Press, 1989

Reinhart, Tanya. "Binding Theory." In Wilson and Keil, *The MIT Encyclopedia of the Cognitive Sciences*, 86–8

Riccoboni, Antonio. "From the Comic Art" (1585). In Lauter, *Theories of Comedy*, 98–107

Richards, Graham. "Emotions into Words – or Words into Emotions." In *Representing Emotions*, edited by Penelope Gowk and Helen Hills, 49–65. Aldershot: Ashgate, 2005

Riche, Barnabe. *The Adventures of Brusanus, Prince of Hungaria*. Edited by Joseph Khoury. Toronto: Centre for Reformation and Renaissance Studies, 2014

– "Of Gonsales and His Virtuous Wife Agatha." In *His Farewell to Military Profession*, edited by Donald Beecher. Ottawa: Dovehouse Editions/Binghamton: MRTS, 1992

Ridley, Matt. *The Agile Gene: How Nature Turns on Nurture*. Toronto: Harper Perennial Canada, 2003

Riley, Bruce T. *The Psychology of Religious Experience in Its Personal and Institutional Dimensions*. New York: Peter Lang, 1988

Roald, Tone. *Cognition in Emotion: An Investigation through Experiences with Art*. Amsterdam: Rodopi, 2007

Robinson, Robert C., et al. "Pathological Laughing and Crying following Stroke: Validation of a Measurement Scale and a Double-Blind Treatment Study." *American Journal of Psychiatry* 150, no. 2 (1993): 286–93

Rosaldo, Michelle. "Toward an Anthropology of Self and Feeling." In *Culture Theory: Essays on Mind, Self, and Emotion*, edited by R. Schweder and R. Le Vine, 137–57. Cambridge: Cambridge University Press, 1984

Rosenwald, G.C., and R.L. Ochberg. *Storied Lives: The Cultural Politics of Self-Understanding*. New Haven: Yale University Press, 1992

Roubine, Jean-Jacques. "La stratégie des larmes au XVIIe siècle." *Littérature* (February 1973): 56–73.

Rowley, William, Thomas Dekker, and John Ford. *The Witch of Edmonton*. Edited by Peter Corbin and Douglas Sedge. Manchester: Manchester University Press, 1999

Rumelhart, David. "The Architecture of Mind: A Connectionist Approach." In Thagard, *Mind Readings*, 207–38

Ruse, Michael. *Darwinism Defended: A Guide to the Evolution Controversies*. Reading, MA: Addison-Wesley, 1982

Salingar, Leo. *Shakespeare and the Traditions of Comedy*. Cambridge: Cambridge University Press, 1974

Samuel, David. *Memory: How We Use It, Lose It and Can Improve It*. London: Weidenfeld & Nicolson, 1999

Sanfey, Alan, et al. "The Neural Basis of Economic Decision-Making in the Ultimatum Game." *Science* 300, no. 5626 (June 2003): 1755–8

Santayana, George. *The Sense of Beauty*. New York: Scribner's, 1896

Sarbin, T.R. "Emotions as Narrative Emplotments." In *Entering the Circle: Hermeneutic Investigation in Psychology*, edited by M.J. Packer et al., 185–201. Albany: State University of New York Press, 1988

Schank, Roger. *Dynamic Memory: A Theory of Reminding and Learning in Computers and People*. New York: Cambridge University Press, 1982

– and Robert Abelson. *Scripts, Plans, Goals, and Understanding: An Inquiry into Human Knowledge Structures*. Hillsdale, NJ: Erlbaum, 1977

Schanzer, Ernest. *The Problem Plays of Shakespeare*. London: Routledge & Kegan Paul, (1963) 1965

Schechtman, Marya. "Personal Identity and the Past." *Philosophy, Psychology and Psychiatry* 12, no. 1 (2005): 9–22

Scheff, T.J. *Catharsis in Healing, Ritual, and Drama*. Berkeley: University of California Press, 1979

Schleiermacher, Friedrich. *On Religion: Speeches to Its Cultured Despisers*. New York: Harper and Brothers, 1958

Schönbrunn-Kölb, Erika. "Markolf in den mittelalterlichen Salomondichtungen und deutscher Wortgeographie." *Zeitschrift für Mundartforschung* 25 (1957): 92–128

Schulz, Kathryn. *Being Wrong: Adventures in the Margin of Error*. New York: HarperCollins, 2010

Scobie, G.E.W. "Types of Christian Conversion." *Journal of Behavioral Science* 1 (1973): 265–71

Scruton, Roger. "Laughter." *Proceedings of the Aristotelian Society*. Supp. 56 (1982): 197–212

Searle, John. *The Rediscovery of Mind*. Cambridge: MIT Press, 1992

Segrè, Carlo. "*Il Secretum* del Petrarca e le *Confessioni* di Sant'-Agostino." *Studi Petrarcheschi* (1911): 1–127

Seigel, Jerrold E. *The Idea of the Self: Thought and Experience in Western Europe since the Seventeenth Century*. Cambridge: Cambridge University Press, 2005

– "Ideas of Eloquence and Silence in Petrarch." *Journal of the History of Ideas* 26 (1965): 147–74

Sellars, Wilfrid. *Science and Metaphysics*. New York: Humanities Press, 1968

Shakespeare, William. *All's Well That Ends Well*. Edited by G.K. Hunter. London: Methuen, 1967

– *Measure for Measure*. Edited by J.W. Lever. London: Methuen, 1967

Sheppard, Robert. "How We Think." *Macleans* 113, no. 18 (May 2000): 42–8

Shermer, Michael. *The Believing Brain: From Ghosts and Gods to Politics and Conspiracies – How We Construct Beliefs and Reinforce Them as Truths*. New York: Times Books, 2011

Shoemaker, Sydney. *Identity, Cause, and Mind: Philosophical Essays*. New York: Cambridge University Press, 1984

– and Richard Swinburne. *Personal Identity*. Oxford: Blackwell, 1984

Shultz, Thomas R. "A Cognitive Developmental Analysis of Humor." In *Humour and Laughter: Theory, Research, and Applications*, edited by Anthony Chapman and Hugh Foot, 11–36. London: Wiley, 1976

Shumaker, Wayne. *Literature and the Irrational*. New York: Washington Square Press, 1960

Sidney, Sir Philip. *An Apology for Poetry or The Defence of Poesy*. Edited by Geoffrey Shepherd. London: Thomas Nelson, 1965

Simon, Herbert. "Some Computer Models of Human Learning." In *How We Know*, edited by Michael Shafto, 89–120. San Francisco: Harper and Row, 1985

Simonov, P.V. *The Emotional Brain: Physiology, Neuroanatomy, Psychology and Emotion*. Translated by Marie J. Hall. New York: Plenum Press, 1986

Slingerland, Edward. "Who's Afraid of Reductionism? The Study of Religion in the Age of Cognitive Science." *Journal of the American Academy of Religion* 76, no. 2 (June 2008): 375–411

Smith, Adam. *The Theory of Moral Sentiments*. New York: Garland, [1759] 1971

Sober, Elliott. *Philosophy of Biology*. Boulder, CO: Westview Press, 1993

Sousa, Ronald de. See de Sousa, Ronald

Spenser, Edmund. *The Faerie Queene: Books Three and Four*. Edited by Dorothy Stephens. Indianapolis: Hackett Publishing, 2006

Sperber, Dan. *On Anthropological Knowledge: Three Essays*. New York: Cambridge University Press, 1985

Speroni, Sperone. *Canace*. Edited by Elio Brancaforte. Toronto: CRRS Publications, 2013

Sperry, R. *Science and Moral Priority: Merging Mind, Brain, and Human Values*. New York: Columbia University Press, 1983

Spilka, Bernard, Ralph W. Hood Jr, and Richard L. Gorsuch. *The Psychology of Religion: An Empirical Approach*. Englewood Cliffs, NJ: Prentice-Hall, 1985

Spolsky, Ellen. *The Contracts of Fiction: Cognition, Culture, Community*. Oxford: Oxford University Press, 2015

Stalnaker, Robert. "Propositional Attitudes." In Wilson and Keil, *The MIT Encyclopedia of the Cognitive Sciences*, 678–9

Steiner, George. *The Death of Tragedy*. New Haven: Yale University Press, 1996

Stich, Stephen. *Deconstructing the Mind*. New York: Oxford University Press, 1996

– "Eliminative Materialism." In Wilson and Keil, *The MIT Encyclopedia of the Cognitive Sciences*, 265–7

Storey, Robert. *Mimesis and the Human Animal: On the Biogenetic Foundations of Literary Representation*. Evanston: Northwestern University Press, 1996

St Victor, Hugo. See Hugo of St Victor

Sully, James. *An Essay on Laughter, Its Forms, Its Causes, Its Development, and Its Value*. London: Longmans, Green & Co., 1902

Sutton, Dana. *The Catharsis of Comedy*. Lanham, MD: Rowman & Littlefield, 1994

Symons, Donald. *The Evolution of Human Sexuality*. Oxford: Oxford University Press, 1979

– "On the Use and Misuse of Darwinism in the Study of Human Behavior." In Barkow, Cosmides, and Tooby, *The Adapted Mind*, 137–59

Tallis, Frank. *Love Sick: Love as a Mental Illness*. New York: Avalon Publishing, 2004

Tambiah, Stanley. "Relations of Analogy and Identity: Toward Multiple Orientations to the World." In Olson and Torrance, *Modes of Thought*, 34–52

Tateo, Francesco. "Interior Dialogue and Ideological Polemic in Petrarch's *Secretum*." Originally published as *Dialogo interiore e polemica ideologica nel "Secretum" del Petrarca*. Florence: Le Monnier, 1965. In *Petrarch's* Secretum, translated by Davy Carozza and H. James Shey, 251–71. New York: Peter Lang, 1989

Taylor, Archer. *The Proverb*. Cambridge: Harvard University Press, 1931

Taylor, Charles. "Self-Interpreting Animals." In *Human Agency and Language*. Philosophical Papers 1, 45–76. Cambridge: Cambridge University Press, 1985

Thagard, Paul. *Mind: Introduction to Cognitive Science*. Cambridge: MIT Press, 1996

– ed. *Mind Readings*. Cambridge: MIT Press, 1998

Thomasson, Amie L. "Speaking of Fictional Characters." *Dialectica* 57, no. 2 (2003): 205–23

Thompson, Richard, and Stephen Madigan. *Memory: The Key to Consciousness*. Princeton: Princeton University Press, 2005

Thouless, Robert Henry. *An Introduction to the Psychology of Religion*. 3rd ed. London: Cambridge University Press, 1971

Tiger, Lionel, and Robin Fox. *The Imperial Animal*. New York: Delta of Dell Publishing, 1971

Tomkins, Sylvan. "Script Theory." In *The Emergence of Personality*, edited by Joel Aranoff et al., 147–216. New York: Springer, 1987

Tooby, John, and Leda Cosmides. "Cognitive Adaptations for Social Exchange." In Barkow, Cosmides, and Tooby, *The Adapted Mind*, 163–228

– "The Psychological Foundations of Culture." In Barkow, Cosmides, and Tooby, *The Adapted Mind*, 19–136

Trevarthen, Colwyn. "Intersubjectivity." In Wilson and Keil, *The MIT Encyclopedia of the Cognitive Sciences*, 415–18

Tripet, Arnaud. *Pétrarque ou la connaissance de soi*. Genève: Droz, 1967

Trivers, Robert. "The Evolution of Reciprocal Altruism." *Quarterly Review of Biology* 46 (1971): 35–57

– *The Folly of Fools: The Logic of Deceit and Self-Deception in Human Life*. New York: Basic Books, 2001

– "Parental Investment and Sexual Selection." In *Sexual Selection and the Descent of Man*, edited by B. Campbell, 136–79. Chicago: Aldine, 1972

– *Social Evolution*. Reading, MA: Benjamin-Cummings, 1985

Tsur, Reuven. *Toward a Theory of Cognitive Poetics*. New York: North-Holland, 1992

Turner, Mark. *Death Is the Mother of Beauty: Mind, Metaphor, Criticism*. Chicago: University of Chicago Press, 1987

– *Reading Minds: The Study of English in the Age of Cognitive Science*. Princeton: Princeton University Press, 1991

Tye, Michael. *Ten Problems of Consciousness: A Representational Theory of the Phenomenal Mind*. Cambridge: MIT Press, 1995

Unger, Peter. *Identity, Consciousness and Value*. New York, Oxford: Oxford University Press, 1990

Updike, John. "The Tried and the Trëowe." *Forbes ASAP* (2 October 2000): 201

van Dijk, Teun A., and Walter Kintsch. *Strategies of Discourse Comprehension*. London: Academic Press, 1983

van Gulick, Robert. "Self-Knowledge." In Wilson and Keil, *The MIT Encyclopedia of the Cognitive Sciences*, 735–6

van Hooff, J.A.R.A.M. "A Comparative Approach to the Phylogeny of Laughter and Smiling." In *Non-Verbal Communication*, edited by R.A. Hinde, 209–41. Cambridge: Cambridge University Press, 1972 van Inwagen, Peter. "Creatures of Fiction." *American Philosophical Quarterly* 14, no. 4 (1977): 299–308

Vega, Lope de. *Castelvins and Monteses*. Translated by Cynthia Rodriguez-Badendyck. Ottawa: Dovehouse Editions, 1998

Vermeule, Blackey. *Why Do We Care about Literary Characters?* Baltimore: Johns Hopkins University Press, 2010

Vincent, Jean-Didier. *Biologie des passions*. Paris: Seuil, 1986

Vives, Juan Luis. *The Passions of the Soul: The Third Book of* De anima et vita. Translated by Carlos G. Noreña. Lewiston: Mellen Press, 1990

Voight, Georg. *Pétrarque, Boccace et les débuts de l'humanisme en Italie*. Paris: Le Monnier, 1894

von Eckardt, Barbara. "Mental Representation." In Wilson and Keil, *The MIT Encyclopedia of the Cognitive Sciences*, 527–9

Vorderer, Peter. "Toward a Psychological Theory of Suspense," In Vorderer, Wulff, and Friedrichesen, *Suspense*, 233–54

– Hans Wulff, and Mike Friedrichsen, eds. *Suspense: Conceptualizations, Theoretical, Analyses, and Empirical Explorations*. Mahwah, NJ: Lawrence Erlbaum, 1996

Wallis, Mary. "Patterns of Wisdom in the Old English *Solomon and Saturn* II." PhD diss., University of Ottawa, 1991

Walsh, Anthony. *The Science of Love: Understanding Love and Its Effects on Mind and Body*. Buffalo: Prometheus Books, 1996

Walton, Kendall. *Mimesis as Make-Believe: On the Foundations of the Representational Arts*. Cambridge: Harvard University Press, 1990

– "Pictures and Make-Believe." *The Philosophical Review* 82, no. 3 (1973): 283–319

Watson, Walter. *The Lost Second Book of Aristotle's* Poetics. Chicago. University of Chicago Press, 2012

Wesling, Donald. *Joys and Sorrows of Imaginary Persons (On Literary Emotions)*. Amsterdam: Rodopi, 2008

Whissell, Cynthia. "Mate Selection in Popular Women's Fiction." *Human Nature* 7 (1996): 427–47

Whiten, Andrew. "Machiavellian Intelligence Hypothesis." In Wilson and Keil, *The MIT Encyclopedia of the Cognitive Sciences*, 495–6

Wilkins, Ernest Hatch. *Life of Petrarch*. Chicago: University of Chicago Press, 1961

Williams, Bernard. *Truth and Truthfulness: An Essay in Genealogy*. Princeton: Princeton University Press, 2002

Williams, George C. *Adaptation and Natural Selection: A Critique of Some Current Evolutionary Thought*. Princeton: Princeton University Press, 1966

Williams, James G. "Proverbs and Ecclesiastes." In Alter and Kermode, *The Literary Guide to the Bible*, 263–82

Willis, Deborah. "Doctor Faustus and the Early Modern Language of Addiction." In *Placing the Plays of Christopher Marlowe: Fresh Cultural Contexts*, edited by Sara Munson Deats and Robert A. Logan, 135–47. Burlington, VT: Ashgate, 2008

Wilson, David S. "Evolution and Social Constructivism." In Gottschall and Wilson, *The Literary Animal*, 20–37

Wilson, James Q. *The Moral Sense*. New York: The Free Press, Macmillan, 1993

Wilson, Robert. "Philosophy." In Wilson and Keil, *The MIT Encyclopedia of the Cognitive Sciences*, xiii–xxxvii

– and Frank Keil, eds. *The MIT Encyclopedia of the Cognitive Sciences*. Cambridge: MIT Press, 1999

Wispé, Lauren. "The History of the Concept of Empathy." In *Empathy and Its Development*, edited by N. Eisenberg and J. Strayer, 17–37. Cambridge: Cambridge University Press, 1987

Wittgenstein, Ludwig. *Philosophical Investigations*. Oxford: Blackwell, 1958

Wolff, S.L. *The Greek Romances in Elizabethan Prose Fiction*. New York: Columbia University Press, 1912

Wright, Robert. *The Moral Animal: Evolutionary Psychology and Everyday Life*. New York: Pantheon Books, 1994

A Yorkshire Tragedy, 1608. Edited by G.R. Proudfoot. Oxford: University Press, The Malone Society Reprints, (1969) 1973

Young, Dudley. *Origins of the Sacred: The Ecstasies of Love and War*. London: Little, Brown and Co., 1991

Young, J.Z. *Philosophy and the Brain*. Oxford: Oxford University Press, 1988

Zajonc, R.B. "Feeling and Thinking: Preferences Need No Inferences." *American Psychologist* 35, no. 2 (February 1980): 151–75

Zamulinski, Brian. *Evolutionary Intuitionism: A Theory of the Origin and Nature of Moral Facts*. Montreal & Kingston: McGill-Queen's University Press, 2007

Zeman, Adam. "Does Consciousness Spring from the Brain? Dilemmas of Awareness in Practice and in Theory." In *Frontiers of Consciousness*, edited by Lawrence Weiskrantz and Martin Davies, 289–321. Oxford: Oxford University Press, 2008

Zillman, Dolf. "Mechanisms of Emotional Involvement with Drama." *Poetics* 23 (1994): 33–51

– "The Psychology of Suspense in Dramatic Exposition." In Vorderer, Wulff, and Friedrichsen, *Suspense*, 199–231

Ziolkowski, Jan, trans. *Solomon and Marcolf*. Cambridge: Department of Classics, Harvard University; Harvard University Press, 2008

Zunshine, Lisa. *Your Head: What Cognitive Science Can Tell Us about Popular Culture*. Baltimore: Johns Hopkins University Press, 2012
– *Why We Read Fiction: Theory of Mind and the Novel*. Columbus: Ohio State University Press, 2006

INDEX